Sustainability in Biofuel Production Technology

Sustainability in Biofuel Production Technology

Pratibha S. Agrawal
Laxminarayan Institute of Technology
Nagpur, India

Pramod N. Belkhode
Laxminarayan Institute of Technology
Nagpur, India

Samuel Lalthazuala Rokhum
National Institute of Technology Silchar
Silchar, India

WILEY

This edition first published 2023

Registered Office
John Wiley & Sons Ltd, The Atrium, Southern Gate, Chichester PO19 8SQ, UK

Editorial Office
The Atrium, Southern Gate, Chichester PO19 8SQ, UK

For details of our global editorial offices, customer services, and more information about Wiley products visit us at www.wiley.com.

Wiley also publishes its books in a variety of electronic formats and by print-on-demand. Some content that appears in standard print versions of this book may not be available in other formats.

Library of Congress Cataloging-in-Publication Data applied for
ISBN 9781119888833 (hardback)

Cover Design: Wiley
Cover Image: © khonkangrua/Shutterstock

Set in 9.5/12.5pt STIXTwoText by Straive, Pondicherry, India

Printed and bound by CPI Group (UK) Ltd, Croydon CR0 4YY

C9781119888833_220822

Contents

Preface

This book *Sustainability in Biofuel Production Technology* highlights the source of renewable energy biofuels to save natural resources and minimise the impact of fossil fuels. It presents detailed information about the challenges and recent trends in biodiesel production which will be useful for future researchers and industrialists at a global level. It includes contributions of leading researchers in the field of biodiesel, and will serve as a valuable source of information for scientists, researchers, graduate students, and professionals alike. This book focusses on several aspects of biodiesel productions, technologies employed, and sustainability. It consists of 14 chapters which gives complete information to the reader on all the aspects of biofuels.

Chapter 1 is an introduction to biofuels covering details about the history, generation, classification, and various technologies involved along with their environmental impact. Chapter 2 is on ethanol which is used as the first-generation biofuel and its development, environmental aspects, and various models are explained. Chapter 3 includes advanced biofuels explaining the next generation of biofuels with reducing cost and technological development.

Chapter 4 details the production technologies of biofuels and covers all production aspects. Chapters 5, 6 and 7 cover the production of biofuels chemically, biologically, and also by agricultural waste treatment along with processing techniques and cost analysis; and present challenges and future requirements are also discussed.

Chapter 8 focusses on the advanced topic of biofuel utilisation in the aviation industry. In Chapter 9, the authors discuss process equipment such as types of reactors, reactor design, safety considerations, performance analysis with cost estimation, used in the production of the biofuels. Investigation of any phenomenon is important and that the best conclusion of a complete process is presented. Based on that principle, the process can be monitored through various techniques such as modelling, analysis and optimisation.

Chapter 10 includes a case study of the investigation of process parameters of biodiesel blends via a mathematical modelling approach, sensitivity analysis, simulation through artificial neural networks followed by its optimisation. The investigation is helpful to control the behaviour of production by identifying the most influencing terms involved in the processing.

Chapter 11 details key issues and challenges in biofuels production pathways, environmental effects and economic impact. Chapter 12 discusses greener catalytic processes in biofuel production and sustainable catalysts. Uses of feedstock, life cycle impact assessment and analysis with ISO life cycle assessment standards are included in Chapter 13, while Chapter 14 discusses the economics, environmental and policy issues regarding the production and assessment of biofuels.

Thus, the book covers the complete information of biofuels which is useful to optimise the natural resources and assure humans about the need for harmony in nature by minimising the environmental pollution. This is only possible by utilising advanced biofuels that are processed by treating the waste and utilising the maximum amount of energy for all the similar types of applications.

1

Introduction to Biofuel

1.1 Introduction

1.1.1 Need of Emerging Technology

Food, water, clothing and shelter are the traditional basic needs of humans. However, today we rely on electricity and transportation like we do on food and water. Electricity and transportation are such an essential part of our modern life that we cannot imagine today what life would be like without it. Even if we radically decrease transportation to that firmly wanted to ensure basic needs, humans can still certainly survive, but this would need a profound re-modelling of anthropoid activities such as economic and social. An upcoming scenario where human actions are not excessively unnatural is likely more tolerable. These amenities come with a need, i.e. fuel to run and this fuel is usually from the fossils of animals and plants buried millions of years ago. Fossil fuels have facilitated our civilisation everywhere. The major energy source for today's world is fossil fuels [1]. Fossil fuels have been driving the world for eras, making them economical and dependable since the infrastructure is previously in place for their sustained use. Since the early nineteenth century, the fossil fuel consumption in the world has almost doubled every 20 years. The advantage of fossil fuels is their ability to produce huge quantities of energy in just a single location. Innovation of the world's energy system must be on the horizon in current worldwide conditions that have led to a collective awareness. The development and advancement of nations worldwide is also dependent on the fossil fuel industry. The significant driver of the fossil fuel industry in its vast expansion over the course of several decades is the political and economic support [2]. The technical advancement in the technologies related to energy conversion to produce heat, electricity and transportation fuels has made a striking impact on society.

Being non-renewable resource, its overconsumption will apparently result in serious environmental issues [2]. Several consequences have already been observed on the habitat from the burning of fossil fuel, which releases noxious gases such as carbon dioxide, carbon monoxide, sulphur dioxide and nitrogen dioxide [3]. Fossil fuel is estimated to account for one-third of global energy utilisation and is responsible for 15 and 31% of global CO_2 and O_3 man-derived emissions, respectively [4]. Acid rain is one of the major consequences of

Sustainability in Biofuel Production Technology, First Edition. Pratibha S. Agrawal, Pramod N. Belkhode, and Samuel Lalthazuala Rokhum.

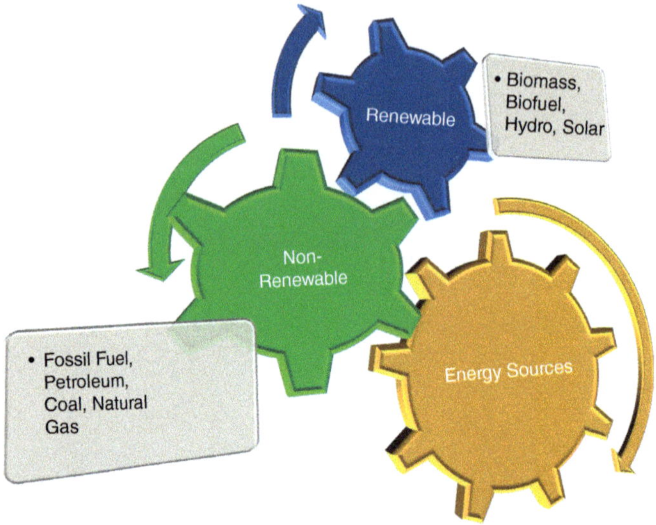

Figure 1.1 Sources of energy.

these gases, which damages the crops, lakes, rivers, trees, wildlife statues and architecture. The consequence of global warming can be reversed and halted with making a shift to clean energy sources. Change to clean can be achieved with the renewable energy fuel (biofuel) that is formed by current processes from biomass rather than geological processes from the formation of fossil fuels (Figure 1.1). Thus, there has been a recent prerequisite to research and thereby to develop advanced energy alternatives that are feasible contenders having the potential to alleviate the matters of climate variation and energy security.

1.1.2 Renewable Energy

Renewable energy sources are the resources that can be replenished as quickly as they are utilised. Some of the main renewable energy sources are solar, biomass and wind. With the use of renewable energy sources, some of the environmental, social and economic problems can be omitted since these resources are environment friendly, with very little release of consumed and toxic gases. Since these resources can be used again and again to produce convenient energy, renewable energy is going to be a significant basis for power generation in the future [5, 6]. A total of 15–20% of the entire world's energy is provided by different renewable energy resources, such as hydropower, wind, solar, biomass, ocean energy, biofuel, geothermal, etc. The chief barriers to the progress of renewable energy are cost, market share and policy [7, 8]. Some of the unique properties of renewable energy resources are that they have the capacity to meet the world's energy demand, offer energy security and protect the atmosphere.

The most utilised non-fossil fuel in today's world which serves as a substitute for engine fuel is bioethanol. Oil being the world's prime source of energy and chemicals, the present demand is about 12 million tonnes per day (84 million barrels a day) with a projection to increase to 16 million tonnes per day (116 million barrels a day) by 2030 [9–11]. The world leaders in biofuel development and usage are United States, Sweden, Brazil, France and Germany. On a worldwide scale, the United Nations (UN) International Biofuels Forum is formed by Brazil, China, India, South Africa, United States and the European Commission [12, 13].

Three percent of the global fuels for road transportation is provided by biofuel [14]. Worldwide biofuel production reached 161 billion litres (43 billion gallons US) in 2019, which was up by 6% from 2018. In the direction of reducing reliability on petroleum by 2050, International Energy Agency (IEA) wants more than a quarter of world demand for transportation fuels to be filled by biofuel [15]. To reach IEA's target annually, global biofuel productivity has to surge by 10% from 2020 to 2030. Only 3% growth/year is expected in the coming few years. However, the production and consumption of biofuels are not on track to meet the IEA's sustainable development scenario. By 2024, nearly 154 billion US dollars is projected to grow in the market for alternative fuels. IEA recorded that the ratio of development and deployment of renewable energy is getting to be deployed, which should be clogged to obtain net zero climate goals using different policies and subsidies [16].

Presently, bioenergy is the major renewable energy source worldwide and more than two-thirds of the renewable energy mix is accounted for with these resources. One can continuously produce biofuels by growing biomass feedstock repeatedly that can withstand indeterminate human exploitation. Biofuels and bioproducts over the previous few years have been produced by biomass, which is one of the renewable resource derivatives of biological precursors [17, 18]. Biofuel is one of civilisation's utmost vital renewable energy sources as well as storable, contrasting conventional fossil energy. Microalgae are being assumed to be the greatest attractive basis for the production of biofuels, amid all these living organisms utilised.

Renewable energy is the leading emergent energy source, which has grown to 90% in the past 20 years [19]. Cost, demand, policy decisions, feedstock availability and public acceptance are the various factors that affect the renewable energy deployment [20].

1.2 History of Biofuel Development

Ever since humans discovered fire, charcoal, woodchips and cattle dung have been used as a source of energy and still today people use these solid fuels for heating and cooking in many parts of the world. In the mid-1700s and early 1800s, oil extracted from whale was broadly used for lighting purposes. The discovery of biofuel is not very recent. In the mid-1700s and early 1800s, oil extracted from whale was largely used for illumination purposes [21]. Since the nineteenth century, transesterification of vegetable oils has been generally identified and employed. In fact, the method presently employed for the production of biofuels from biomass is the same inherited from the ancient times. The feedstock used for their production was also very analogous [22]. Peanut, hemp, and corn oil and animal tallow were conventionally utilised and have been partly substituted by soya bean, rapeseed, recycled oil, forest wastes, and trees and sugar cane. The history of biofuel is considered to be more political and economic than being technical. Ever since humans have explored fire, feedstock like charcoal, woodchips, and cattle dung have been utilised as a basis of energy, and still in many countries around the world these solid biofuels are used for heating and cooking [23].

German named Nikolaus August Otto was the first discoverers to influence public for the usage of ethanol. In the 1860s, he ran his initial engines on ethanol, which is a fermented product of yeasts [24]. This car was completely designed to use hemp derived biofuel as fuel. In the 1880s, Henry Ford's first sample automobile could be functioned with ethanol as fuel, the 'Quadricycle' and his "Model T", the most prevalent car produced between 1908 and 1927 [25]. The first communal demo of vegetable-oil-based diesel fuel was at the 1900 World's Fair in Paris, when the diesel engine was built by Otto Company to run on peanut oil when the French government commissioned it. Diesel engine was invented by another German scientist Rudolf Diesel. His diesel engine was designed to run on peanut oil.

However, during the 1920s, due to inexpensive prices and low viscosity of the diesel engine, producers reformed their engines to petroleum-derived diesel fuel, leading to the improved atomisation of the fuel in the engine's combustion chamber [26, 27]. This problem was resolved when the Belgian patent 422,877 was granted on 31 August 1937 to George Chavanne of the University of Brussels. It designates the usage of methyl and ethyl esters of vegetable oil, attained by acid-catalysed transesterification [28], as diesel fuel, being the first report on what is today known as biodiesel.

1.3 Generation in Biofuel

Based on feedstock and method of production, biofuels are classified in different groups named as first-, second-, third- and fourth-generation biofuels [29] (Figure 1.2). The composition and calorific content of biofuel depend on the kind of biomass and process used.

In first-generation biofuels (Figure 1.3), edible biomass is used for starch and sugar, which leads to increased production cost; the utilisation of resources are inefficient; and energy is consumed in cultivating crops. Specifically, edible biomass uses a large area of crop fields for its production and requires a large quantity of fertilisers and water which leads it to compete with food crops.

The second generation of biofuels is built on more resourceful renewable substituents by employing switch grass, sawdust, low-priced woods, crop wastes and municipal wastes that are categorised under inedible lignocellulose biomass [30]. Although this generation requires more phases to generate acceptable biofuels at a viable cost, it overcomes the disadvantages of the first generation (Figure 1.4) [32].

Algae biomass is used in third-generation biofuels, which is an aquatic feedstock [33, 34]. Seaweed is the example of algae that are photosynthetic plants that capture large amounts of CO_2 and produce O_2 and oil as well. Biofuels are less stable when produced from algae than from other sources because the core reason is that oil produced is highly unsaturated making them more volatile specifically at higher temperatures. However, this kind of biomass has some disadvantages such as its huge price and the fact that they are more likely to degrade [35]. Bioengineered microorganisms such bioengineered algae or crops conquered fourth-generation biofuel. They are still in an early stage of development and are genetically modified such that they can consume a large amount of CO_2 than they emit when burned in the environment [36].

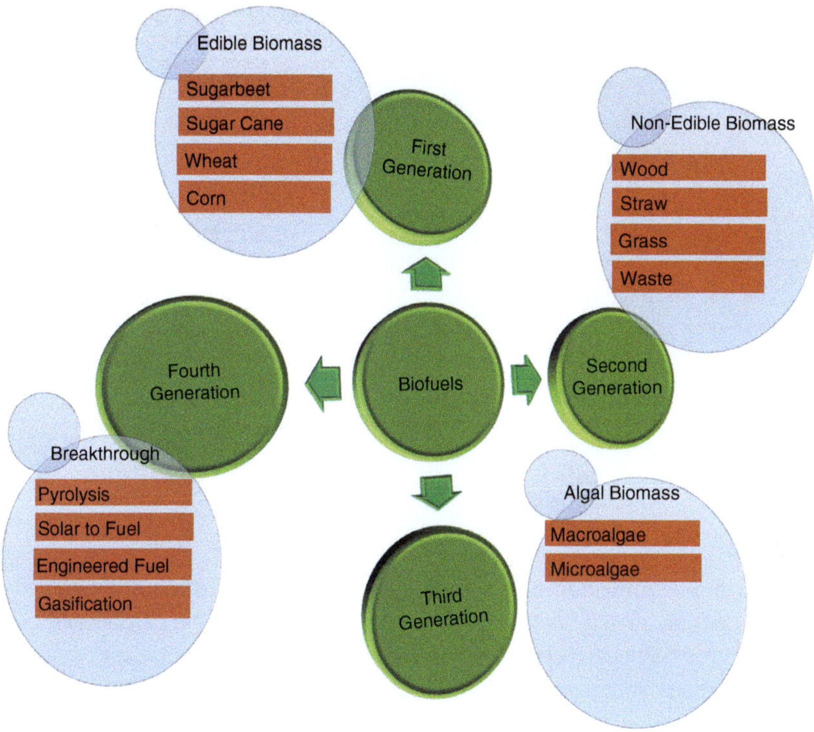

Figure 1.2 Different generation biofuels.

Figure 1.3 First-generation biofuels.

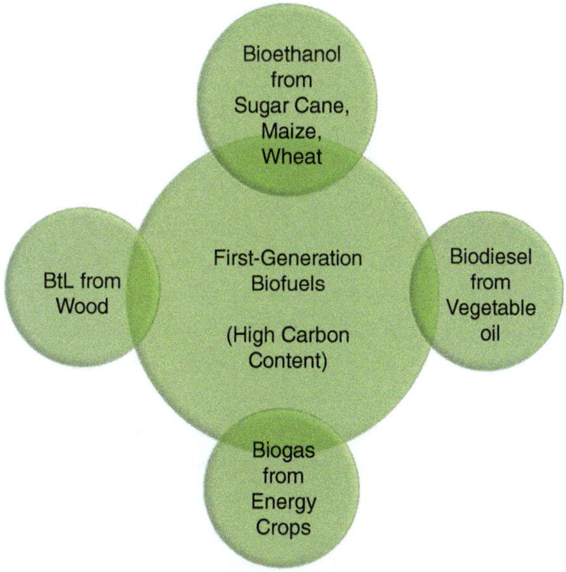

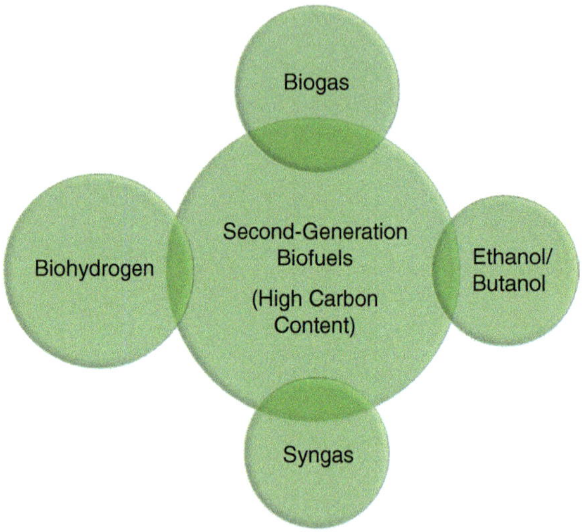

Figure 1.4 Second-generation biofuels. *Source:* Adapted from [31].

First-generation biofuels include ethanol and biodiesel and are directly related to a biomass that is more than often edible or in other terms food-related items. Concerns arose about using edible crops as feedstock and the impacts on croplands, biodiversity and food supply [37]. The emergent problems that can be observed directly were as follows:

a) the biomass chemical composition,
b) energy balance,
c) availability of croplands and the contribution to biodiversity and cropland value losses,
d) competition with food needs,
e) cultivation practices,
f) emission of pollutant gases,
g) impact of mineral absorption on water resources and soil,
h) use of pesticides,
i) cost of the biomass and its transport and storage,
j) soil erosion,
k) economic evaluation considering both the coproducts and feedstock,
l) creation or maintenance of employment and
m) resource availability such as water.

Second-generation biofuels are defined as fuels produced from a wide array of different feedstock, [38] especially but not limited to non-edible lignocellulose biomass or non-food sources. A wide variety of abandoned materials can be used as biofuel feedstock such as agricultural waste [39], poplar trees [40], willow and eucalyptus [41], miscanthus [42], switch grass [43], reed canary grass [44] and wood [45], and they mostly consist of plant cell walls whose primary components is polysaccharides [46]. Second-generation biofuels are bio-ethanol, bio-methanol, Fischer–Tropsch (FT) diesel, dimethyl ethanol (DME), bio-hydrogen, which are a few examples of second-generation biofuels. Second-generation biofuels also generate higher energy yields per acre than first-generation biofuels [47]. The general pathway for the production of second-generation biofuel is biochemical or

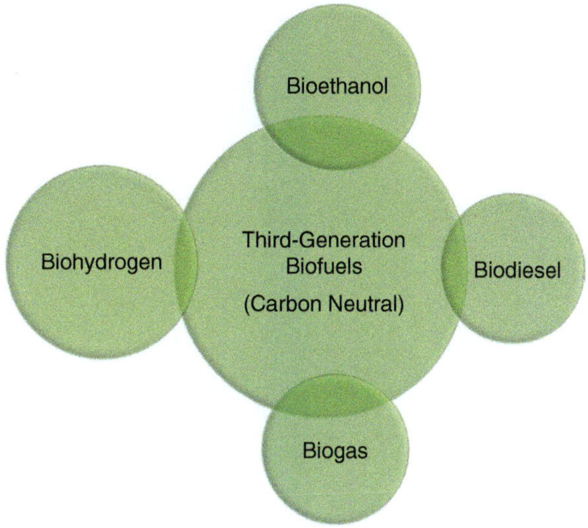

Figure 1.5 Third-generation biofuels.

thermochemical [48], but as far as sustainability is considered, the following problems are incurred:

a) feedstock is not economically and practically viable for stable energy supply due to their low conversion rates,
b) lack of feedstock and
c) with the ever-growing demand, various algae-based biofuels have been augmented.

The cost effectiveness of this generation of biofuels still needs development because there are several technical barriers that need to be overcome.

The third-generation biofuels are usually made up of algae or microbial feedstock (Figure 1.5). Third-generation biofuels are more energy dense than first- and second-generation biofuels per area of harvest [49]. There are two main classifications for algae based on their size and morphology: macroalgae and microalgae [50]. Microalgae have several important properties such as requiring less space to grow, high oil content, the ability to grow in both artificial and natural environments, and being eco-friendly. They also possess a unique advantage that is the capability for both oxygenic photosynthesis and hydrogen production. In addition, their growth requirements are simple and limited to light, carbon dioxide and other inorganic nutrients [51, 52].

It is considered the most energy-intensive fuel. Third-generation biofuels still have the following disadvantages:

a) requirement of complex structure, storage and content,
b) capital intensity of the process of third-generation biofuels,
c) cultivation of cultures and
d) lack of preferences during cultivation.

For a country to adopt alternative fuels, it must be able to avoid a food crisis and control measures regulating the fuel markets [53].

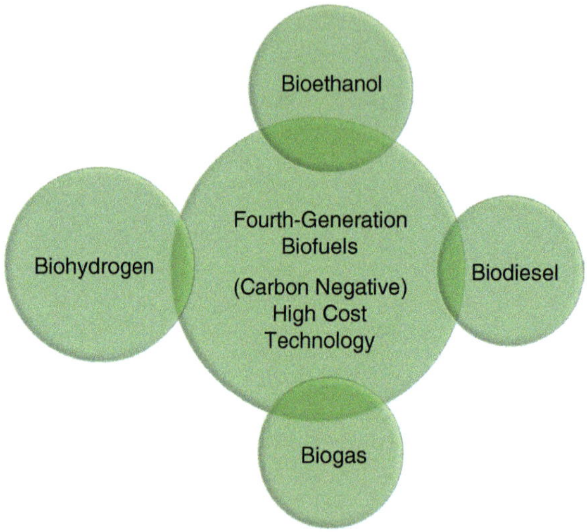

Figure 1.6 Fourth-generation biofuels.

The fourth category would include biofuels produced from biomass (Figure 1.6), whose genetic modification would additionally increase the absorptivity of carbon dioxide in the photosynthesis process [54].

This generation approach utilises metabolic engineering of algae for generating biofuels from oxygenic photosynthetic microbes and creating artificial carbon sinks. Figure 1.7 demonstrates various pathways for the production of biofuels.

1.4 Classification of Biofuels

Biofuels, as discussed, are produced from the organic materials. They can be present in any type of state and form. Thus, on the basis of the state, the biofuels can be classified as the solid, liquid and gaseous biofuels [55].

Solid biofuel includes wood, coal, dried plant material and manure. Improvement in the physical and chemical properties, i.e. particle size, moisture content and energy content, is implemented time to time. The use of solid biofuels is renewable, which can replace energy generated from fossil fuels and help to displace greenhouse gas (GHG) emissions from fossil fuels. It can also lower the risk of forest fires by managing the forest floor debris.

The main drawbacks associated with these fuels are their variable composition, high moisture content, low energy density and availability of the related resources. However, the high content of volatile matter, water-soluble nutrients, low maintenance cost and reduced harmful emission make it a preferable candidate in biofuel market [56]. The technological and other barriers can be dealt with due care in progressive modifications. Thus, renewable standardised solid biofuels offer consistent quality, leading to improved performance, lower maintenance costs and reduced emissions.

Liquid biofuels include bioethanol, dimethyl ether, bio-oil and biodiesel. A wide range of feedstock, process technologies, and field of applications are opened in this area [57].

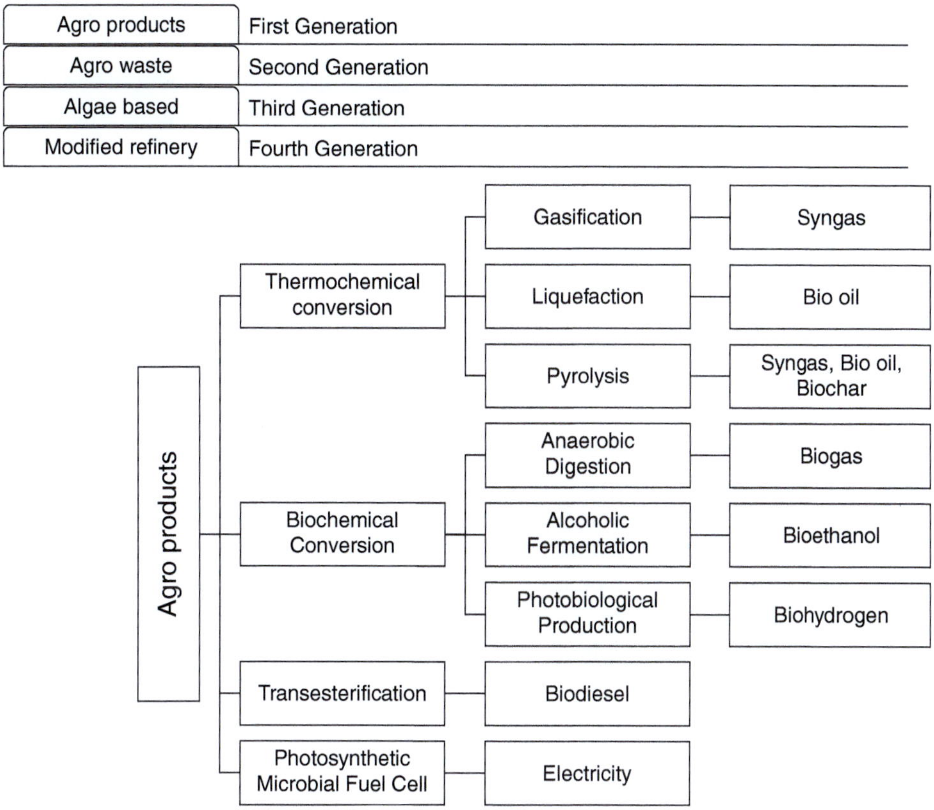

Figure 1.7 Production of biofuels.

Ethanol is a type of alcohol that can be produced by fermentation using any feedstock containing significant amounts of sugar. Ethanol can be blended with petrol or burned in nearly pure form in slightly modified spark-ignition engines. Bioethanol suffers from the disadvantage that it can degrade certain elastomers and corrode certain metals inside the vehicle, leading to the conclusion that continual replacement is needed. Most existing car petrol engines can run on blends of up to 15% bioethanol with petroleum/gasoline. Biodiesel can be derived from a wide range of oils, including rapeseed, soya bean, palm, coconut or jatropha oils using transesterification process [58–60]. Biodiesel can be used in regular diesel engines in pure form or may be blended with petro-diesel in any proportion. Sixty percent of the emission is reduced with the use of biodiesel instead of diesel. One of the problems associated with the biodiesel is that its viscosity increases with a decrease in the temperature. Thus, its use in countries with lower climate requires extra precautions while using biodiesel [61]. Industry, government and research organisations are also supporting the research activities and the related growth of liquid biofuels [62].

Major examples of gaseous biofuel include biogas and syngas. Biogas and bio-hydrogen are companionable gaseous biofuels for blending with natural gas. Usually, woody biomass is a suitable feedstock used in the production of biofuel due to its easy availability and use of non-arable lands. The cellulose and hemicellulose divisions of wood can be used for the fermentative production of gaseous biofuels [63].

Biogas is a gas composed principally of methane and carbon dioxide produced by anaerobic digestion of biomass or by thermal processes from biomass, including biomass in waste [64]. Biogas is a blend of gases produced by the decomposition of organic matter in the absence of oxygen. Anaerobic biogas, which is comparable to landfill gas, comprises 60% methane and 40% carbon dioxide [65]. Biogas is used in power generation systems such as combined-cycle power plants. Syngas, also called a synthesis gas, is a mix of molecules containing hydrogen, methane, carbon monoxide, carbon dioxide, water vapours as well as other hydrocarbons and condensable compounds [66]. It is the main product of gasification, and the majority of products of high-temperature pyrolysis are carried on any biomass, residues and waste. Thermochemical process called gasification converts carbonaceous materials, such as biomass, municipal wastes, coal, petroleum and tyres under controlled amounts of oxidants such as oxygen, air and CO_2 inside a gasifier to obtain syngas [67]. Syngas is mainly used for the generation of heat and power in both stand-alone combined heat and power plants and also in large-scale power plants. It can also be used in internal combustion engines as fuel.

The characterisation of the biomass and selection of the feedstock have a major effect on the technology used and the application field. Thermochemical processes involved such as pyrolysis, gasification and combustion involve several steps, chemical species reactions, catalysts, and routes. To lessen the efforts, simulations and modelling need to be simplified, which will allow optimal conditions for biofuel productions [68].

1.5 Technologies Involved in Biofuel Production

The biofuel is nothing but the monoalkali ester that is derived from the animal fat or vegetable oil. These monoalkali esters are to be extracted from the feedstock, and to do so different technologies are to be employed. Some of the technologies involved in the production of biofuels are transesterification, thermochemical process, microwave-assisted synthesis [69–71], etc.

Pyrolysis of oils is a chemical change using heat in the presence of nitrogen (without participation of oxygen) to produce a wide range of different products such as alkanes, alkenes, alkadienes, aromatics, carboxylic acids, etc. The purpose of this route is to obtain high-value fuel products from biomass by thermal and catalytic methods. Pyrolysis of triglycerides was found to be a potential option for the production of biodiesel [72]. This process is not widely accepted due to huge ash and carbon residue content (79% carbon in the case of soya bean oil), large input energy and high pour point of the final fuel product [73].

Microemulsions are defined as transparent, thermodynamically stable colloidal dispersions, in which the diameter of the dispersed-phase particles is less than one-fourth of the wavelength of visible light [74]. The flow behaviour of biofuel in the application of a diesel engine needs to be simplified, which was successfully solved with this technique.

Transesterification of oil to its corresponding fatty ester is one of the most promising techniques involved in the synthesis of biofuel. Basically, in transesterification reaction oil reacts with alcohol in the presence of some catalysts to yield fatty acid alkyl ester and glycerol [58, 75, 76]. The catalysts that are primarily used in transesterification reaction in biofuel synthesis are alkali, acid and enzymes. Alkali catalysts are considered suitable catalysts for the transesterification reaction.

Microwaves are electromagnetic radiations that represent a non-ionising radiation that influences molecular motions such as ion migration or dipole rotations, but does not alter the molecular structure. The frequencies of microwave range from 300 MHz to 30 GHz, generally a frequency of 2.45 GHz is preferred in laboratory applications. The microwave process can be explained for the biodiesel production with transesterification reaction: the oil, methanol and base catalyst contain both polar and ionic components [77]. Microwaves activate the smallest degree of variance of polar molecules and ions, leading to molecular friction, and therefore the initiation of chemical reactions is possible. When the reaction is carried out under microwaves, transesterification is efficiently accelerated in a short reaction time. As a result, a drastic reduction in the quantity of by-products and a short separation time are obtained with high yields of highly pure products [78].

Ultrasonic waves are energy applications of sound waves that are vibrated more than 20 000 per second. Ultrasonic processing of biodiesel involves the following steps: (i) mixing vegetable oil with the alcohol (methanol or ethanol) and catalyst, (ii) heating the mixture, (iii) the heated mixture is sonicated inline and (iv) glycerine separation by using centrifugation [79].

In all the above methods, four steps are involved for the biofuel synthesis in general. In the first step, transesterification reaction takes place between reagents and reactants under the controlled conditions. The resultant slurry is settled and centrifuged for phase separation in the second step. Thirdly, in an evaporator or a flash unit unreacted alcohol, the formed biofuel is sent to separate alcohol. Neutralisation and distillation of biofuel from other unwanted components such as catalyst and unreacted triglyceride is carried out in the fourth step.

The two most commonly used reactors for the commercial production of biofuel are batch-mode process and continuous flow reactor. In the batch-mode process, there is no flow of reactant and product of the reagents in and out of the reactor in a specific period of time, [80] while in the continuous flow process, feedstock is continuously fed into the continuous-mode reactor while product stream leaving the system.

Many reactors for the biodiesel production are available. These reactors are based on the operational parameter with respect to chemical properties of reactants, reagents and products as well as physical parameters of procedure [81]. Thus, cost-effective and eco-friendly biofuel production technologies should be adopted to make this biofuel more competitive against the traditional fuels used. Different transesterification reactors are listed in Figure 1.8, which are finally analysed from the various angles. Merits and limitations of each reactor open up a new area of research.

1.6 Biofuel Properties

Properties of biofuel will indicate whether it is suitable or not for the performance, emission and life for the desired application. The major deviation in the range of standard parameters of biofuel could damage the engine in which it is used. An acceptable and

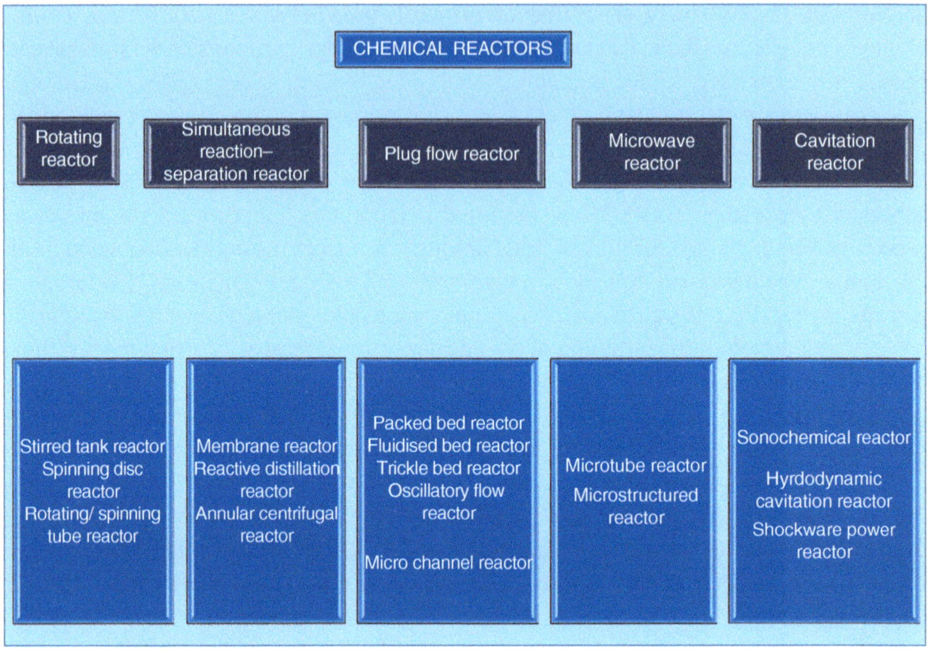

Figure 1.8 Types of reactors.

persistent properties of biodiesels can only be guaranteed by relating to the biodiesel quality standards. It is essential to watch the quality during the biodiesel manufacturing procedure, from the feedstock to the delivery stations to accomplish this goal. The composition and nature of the feedstock used for the production strongly impact the physicochemical properties of biofuels. Blending, testing, storage and distribution should be involved in quality assurance and monitoring. Usually, the differences in biofuel properties listed in Table 1.1 are seen in oxidation stability, cetane number, iodine value, viscosity and density [105, 106].

Acceptability of biodiesel by various sectors depends on its socioeconomic and environmental impacts. Adverse cold flow properties, poor oxidation stability, corrosive and acidic nature, and degradation tendency of biodiesel present it as a non-compatible system. A vast area is opened for the R and D sector to further improve its properties and make it compatible with automotive materials.

1.7 Socioeconomic and Environmental Impact

The development of world economy is restricted because of one of the important factors, i.e. shortage of energy. Public awareness, opinion and knowledge could contribute to the social acceptance of biofuels [107]. The current surge in biofuel investments and manufacturing capacities is determined by the potential of multiple social, economic, ecological, and geopolitical aids. From a broader view, there are mainly three motives behind the promotion of biofuel in the society (Figure 1.9). The socioeconomic and environmental impact

Table 1.1 Specifications of biodiesel.

Property	ASTM D975-08a		ASTM D6751-12			EN 590:2004		EN 14214:2012	
			2-B	1-B	Test				
Flash point [82], min	No 1D 38 °C No 2D 52 °C	D93	93 °C		D93	55 °C	EN 22719	101 °C	EN ISO 2719
Water and sediment [83], max	0.05% vol.	D2709	0.050% vol.		D2709				
Water, max						200 mg/kg	EN ISO 12937	500 mg/kg	EN ISO 12937
Total contamination, max						24 mg/kg	EN 12662	24 mg/kg	EN 12662
Distillation temperature [84] (% vol. recovered)	90%: 1D 288 °C max 2D 282–338 °C	D86	90%: 360 °C max		D1160	65%: 250 °C min 85%: 350 °C max	EN ISO 3405		
Kinematic viscosity [85]	1D 1.3–2.4 mm^2/s 2D 1.9–4.1 mm^2/s	D445	1.9–6.0 mm^2/s		D445	2.0–4.5 mm^2/s	EN ISO 3104	3.5–5.0 mm^2/s	EN ISO 3104
Density [86]						820–845 kg/m^3	EN ISO 3675 EN ISO 12185	860–900 kg/m^3	EN ISO 3675 EN ISO 12185
Ester content [87]	5% vol. max	EN 14078				5% vol. max FAME	EN 14078	96.5% min	EN 14103
Ash [88], max	0.01% wt.	D482				0.01% wt.	EN ISO 6245		
Sulfated Ash [89], max			0.020% mass		D874			0.02% mass	ISO 3987
Sulphur [90], max (by mass)	1D and 2D: S15 15 mg/kg S500 0.05% S5000 0.50%	D5453 D2622 D129 [2]	Two grades: S15 15 ppm S500 0.05%		D5453	Two grades: 50 mg/kg 10 mg/kg	EN ISO 14596 EN ISO 8754 EN ISO 24269	10.0 mg/kg	EN ISO 20846 EN ISO 20884 EN ISO 13032

(Continued)

Table 1.1 (Continued)

Property	ASTM D975-08a		ASTM D6751-12			EN 590:2004		EN 14214:2012	
			2-B	1-B	Test				
Copper strip corrosion [91–93], max	No 3	D130	No 3		D130	class 1	EN ISO 2160	class 1	EN ISO 2160
Cetane number [94], min	40	D613	47		D613	51.0	EN ISO 5165	51.0	EN ISO 5165
Cetane index [95], min						46.0	EN ISO 4264		
One of [3]: • cetane index • aromaticity	40 min 35% vol. max	D976–80 D1319							
PAH, max						11% wt.	IP 391 EN 12916		
Operability [96] • cloud point • LTFT/CFPP	Report	D2500 D4539 D6371							
Cloud point			Report		D2500	Location and season dependent	EN 23015	Location and season dependent	EN 23015
CFPP [97]						Location and season dependent	EN 116	Location and season dependent	EN 116
Carbon residue on 10% distillation residue, max	1D: 0.15% wt. 2D: 0.35% wt.	D524	0.050% wt. [5]		D4530	0.30% wt.	EN ISO 10370		
Acid number [91], max			0.50 mg KOH/g		D664			0.50 mg KOH/g	EN 14104
Oxidation stability [98]			3 h min		EN 14112	25 g/m^3 max	EN ISO 12205	8 h min	EN 14112

Iodine value [99], max							120 g Iod/100 g [1]	EN 14111 EN 16300
Linolenic acid methyl ester [100], max							12.0% wt.	EN 14103
Polyunsaturated methyl esters, max							1.00% wt.	EN 15779
Alcohol control			0.2% wt. methanol max, or	EN14110			0.20% wt. methanol max	EN 14110
			130 °C flash point min	D93				
Monoglycerides, diglycerides and triglycerides, max			MG 0.40% wt.	D6584			MG 0.70% wt. DG 0.20% wt. TG 0.20% wt.	EN 14105
Group I metals (Na + K), max			5 mg/kg	EN 14538			5.0 mg/kg	EN 14108 EN 14109 EN 14538
Group II metals (Ca + Mg), max			5 mg/kg	EN 14538			5.0 mg/kg	EN 14538
Free glycerine [101], max			0.020% wt.	D6584			0.02% wt.	EN 14105 EN 14106
Total glycerine [102], max			0.240% wt.	D6584			0.25% wt.	EN 14105
Phosphorous, max			0.001% wt.	D4951			4.0 mg/kg	EN 14107 prEN 16294
Lubricity [103], max	520 μm	D6079			460 μm	ISO 12156-1		
Conductivity [104], min	25 pS/m	D2624 D4308						
Cold soak filtration time (CSFT), max			360 s [4] 200 s	D7501				

of biofuel varies widely based on the specific condition of the country. Political support or strategy is evolved out of or by the combination of those three areas.

The profit and cost for biofuel have a tendency to differ across commodities, landscape and business models. The main positive influence of the countries is income, rural development, employment and energy security [108]. Energy security refers to the capacity of nation to access the energy resources desired to sustain its national power. The evaluation of the economics of renewable energy development included the estimation of the resulting economic costs and benefits to any country including (i) cost of electricity, (ii) direct and indirect impacts on jobs, income and economic output, and (iii) renewable liquid fuel price impacts. The benefits of bioenergy comprise support of traditional trades, the financial expansion of rural civilisations and rural diversification. The cost of biomass conversion includes factors such as the scale of operation, types of biomasses, conversion process and the location of feedstock. The relative cost of biofuel will also depend on the accessibility of the alternative energy possibilities. In the development of biofuels, trade barriers, price interventions and financial support played a critical role. Biofuel chain (harvesting, production, transportation, etc.) will generate employment opportunities and increase people's income [109]. The socioeconomic effect of biofuel includes its impact on food security, supply and accessibility. Moreover, the reuse of unrestrained crops imitates the idea of a recycling economy, advances the efficacy of resource usage, and thus protects resources associated with the production method. The first-generation biofuels are considered to be cost-effective, while the second-generation biofuels cost is indeterminate and differ with the production and conversion process. Feedstock for second-generation biofuel would reduce the cost by 50% for production than the use of corn and sugar-based feedstock.

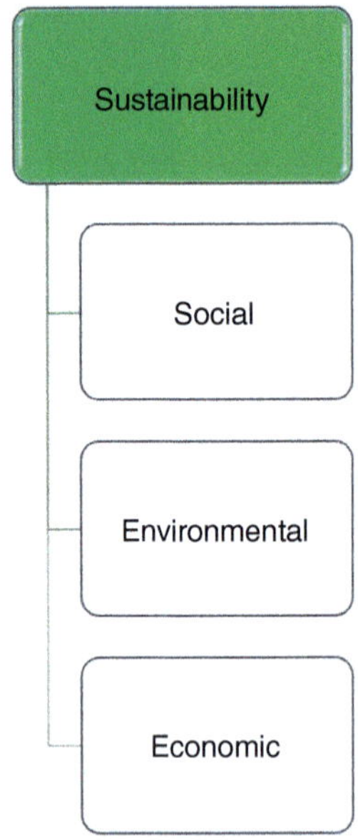

Figure 1.9 Socioeconomic and environmental impacts of biofuels.

Biofuels have the strongest association with agricultural markets and residues as they are formed from agricultural supplies and have the utmost ability to influence food production and values [110], while next-generation biofuels derived from lignocellulosic biomass and photosynthetic algae may have rarer straight relations to food production systems. Biofuel is explored so much because of its advantage on environment. The reason being biofuel is carbon neutral, i.e. the amount of carbon released when it is burned is equal to the amount of carbon utilised in the photosynthesis of plants, thereby reducing the GHG gas emission in the environment. Environmental impact includes issues such as loss of biodiversity, degradation of soil and water resources quality. Environmentally benign biofuel-making procedure is projected to have little energy involvements in all steps starting from agriculture to processing of plants and manufacturing of biofuels with higher energy balance and better engine performance. Ecosystem is affected significantly with feedstock production for biofuel, either by enhancing biodiversity or intimidating the natural habitat and species. Some feedstock like sugar beet has higher impact on aquatic ecosystem than maize or

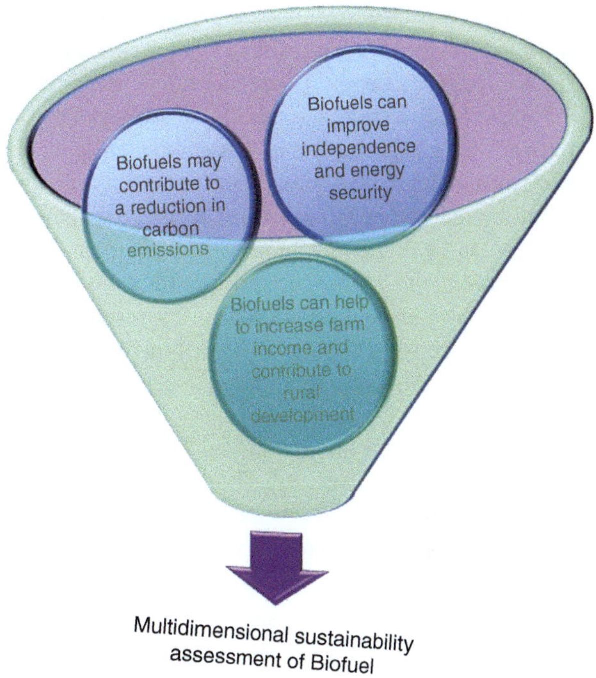

Figure 1.10 Multidimensional sustainability assessment.

wheat so that its monoculture is preferred compared to others. A number of tasks are placed for biofuel and managing the soil fertility. First is the possibility of recycling for small organic and plant nutrients. Current agricultural practices (in particular in developing countries) for soil management depend on the crop wasted. Secondly, feedstock nutrients can be retrieved during land conversion processes and applied to the crop field for biofuel production. Finally, hydrological effects are also important. Some bioenergy crops require the same amount of water irrigation as food crops (i.e. sugar cane). Best agricultural practices should avoid water infiltrations of water wastes to guarantee an efficient growth of bioenergy crop. Economic and environmental credibility of coproducts is essential to completely justify that investments in biofuel sectors are profitable and will meet the present energy demands and also curb greenhouse emissions (Figure 1.10). Thus, multidimensional sustainability assessment is very essential for a sustainable energy future.

1.8 Conclusion

Fossil fuel being limited, i.e. non-renewable energy source, its combustion causes side effects on the environment; thus, an alternative has to be considered. The usage of biofuels is projected to contribute to the energy maintenance and global warming reduction [111]. Biofuel is a renewable energy that can be used again and again. More than two-thirds of the total renewable energy mix is accounted by the bioenergy. The centre stage is gained by biofuel as human activities are rising. The high reliance and burden on limited fossil fuels will be shifted due to the worldwide call to look into renewable eco-friendly fuel carriers.

Biofuel is expected to increase the overall economy of the country. It benefits the society by providing employment, rural development and energy security. The second-generation cost is indeterminate, but the production cost can be reduced by 50% as compared to that of first generation. Biofuel is used because of its environmental benefits. It has the capacity to reduce the GHSs and can mitigate many environmental issues. The production of feedstock for biofuel has an impact on the fertility of soil, water usage and biodiversity. It is important to have the biofuel of the desired properties, which will make it efficient for use. There are some standard values for the properties of biofuel, which are given by the agency that makes its accomplished for use. With a diverse range of potentially renewable energy resources, the idea and expected benefits developing from the usage of biofuels are inspiring; subsequently, in demand to contribute to a sustainable and energy secure future, they must be taken into consideration.

References

1 Yildiz, L. (2018). Fossil fuels. *Compr. Energy Syst.* 1: 521–567. https://doi.org/10.1016/B978-0-12-809597-3.00111-5.

2 Inumaru, J., Hasegawa, T., Shirai, H. et al. (2021). Fossil fuels combustion and environmental issues. *Adv. Power Boil.* 1–56. https://doi.org/10.1016/b978-0-12-820360-6.00001-1.

3 Verma, R.K., Sankhla, M.S., Parihar, K. et al. (2021). The study of assessing the impact on environment by the noxious airborne chemicals: A review. *Biointerface Res. Appl. Chem.* 11 (3): 10844–10863. https://doi.org/10.33263/BRIAC113.1084410863.

4 Othman, M.F., Adam, A., Najafi, G., and Mamat, R. (2017). Green fuel as alternative fuel for diesel engine: a review. *Renew. Sust. Energ. Rev.* 80: 694–709. https://doi.org/10.1016/j.rser.2017.05.140.

5 Searchinger, T.D. (2010). Biofuels and the need for additional carbon. *Environ. Res. Lett.* 5: 024007. (10pp), https://doi.org/10.1088/1748-9326/5/2/024007.

6 Leng, R.A. (2008). Decline in available world resources; implications for livestock production systems in Asia. *Livest. Res. Rural. Dev.* 20 (1): 2008.

7 Lee, D.G. (2017). Renewable energy: power for a sustainable future. *Australas J. Environ. Manag.* 25 (2): 248. https://doi.org/10.1080/14486563.2017.1378135.

8 Ellis Paine, A., Kamerade, D., Mohan, J., and Davidson, D. (2019). Communities as 'renewable energy' for healthcare services? a multimethods study into the form, scale and role of voluntary support for community hospitals in England. *BMJ Open* 9 (10): e030243. https://doi.org/10.1136/bmjopen-2019-030243.

9 Van Gunawan, L. and Effendy, M. (2020). Engine performance and exhaust gas emission of a gasoline engine using ethanol-gasoline blended from bioethanol of duriozibethinus seed and salaccazalacca. *Int. J. Emerg. Trends Eng. Res.* 8 (4): 1129–1134. https://doi.org/10.30534/ijeter/2020/29842020.

10 Teoh, Y.H., Yu, K.H., How, H.G., and Nguyen, H.T. (2019). Experimental investigation of performance, emission and combustion characteristics of a common-rail diesel engine fuelled with bioethanol as a fuel additive in coconut oil biodiesel blends. *Energies* 12 (10): 1954. https://doi.org/10.3390/en12101954.

11 Sayin Kul, B. and Ciniviz, M. (2020). Assessment of waste bread bioethanol-gasoline blends in respect to combustion analysis, engine performance and exhaust emissions of a SI engine. *Fuel* 277: 118237. https://doi.org/10.1016/j.fuel.2020.118237.

12 Hyrchenko, Y., Skibina, T., Us, Y., and Veckalne, R. (2021). World market of liquid biofuels: trends, policy and challenges. *E3S Web Conf.* 280: 05005. https://doi.org/10.1051/e3sconf/202128005005.

13 Kapoor, N., Jamwal, V.L., Shukla, M.R., and Gandhi, S.G. (2020). The rise of nutraceuticals: Overview and future. *Biotechnol. Bus. Concept Delivery* 67–74. https://doi.org/10.1007/978-3-030-36130-3_4.

14 Ganev, E.I., Ivanov, B.B., Dobrudzhaliev, D.G. et al. (2018). Adopting environmental transportation practice in biodise lproduction as key factor for sustainable development: a Bulgarian case study. *Bulg. Chem. Commun.* 50: 106–111.

15 Shi, E. and Hanson, S. (2020). EIA projects U.S. biofuel production to slowly increase through 2050 – Today in Energy – U.S. Energy Information Administration (EIA). Today. *Energy* 9.

16 Kuzemko, C. and Britton, J. (2020). Policy, politics and materiality across scales: a framework for understanding local government sustainable energy capacity applied in England. *Energy Res Soc. Sci.* 62: 101367. https://doi.org/10.1016/j.erss.2019.101367.

17 Abdeshahian, P., Dashti, M.G., Kalil, M.S., and Yusoff, W.M.W. (2010). Production of biofuel using biomass as a sustainable biological resource. *Biotechnology* 9 (3): 274–282. https://doi.org/10.3923/biotech.2010.274.282.

18 Ali, S., Shafique, O., Mahmood, S. et al. (2020). Biofuels production from weed biomass using nanocatalyst technology. *Biomass Bioenergy* 139: 105595. https://doi.org/10.1016/j.biombioe.2020.105595.

19 Kapitonov, I.A. and Batyrova, N.T. (2021). Economic prospects of replacing distributed generation with biofuel. *Int. J. Energy Res.* 45 (12): 17502–17514. https://doi.org/10.1002/er.7105.

20 De Laurentis, C. and Pearson, P.J.G. (2021). Policy-relevant insights for regional renewable energy deployment. *Energy Sustain. Soc.* 11 (1): 1–21. https://doi.org/10.1186/s13705-021-00295-4.

21 Guo, M., Song, W., and Buhain, J. (2015). Bioenergy and biofuels: history, status, and perspective. *Renew. Sust. Energ. Rev.* 42: 712–725. https://doi.org/10.1016/j.rser.2014.10.013.

22 Songstad, D.D., Lakshmanan, P., Chen, J. et al. (2009). Historical perspective of biofuels: learning from the past to rediscover the future. *in vitro Cell Dev. Biol. Plant* 45 (3): 189–192. https://doi.org/10.1007/s11627-009-9218-6.

23 Kumar, K.A., Pinto, P., Hawaldar, I.T., and Kumar, B.R.P. (2020). Biogas from cattle dung as a source of sustainable energy: a feasibility study. *Int. J. Energy Econ. Policy* 10 (6): 370. https://doi.org/10.32479/ijeep.10135.

24 Hatanaka, R.R., Sequinel, R., Gualtieri, C.E. et al. (2013). Development and validation of an environmentally friendly attenuated total reflectance in the mid-infrared region method for the determination of ethanol content in used engine lubrication oil. *Talanta* 109: 191–196. https://doi.org/10.1016/j.talanta.2013.02.047.

25 Report, S. (2005). Shaping the future of mobility. *Environment* 35–63. https://doi.org/10.1007/978-3-030-75590-4_3.

26 Knothe, G. (2012). Historical perspectives on biofuels. *Compr. Renew. Energy* 5: 11–14. https://doi.org/10.1016/B978-0-08-087872-0.00502-3.

27 Ayadi, M., Sarma, S.J., Pachapur, V.L. et al. (2016). History and global policy of biofuels. *Green Energy Technol.* 1–14. https://doi.org/10.1007/978-3-319-30205-8_1.

28 Stamenković, O.S., Veličković, A.V., and Veljković, V.B. (2011). The production of biodiesel from vegetable oils by ethanolysis: current state and perspectives. *Fuel* 90 (11): 3141–3155. https://doi.org/10.1016/j.fuel.2011.06.049.

29 Alalwan, H.A., Alminshid, A.H., and Aljaafari, H.A.S. (2019). Promising evolution of biofuel generations. Subject review. *Renew Energy Focus* 28: 127–139. https://doi.org/10.1016/j.ref.2018.12.006.

30 Mohr, A. and Raman, S. (2015). Lessons from first generation biofuels and implications for the sustainability appraisal of second generation biofuels. *Effic. Sustain. Biofuel Prod. Environ. Land-Use Res.* 63: 114–122. https://doi.org/10.1016/j.enpol.2013.08.033.

31 Carriquiry, M.A., Du, X., and Timilsina, G.R. (2011). Second generation biofuels: economics and policies. *Energy Policy* 39 (7): 4222–4234. https://doi.org/10.1016/j.enpol.2011.04.036.

32 Eggert, H. and Greaker, M. (2014). Promoting second generation biofuels: does the first generation pave the road? *Energies* 7 (7): 4430–4445. https://doi.org/10.3390/en7074430.

33 Andersson, V., Broberg Viklund, S., Hackl, R. et al. (2014). Algae-based biofuel production as part of an industrial cluster. *Biomass Bioenergy* 71: 113–124. https://doi.org/10.1016/j.biombioe.2014.10.019.

34 Chhandama, M.V.L., Satyan, K.B., Changmai, B. et al. (2021). Microalgae as a feedstock for the production of biodiesel: a review. *Bioresour. Technol. Rep.* 15: 100771, 15, 100771, https://doi.org/10.1016/j.biteb.2021.100771.

35 Vassilev, S.V. and Vassileva, C.G. (2016). Composition, properties and challenges of algae biomass for biofuel application: an overview. *Fuel* 181: 1–33. https://doi.org/10.1016/j.fuel.2016.04.106.

36 Shokravi, H., Shokravi, Z., Heidarrezaei, M. et al. (2021). Fourth generation biofuel from genetically modified algal biomass: challenges and future directions. *Chemosphere* 285: 131535. https://doi.org/10.1016/j.chemosphere.2021.131535.

37 Kesharwani, R., Sun, Z., Dagli, C., and Xiong, H. (2019). Moving second generation biofuel manufacturing forward: investigating economic viability and environmental sustainability considering two strategies for supply chain restructuring. *Appl. Energy* 242: 1467–1496. https://doi.org/10.1016/j.apenergy.2019.03.098.

38 Antizar-Ladislao, B. and Turrion-Gomez, J.L. (2008). Second-generation biofuels and local bioenergy systems. *Biofuels, Bioprod. Biorefin.* 2 (5): 455–469. https://doi.org/10.1002/bbb.97.

39 Srivastava, S., Dafale, N.A., Tulsani, N. et al. (2021). Evaluation of designed consortium SNH-1 for efficient hydrolysis of agriculture waste to benefit bioethanol production. *J. Clean. Prod.* 254: 120073. https://doi.org/10.1016/j.jclepro.2020.125601.

40 An, Y., Liu, Y., Liu, Y. et al. (2021). Opportunities and barriers for biofuel and bioenergy production from poplar. *GCB Bioenergy* 13 (6): 905–913. https://doi.org/10.1111/gcbb.12829.

41 Searle, S.Y. and Malins, C.J. (2014). Will energy crop yields meet expectations? *Biomass Bioenergy* 65: 3–12. https://doi.org/10.1016/j.biombioe.2014.01.001.

42 Bhatia, R., Lad, J.B., Bosch, M. et al. (2021). Production of oligosaccharides and biofuels from miscanthus using combinatorial steam explosion and ionic liquid pretreatment. *Bioresour. Technol.* 307: 123245. https://doi.org/10.1016/j.biortech.2020.124625.

43 Schmer, M.R., Vogel, K.P., Mitchell, R.B. et al. (2012). Temporal and spatial variation in switch grass biomass composition and theoretical ethanol yield. *Agron. J.* 104 (1): 54–64. https://doi.org/10.2134/agronj2011.0195.

44 Šiaudinis, G., Jasinskas, A., Šarauskis, E. et al. (2021). The influence of lime material and nitrogen fertilization on reed canary grass productivity, plant quality and environmental impact of using biomass for energy purposes. *Agronomy* 11 (5): 895. https://doi.org/10.3390/agronomy11050895.

45 Akhtari, S., Malladi, K.T., Sowlati, T., and Mirza, F. (2021). Incorporating risk in multi-criteria decision making: the case study of biofuel production from construction and demolition wood waste. *Resour. Conserv. Recycl.* 167: 105233. https://doi.org/10.1016/j.resconrec.2020.105233.

46 Glithero, N.J., Wilson, P., and Ramsden, S.J. (2013). Straw use and availability for second generation biofuels in England. *Biomass Bioenergy* 55: 311–321. https://doi.org/10.1016/j.biombioe.2013.02.033.

47 Groves, C., Sankar, M., and Thomas, P.J. (2018). Second-generation biofuels: exploring imaginaries via deliberative workshops with farmers. *J. Responsible Innov.* 5 (2): 149–169. https://doi.org/10.1080/23299460.2017.1422926.

48 Valdez-Vazquez, I. and Sanchez, A. (2018). Proposal for biorefineries based on mixed cultures for lignocellulosic biofuel production: a techno-economic analysis. *Biofuels, Bioprod. Biorefin.* 12 (1): 56–67. https://doi.org/10.1002/bbb.1828.

49 Mohan, S.V., Rohit, M.V., Chiranjeevi, P. et al. (2016). Integrating microalgae cultivation with wastewater treatment for biodiesel production. *Algal Biorefinery An Integr. Approach* 321–337. https://doi.org/10.1007/978-3-319-22813-6_15.

50 Jalilian, N., Najafpour, G.D., and Khajouei, M. (2020). Macro and micro algae in pollution control and biofuel production – a review. *Chem. Bio. Eng. Rev.* 7 (1): 18–33. https://doi.org/10.1002/cben.201900014.

51 Chong, T.Y., Cheah, S.A., Ong, C.T. et al. (2020). Techno-economic evaluation of third-generation bioethanol production utilizing the macroalgae waste: a case study in Malaysia. *Energy* 210. https://doi.org/10.1016/j.energy.2020.118491.

52 Lackner, M. (2016). 3rd-generation biofuels: bacteria and algae as sustainable producers and converters. *Handb. Clim. Chang. Mitig. Adapt. Second Ed.* 1: 4. https://doi.org/10.1007/978-3-319-14409-2_90.

53 Junginger, M., van Dam, J., Zarrilli, S. et al. (2011). Opportunities and barriers for international bioenergy trade. *Energy Policy* 39 (4): 2028–2042. https://doi.org/10.1016/j.enpol.2011.01.040.

54 Ganesan, R., Manigandan, S., Samuel, M.S. et al. (2020). A review on prospective production of biofuel from microalgae. *Biotechnol. Rep.* 27: e00509. https://doi.org/10.1016/j.btre.2020.e00509.

55 Debiagi, P., Faravelli, T., Hasse, C., and Ranzi, E. (2020). Kinetic modeling of solid, liquid and gas biofuel formation from biomass pyrolysis. In: *Production of Biofuels and Chemicals with Pyrolysis. Biofuels and Biorefineries*, vol. 10 (ed. Z. Fang, R.L. Smith Jr. and L. Xu). Singapore: Springer https://doi.org/10.1007/978-981-15-2732-6_2.

56 Vassilev, S.V., Vassileva, C.G., and Vassilev, V.S. (2015). Advantages and disadvantages of composition and properties of biomass in comparison with coal: an overview. *Fuel* 158: 330–350. https://doi.org/10.1016/j.fuel.2015.05.050.

57 Littlejohns, J., Rehmann, L., Murdy, R. et al. (2018). Current state and future prospects for liquid biofuels in Canada. *Biofuel Res. J.* 5 (1): 759–779. https://doi.org/10.18331/BRJ2018.5.1.4.

58 Changmai, B., Vanlalveni, C., Ingle, A.P. et al. (2020). Widely used catalysts in biodiesel production: a review. *RSC Adv.* 10 (68): 41625–41679. https://doi.org/10.1039/d0ra07931f.

59 Changmai, B., Rano, R., Vanlalveni, C., and Rokhum, L. (2021). A novel Citrus sinensis peel ash coated magnetic nanoparticles as an easily recoverable solid catalyst for biodiesel production. *Fuel* 286: 119447. https://doi.org/10.1016/j.fuel.2020.119447.

60 Laskar, I.B., Changmai, B., Gupta, R. et al. (2021). A mesoporous polysulfonic acid-formaldehyde polymeric catalyst for biodiesel production from Jatropha curcas oil. *Renew. Energy* 173: 415–421. https://doi.org/10.1016/j.renene.2021.04.004.

61 Fivga, A., Speranza, L.G., Branco, C.M. et al. (2019). A review on the current state of the art for the production of advanced liquid biofuels. *AIMS Energy* 7 (1): 46–76. https://doi.org/10.3934/ENERGY.2019.1.46.

62 Enagi, I.I., Al-attab, K.A., and Zainal, Z.A. (2018). Liquid biofuels utilization for gas turbines: a review. *Renew. Sust. Energ. Rev.* 90: 43–55. https://doi.org/10.1016/j.rser.2018.03.006.

63 Wijeyekoon, S.L.J. and Vaidya, A.A. (2021). Woody biomass as a potential feedstock for fermentative gaseous biofuel production. *World J. Microbiol. Biotechnol.* 37 (8): 1–15. https://doi.org/10.1007/s11274-021-03102-6.

64 Safieddin Ardebili, S.M. and Khademalrasoul, A. (2020). An assessment of feasibility and potential of gaseous biofuel production from agricultural/animal wastes: a case study. *Biomass Convers. Biorefinery* 1–10. https://doi.org/10.1007/s13399-020-00901-z.

65 Defries, R.S., Rudel, T., Uriarte, M., and Hansen, M. (2010). Deforestation driven by urban population growth and agricultural trade in the twenty-first century. *Nat. Geosci.* 3 (3): 178–181. https://doi.org/10.1038/ngeo756.

66 Peres, S., Loureiro, E., Santos, H. et al. (2020). The production of gaseous biofuels using biomass waste from construction sites in Recife, Brazil. *Processes* 8 (4): 457. https://doi.org/10.3390/PR8040457.

67 Frazão, C.J.R. and Walther, T. (2020). Syngas and methanol-based biorefinery concepts. *Chemie. Ingenieur. Technik.* 92 (11): 1680–1699. https://doi.org/10.1002/cite.202000108.

68 Pavlenko, A. and Slowak, A.M. (2020). Mathematical modelling of gasification processes of the biofuel in the conditions of chemical equilibrium. *IOP Conf. Ser. Earth Environ. Sci.* 467 (1): 012094. https://doi.org/10.1088/1755-1315/467/1/012094.

69 Agarwal, A., Rana, M., and Park, J.H. (2018). Advancement in technologies for the depolymerization of lignin. *Fuel Process. Technol.* 181: 115–132. https://doi.org/10.1016/j.fuproc.2018.09.017.

70 Sekar, M., Mathimani, T., Alagumalai, A. et al. (2021). A review on the pyrolysis of algal biomass for biochar and bio-oil – Bottlenecks and scope. *Fuel* 283: 119190. https://doi.org/10.1016/j.fuel.2020.119190.

71 Selvam, S.M. and Paramasivan, B. (2022). Microwave assisted carbonization and activation of biochar for energy-environment nexus: a review. *Chemosphere* 286: 131631. https://doi.org/10.1016/j.chemosphere.2021.131631.

72 Maher, K.D. and Bressler, D.C. (2007). Pyrolysis of triglyceride materials for the production of renewable fuels and chemicals. *Bioresour. Technol.* 98 (12): 2351–2368. https://doi.org/10.1016/j.biortech.2006.10.025.

73 Demirbas, A. (2008). Biodiesel: a realistic fuel alternative for diesel engines. In: 1–23. Switzerland: Springer Nature https://doi.org/10.1007/978-1-84628-995-8.

74 Kale, S.N. and Deore, S.L. (2016). Emulsion micro emulsion and nano emulsion: a review. *Syst. Rev. Pharm.* 8 (1): 39. https://doi.org/10.5530/srp.2017.1.8.

75 Rajkumari, K. (2020). A sustainable protocol for production of biodiesel by transesterification of soybean oil using banana trunk ash as a heterogeneous catalyst. *Biomass Conversion and Biorefinery* 10 (4): 839–848. https://doi.org/10.1007/s13399-020-00647-8.

76 Moser, B.R. (2009). Biodiesel production, properties, and feedstocks. *in vitro Cell Dev. Biol. Plant* 45: 229–266. https://doi.org/10.1007/s11627-009-9204-z.

77 Aziz, N.A.M., Yunus, R., Hamid, H.A. et al. (2020). An acceleration of microwave-assisted transesterification of palm oil-based methyl ester into trimethylolpropane ester. *Sci. Rep.* 10 (1): 1–17. https://doi.org/10.1038/s41598-020-76775-y.

78 Koech, A.K., Kumar, A., and Siagi, Z.O. (2020). In situ transesterification of spirulina microalgae to produce biodiesel using microwave irradiation. *J Energy* 2020: https://doi.org/10.1155/2020/8816296.

79 Colucci, J.A., Borrero, E.E., and Alape, F. (2005). Biodiesel from an alkaline transesterification reaction of soybean oil using ultrasonic mixing. *J. Am. Oil Chem. Soc.* 82 (7): 525–530. https://doi.org/10.1007/s11746-005-1104-3.

80 Bonet, J., Plesu, V., Ruiz, A.E.B. et al. (2014). Thermodynamic study of batch reactor biodiesel synthesis. *Rev. Chim.* 65 (3): 358.

81 Tabatabaei, M., Aghbashlo, M., Dehhaghi, M. et al. (2019). Reactor technologies for biodiesel production and processing: a review. *Prog. Energy Combust. Sci.* 74: 239–303. https://doi.org/10.1016/j.pecs.2019.06.001.

82 Fu, J. (2019). Flash points measurements and prediction of biofuels and biofuel blends with aromatic fluids. *Fuel* 241: 892–900. https://doi.org/10.1016/j.fuel.2018.12.105.

83 Zakaria, H., Khalid, A., Sies, M.F., and Mustaffa, N. (2014). Overview effect of biodiesel storage on properties and characteristic. *Appl. Mech. Mater.* 465: 260–264. https://doi.org/10.4028/www.scientific.net/AMM.465-466.260.

84 Poyadji, K., Stylianou, M., Agapiou, A. et al. (2018). Determination of quality properties of low-grade biodiesel and its heating oil blends. *Environ - MDPI* 5 (9): 96. https://doi.org/10.3390/environments5090096.

85 Valente, O.S., Pasa, V.M.D., Belchior, C.R.P., and Sodré, J.R. (2011). Physical-chemical properties of waste cooking oil biodiesel and castor oil biodiesel blends. *Fuel* 90 (4): 1700–1702. https://doi.org/10.1016/j.fuel.2010.10.045.

86 Zahran, m., Negm, N., Mahmoud Abbas, W., and Abd El-Raouf, M. (2020). Physicochemical characteristics and fuel performance of (Castor-Linseed biofuel)-petroleum fuel mixture. *Egypt. J. Chem.* 64 (2): 1035–1039. https://doi.org/10.21608/ejchem.2020.40373.2817.

87 Anguebes-Franseschi, F., Bassam, A., Abatal, M. et al. (2019). Physical and chemical properties of biodiesel obtained from Amazon sailfin catfish (Pterygoplichthys pardalis) biomass oil. *J. Chem.* 2019: https://doi.org/10.1155/2019/7829630.

88 Vassilev, S.V., Vassileva, C.G., Song, Y.C. et al. (2017). Ash contents and ash-forming elements of biomass and their significance for solid biofuel combustion. *Fuel* 208: 377–409. https://doi.org/10.1016/j.fuel.2017.07.036.

89 Hanifa, A., Nadia, A., Saputri, W.D. et al. (2021). Performance of ni-mo sulfated nanozirconia catalyst for conversion of waste cooking oil into biofuel via hydrocracking process. *Mater. Sci. Forum* 1045: 79–89. https://doi.org/10.4028/www.scientific.net/MSF.1045.79.

90 Abubakar, H.G., Abdulkareem, A.S., Jimoh, A. et al. (2016). Optimization of biodiesel production from waste cooking oil. *Energy Sources Part A* 38 (16): 2355–2361. https://doi.org/10.1080/15567036.2015.1040899.

91 Coronado, M.A., Montero, G., García, C. et al. (2017). Quality assessment of biodiesel blends proposed by the new Mexican policy framework. *Energies* 10 (5): 631. https://doi.org/10.3390/en10050631.

92 Kalantar, A. and Levin, M. (2008). Factors affecting the dissolution of copper in transformer oils. *Lubr. Sci.* 20 (3): 223–240. https://doi.org/10.1002/ls.63.

93 Achenbach, E., Ag, B., Almeida, E.S. et al. (2010). Standard test method for corrosiveness to copper from petroleum products by copper strip test 1. *Fuel Process. Technol.* 05 (01): https://doi.org/10.1520/D0130-19.

94 Gangadhar, K.N., Pereira, H., Diogo, H.P. et al. (2016). Assessment and comparison of the properties of biodiesel synthesized from three different types of wet microalgal biomass. *J. Appl. Phycol.* 28 (3): 1571–1578. https://doi.org/10.1007/s10811-015-0683-5.

95 Da Silva, N.D.L., Santander, C.M.G., Rueda, S.M.G. et al. (2013). Characterization of blend properties of castor biodiesel and bioethanol. *Ind. Eng. Chem. Res.* 52 (44): 15504–15508. https://doi.org/10.1021/ie400680t.

96 Altalhi, A.A., Mohamed, E.A., Morsy, S.M. et al. (2021). Catalytic manufacture and characteristic valuation of biodiesel-biojet achieved from Jatropha curcas and waste cooking oils over chemically modified montmorillonite clay. *J. Mol. Liq.* 340: 117175. https://doi.org/10.1016/j.molliq.2021.117175.

97 Bejan, C.C.C., Celante, V.G., De Castro, V.R., and E Pasa, VMD. (2014). Effect of different alcohols and palm and palm kernel (palmist) oils on biofuel properties for special uses. *Energy Fuel* 28 (8): 5128–5135. https://doi.org/10.1021/ef500776u.

98 Wierzbicki, V. (2010). Determining the oxidation stability of biodiesel and blends using a new rapid small scale oxidation test (RSSOT) – The PetroOXY. *J. ASTM Int.* 7 (4): 1–8. https://doi.org/10.1520/JAI102578.

99 Schober, S. and Mittelbach, M. (2007). Iodine value and biodiesel: is limitation still appropriate? *Lipid Technol.* 19 (12): 281–284. https://doi.org/10.1002/lite.200700091.

100 Munari, B.F., Cavagnino, D., and Scientific, T.F. (2009). Determination of total FAME and linolenic acid methyl ester in pure biodiesel (B100) by GC in compliance with EN 14103. *Test* 10212.

101 Louzeiro, H.C., Soledade, L.E.B., Souza, A.G. et al. (2017). Free glycerin determination in ethanol biodiesel utilizing spectrophotometric and chromatographic (gc/fid) methods. *Eclética Química J.* https://doi.org/10.26850/1678-4618eqj.v38.1.2013.p25-36.

102 Rathore, V., Tyagi, S., Newalkar, B., and Badoni, R.P. (2015). Jatropha and Karanja oil derived DMC-biodiesel synthesis: a kinetics study. *Fuel* 140: 597–608. https://doi.org/10.1016/j.fuel.2014.10.003.

103 Weinebeck, A., Kaminski, S., Murrenhoff, H., and Leonhard, K. (2017). A new QSPR-based prediction model for biofuel lubricity. *Tribol. Int.* 115: 274–284. https://doi.org/10.1016/j.triboint.2017.05.005.

104 de Oliveira, E.C., de Santana Maia, L.A., da Costa, L.G., and Lourenço, T.C. (2018). Validation of new methodology for the definition of tolerance limits of critical properties in fuels handled in terminals and pipelines – giveaway: compliance with the specification. *Accred. Qual. Assur.* 23 (6): 365–369. https://doi.org/10.1007/s00769-018-1353-5.

105 Hoekman, S.K., Broch, A., Robbins, C. et al. (2012). Review of biodiesel composition, properties, and specifications. *Renew. Sust. Energ. Rev.* 16 (1): 143–169. https://doi.org/10.1016/j.rser.2011.07.143.

106 Sorate, K.A. and Bhale, P.V. (2015). Biodiesel properties and automotive system compatibility issues. *Renew. Sust. Energ. Rev.* 41: 777–798. https://doi.org/10.1016/j.rser.2014.08.079.

107 Lai, B., Yi, P., Sui, Y., and Zhang, Q. (2021). Energy distribution in EV energy network under energy shortage. *Neurocomputing* 444: 179–188. https://doi.org/10.1016/j.neucom.2020.08.090.

108 Senatore, A., Dalena, F., Sola, A. et al. (2019). First-generation feedstock for bioenergy production. *Second Third Gener. Feed. Evol. Biofuels* 35–37. https://doi.org/10.1016/B978-0-12-815162-4.00002-1.

109 Taheripour, F., Cui, H., and Tyner, W.E. (2019). The economics of biofuels. Routledge. *Handb. Agric. Econ.* 638–657. https://doi.org/10.4324/9781315623351-34.

110 Bryan, B.A., King, D., and Wang, E. (2010). Biofuels agriculture: landscape-scale trade-offs between fuel, economics, carbon, energy, food, and fiber. *GCB Bioenergy* 2 (6): 330–345. https://doi.org/10.1111/j.1757-1707.2010.01056.x.

111 Levasseur, A., Lesage, P., Margni, M. et al. (2010). Considering time in LCA: dynamic LCA and its application to global warming impact assessments. *Environ. Sci. Technol.* 44 (8): 3169–3174. https://doi.org/10.1021/es9030003.

2

Ethanol as the Leading 'First-Generation' Biofuel

2.1 Introduction

Over the past few decades, countries all over the world have been continuously engaged in environmental protection, securing global energy sources and new income opportunities in the transportation sector (Figure 2.1). Sustainable energy sources and cleaner environment are the primary focus of all countries [2, 3]. Countries with low carbon footprint drew the attention of the world from a technological revolution point of view. Finite reserves of fossil fuels, high energy prices, increased population and global industrialisation have spurred the search for alternative, renewable biofuels [4]. These substitutes for biofuels must meet the demand of net energy output, environmental health, should be economically viable and should be producible in a large amount. Biofuels play distinct roles amongst all energy sources due to strong interrelations between agriculture, energy production, social development and green environment [5, 6]. Amongst all natural energy sources, such as wind, solar, hydroelectric power and geothermal power, biomass created its own space in the revolution successfully substituting the fossil fuels to a certain extent.

Biofuels produced from biomass or organic parts of the plants could reduce world's dependence on fossil fuel and reduce CO_2 emission because the feedstock has already used CO_2 for the growth [7]. It has been believed that it has potential to fulfil 25% world's fuel demand by 2035, which will also be helpful from a social, economical and environmental point of view.

Although this is a well-accepted fact all over the world, the use of fossil fuel has been observed to be increasing by approximately 7–10% since 1980. The lowest initial investment in technology part for using fossil-based fuels may be the reason behind this [8]. Fortunately, awareness regarding environment and cost-effective studies keeping this number minimum, and constant research is going on minimising this number in the future, increasing the energy consumption and utilisation simultaneously.

The world's transport heavily relies on crude oil and the availability and supply of gasoline [9] (Figure 2.2). Over the past 30 years, government and industry have invested heavily in the research and development of the new green fuel technology. It is not surprising that investment and development have been increased considerably. This dependence also generated tremendous socioeconomic impacts. During the 1970s, major countries such as

Sustainability in Biofuel Production Technology, First Edition. Pratibha S. Agrawal, Pramod N. Belkhode, and Samuel Lalthazuala Rokhum.

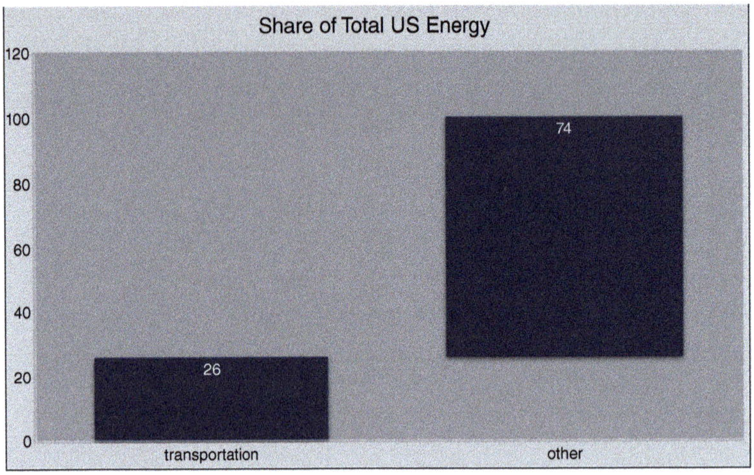

Figure 2.1 Share of the total US energy. *Source:* Adapted from [1].

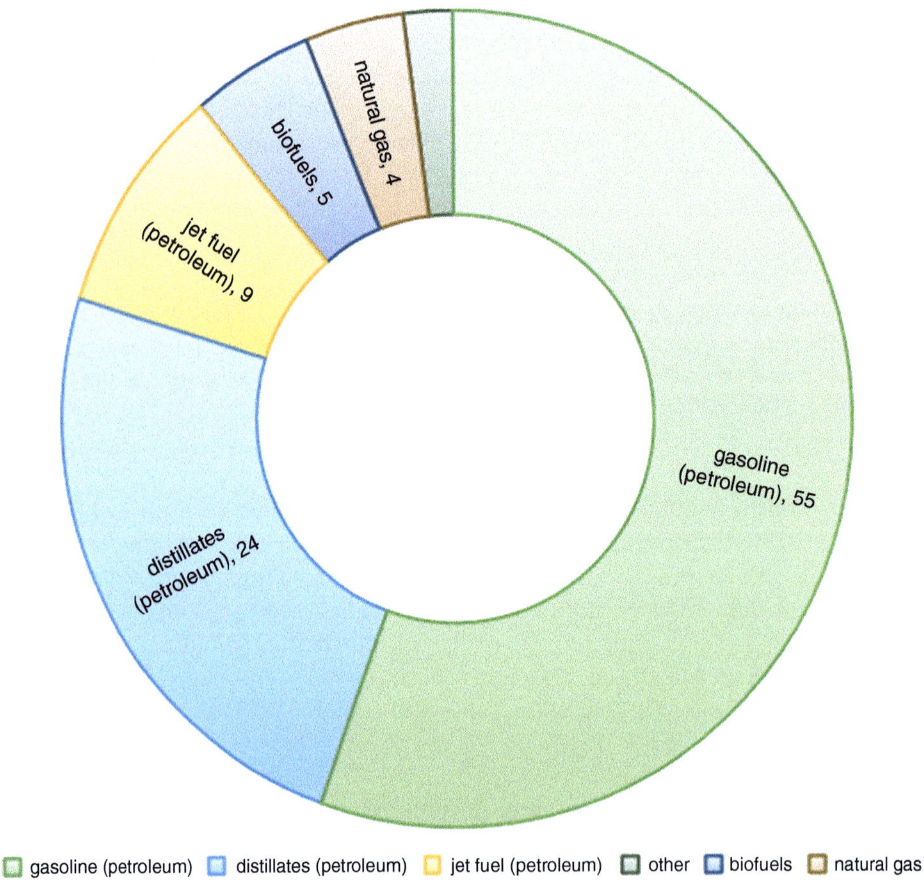

Figure 2.2 US transportation energy sources.

United States, Western Europe and Australia also faced fuel shortage problems due to Iranian Revolutions, and citizens' growing environmental concern [10]. Therefore, it has become urgent to develop a substitute fuel that can fit to the gasoline properties without any change in the existing engine set-up.

The combustion of fossil fuels has been responsible for an increase in the carbon dioxide emission in the atmosphere and thus contributing to global warming. Many countries started directing the use of biomass and decreasing the use of fossil fuel as stated in the Kyoto Protocol [11]. Government agencies, NGOs, research departments and ministry of renewable energy impose various policies and regulations in the field of biofuel technology. Biofuel has become the thirst area of research satisfying the future needs [12, 13]. Government policies encourage the suppliers of agricultural feedstock for biofuels, the industries for the production of biofuels and its consumers by setting blending instructions. However, depending on the development in this technology, various generations of biofuel have come into existence. Bioethanol – biodiesel from plant – was hence developed as a first-generation biofuel. The first-generation biodiesel [14] is produced by conventional pressing using the feedstock of different plants such as sugar-rich crops, starch-containing crops, vegetable oil crops or non-edible oil crops also known as straight vegetable oil. It mainly includes bioethanol, biodiesel, biogas, oils from plant products, agro waste, etc. The second-generation biofuels include lignocellulosic-based biofuels produced from agro waste vegetables, third-generation biofuels include algae-based feedstock, and more advanced biofuels will be termed as green diesel, which is used as aviation fuel [15]. Thus, biofuels are broadly classified into different groups as shown in Figure 2.3.

As compared to fossil fuels, biofuels have the following benefits:

a) they are renewable [16] sources,
b) feedstock for biofuels are plentifully available [17],
c) it limits carbon dioxide emission [18, 19],
d) they are responsible for negligible SOx emission [20],
e) they minimise NOx emissions [21, 22],
f) they are termed as environmentally friendly fuels [23],

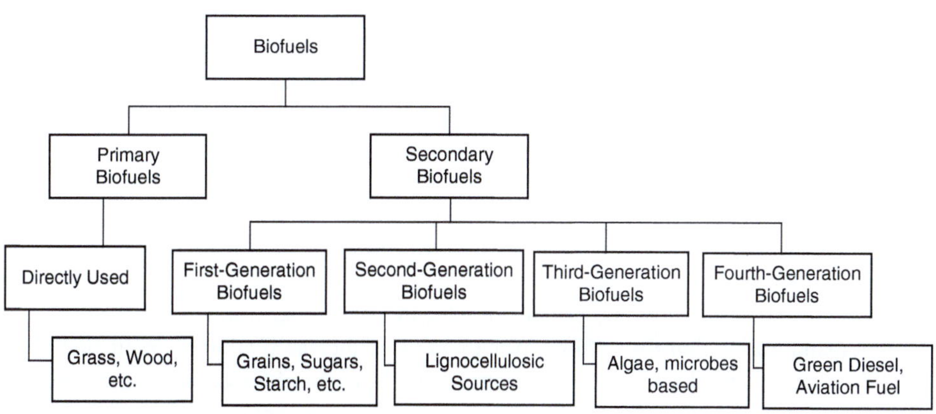

Figure 2.3 Classification of biofuels.

g) they are biodegradable and sustainable [24],
h) their production process is fast and safe,
i) they help in rural developments and employments [25] and
j) they reduce dependence on energy imports [26].

In addition to the above advantages, biofuels in transport show numerous benefits. It can be used with reduced carbon footprints compared to fossil fuels [27, 28]. As compared to electric or other fuel cells requiring novel infrastructure, biofuels proved to be an economical one [29, 30]. In addition, this technology can foster numerous employments, and thus it can help rural area people in their regular struggle for survival.

There are numerous definitions available for biofuels depending on the sources (living organism, plant materials and biomass) from which they have been produced. Biofuel is the fuel that is specifically produced from renewable living organism in global meaning. The basic requirement is that it should be renewable, produced from living parts and can be used as energy [31]. Technically, biofuels can only replace part of the liquid fuel without any changes in the engine technology. However, whole replacement is still not feasible and needs further development in innovative technologies. Taking into account all situations, the biofuels are necessary to be tackled carefully to fulfil the next-generation demand. This chapter gives a comprehensive introduction to bioethanol.

2.2 Historical Development of Ethanol as Biofuels

Bioethanol, CH_3–CH_2–OH, is a liquid biofuel with the same molecular formula as that used in alcoholic drinks [32]. The history of ethanol has been deep routed before human civilization, although it is well described in the Energy Information Agency (2005). Biofuels always served the human need for lightening, burning, heating and cooking in the form of wood and grass. It was thought that ethanol was discovered when the oil prices hacked in the 1970s [33], but the fact is that ethanol (blended with turpentine) was first used to power an internal combustion engine in 1826 by Samuel Morey. He was able to drive the boat at a speed of 7–8 mph and also filed a patent, but the engine still needs further research [34]. Earlier, it was used in 1850 as a lightening fuel blended with turpentine oil replacing the whale oil, but its use was curtailed due to high taxes during war time.

It was Nicolaus Otto, in 1876, who used elixir containing ethanol in a combustion power engine. He was funded with a government grant and owned a sugar refining company putting ethanol at a good place in the European market [35].

After that phase, ethanol was re-used again in 1908 in Model T, especially for designed biofuel vehicles. Again, the world faced a fuel shortage problem, and ethanol-blended fuels have been used as an alternative solution [36]. During the 1920s–1930s, a new programme ‘Chemurgy’ was supported by Henry Ford, which focused on the use of crop for bio-based fuel production [37]. In addition, the production of synthetic rubber from such ethanol was on high demand during that time. World War I witnessed increased ethanol production above 50 million from natural resources, which was further also used for domestic purpose. This increase was raised 10-fold during World War II. The realisation of senses of using vegetable oil in diesel engine motivated oil-crop-producing countries from 1940 onwards. In 1942, Ford invented soya bean car by combining and strengthening the automobile and agricultural developments. Thereafter, depending upon the types of feedstock available, geographical and economic concern, and industrial growth in any country, bioethanol production started [38]. The USA produced soya bean oil, whereas European countries produced canola-based oil.

In 1970, ethanol production industries started due to the abundance of feedstock and the ease of transformation of production of ethanol from corn. These attempts also strengthened rural communities. This demand drastically increased further. In 1978, South Dakota State University provided funds for the farm-scale ethanol production plant [39]. A notable contribution was provided by Dr Middaugh (1979) from the USA, who opened the first operating dry mill ethanol plant. He received a research grant of approximately one lakh dollars from government for developing the training and research centre, leading to various innovations that are still in existence [40].

Meanwhile, other companies also started to grow their stocks in the Midwest region. Jeff Broin, in 1986, developed an ethanol plant producing 100000 gallons/year of ethanol. He extended his efforts to develop 26 more ethanol plants in seven states [41, 42]. VeraSun by Don Endres in the USA (2001) became the largest ethanol producer, producing 1.64 billion gallons per year across different states [43]. It increased to more than 200 until 2015. Further development in biorefineries at the global level is presented in Table 2.1 and Figures 2.4 and 2.5 [44].

Table 2.1 Overview of the world ethanol market in the past 10 years.

Year	World ethanol trade in billions of litres	World ethanol production in billions of litres
2010	6	104
2011	10	103
2012	9	101
2013	8	109
2014	7	115
2015	7	119
2016	9	118
2017	10	120
2018	9	123
2019	9	124
2020	9	125
2021	9	126
2022	9	127
2023	9	128
2024	9	129
2025	9	130
2026	9	130
2027	9	131

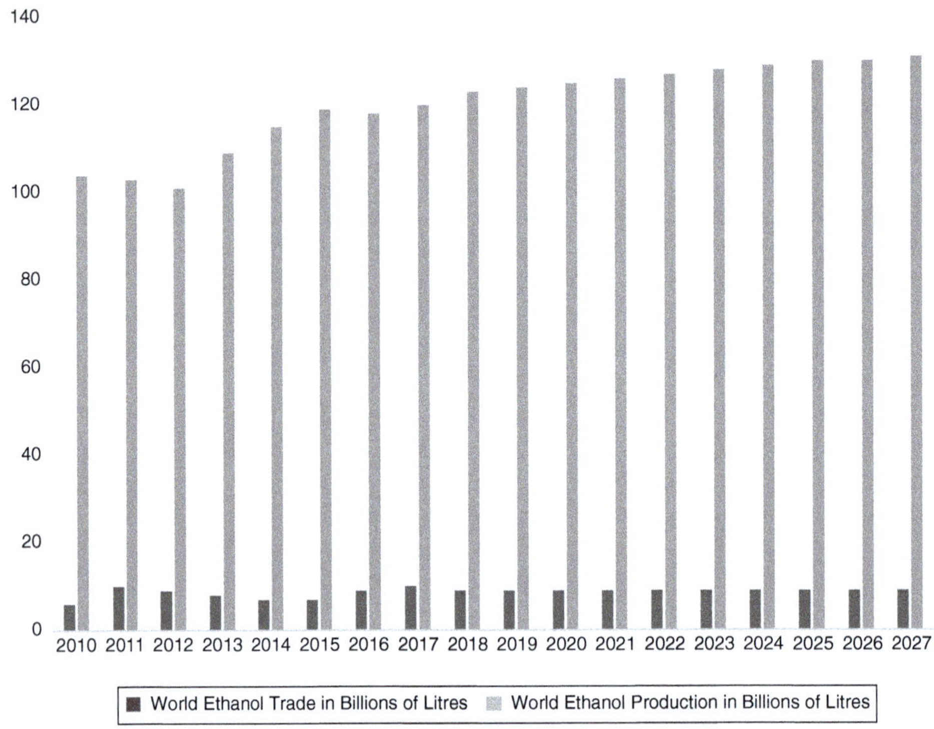

Figure 2.4 Development of the world ethanol market. *Source:* Adapted from [44].

Restrictions due to the COVID-19 epidemic have also negatively affected the bioethanol market. The worldwide spread of COVID-19 in late 2019 sharply affected the demand and production of fuels due to restricted transportation, which has resulted in bringing down the price of fuel as shown in Figure 2.6. The global-level biofuel industry has been strongly impacted by the COVID-19 pandemic, when compared to previous years. Most of the reduction is expected in European markets. If transportation and policy can be well managed in pandemic situations, then the global production of biofuels used for transport can again reach that of 2019 in 2021, approximately 165 billion litres. This growth is expected in Asian and South American countries with Brazil, China and India making a strong place in growth markets [46]. However, many countries have to opt for blended biofuels for solving the conflict of 'food versus fuel' conflicts.

The sharp decrease in the price of ethanol and gasoline affected ethanol producers, thereby also shifting the profit margin. In addition, social distancing has negatively affected the transportations that adversely affected the demand for fuels.

Actually, biofuel refers to the liquid fuel extracted from plant sources. Thus, the production of biodiesel relates agricultural as well as processing industrial sector. Liquid biofuels are mainly used in three major areas – heat production, electric generation and transportations. In the first two areas, biofuels are used in the stationary diesel engine but in the third area two types of engines, viz constant volume – spark ignition engine (used in lighter

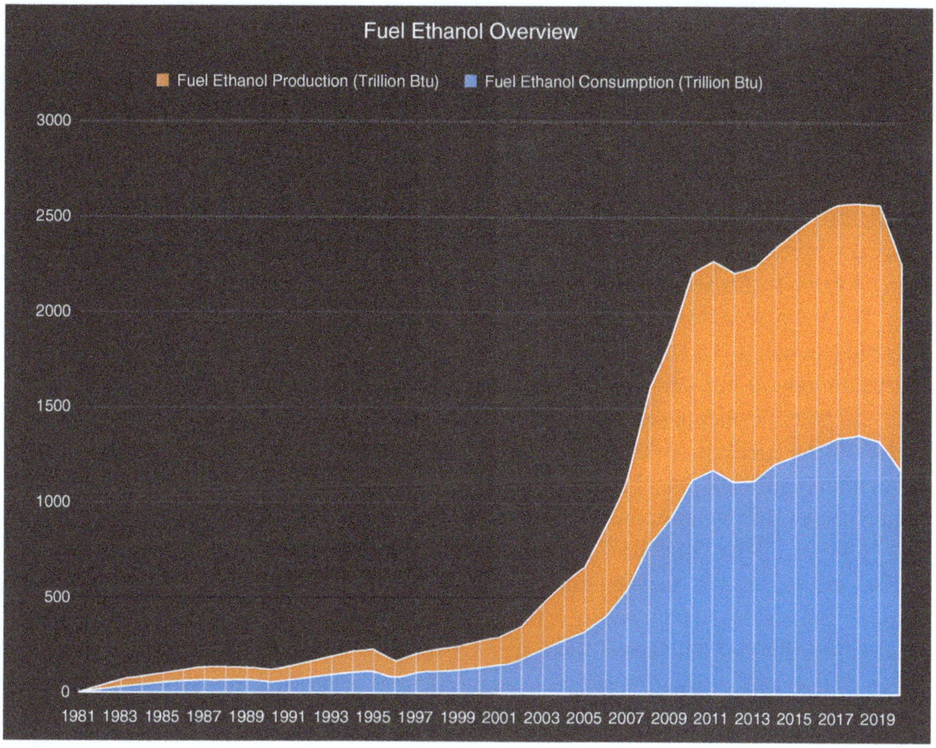

Figure 2.5 Ethanol production and consumption review.

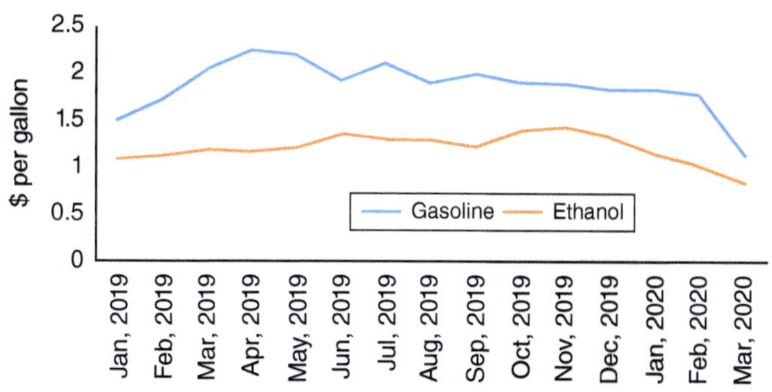

Figure 2.6 Unleaded gasoline and ethanol prices, January 2019 to March 2020. *Source:* Adapted from [45].

vehicles) – and constant pressure – compression ignition engine (used for heavier vehicles), exist. Thus, once again biofuels have been brought to the spotlight. Bioethanol has already been used in Brazil as a gasoline blend, but thereafter it gained its own space into the international market [47]. Though hydrous bioethanol (95–96%) is not very much applicable, but anhydrous bioethanol (99%) stands good with all mixtures of gasoline. Bioethanol proved to have better combustion efficiency with a higher octane number, increased flame speed and heat over conventional fuels [48].

The first-generation biofuel was produced similar to that of commercially practiced alcoholic beverages by a fermentation process that was used for berries, grapes, honey and cereals. However, the different taxation scheme enacted to fund civil war made it costly, and hence a slight change in production was practised. Bioethanol can be produced from a number of agricultural origins such as sugar-based plants (sugar cane), cereal-based plants (maize and wheat) and starch-based plants (potatoes) [49].

Ethanol for fuel was treated with a denaturation process by adding gasoline hydrocarbons, making it non-edible. If feedstock contains complex carbohydrates, it can be easily cracked using a catalyst. An acid or enzymatic hydrolysis process is used before the fermentation step. The main process of production was the biochemical fermentation process changing sugar to alcohol. It can be carried out using bioenzymes, microorganism like yeast [50, 51]. Fermentation can be carried out by batch, fed batch or continuous method.

In the batch fermentation process, the liquid fraction of the feedstock juice is directly added to the medium at the beginning of the fermentation process. In the fed-batch fermentation process, one or more ingredients are added as the fermentation progresses. In the continuous fermentations process, ingredients are constantly added at a specific flow rate and time interval; the products are also removed continuously from the reactors.

The chemical and physical technology involved in ethanol production from corn is more complicated and multistep than that produced from sugar cane (in which the sucrose part is already available). The ethanol produced from corn is five times higher in volume per tonne of feedstock, while the ethanol produced from sugar cane has greater productivity per hectare [52]. The chemical reaction pathway is presented in Figure 2.7.

Thus, theoretically, parts of any plant can be used as feedstock for bioethanol production. Lastly, the distillation process can separate bioethanol to 85–95%, producing hydrous bioethanol. Azeotropic distillation under 0.1 atm can be costly but useful in producing anhydrous bioethanol [53]. Instead, some solid potential adsorbents like starchy flours can also be used for the above purpose. Such bioethanol can be directly used instead of gasoline in engines. However, the lower heating value (LHV) of ethanol (21.1 MJ/l) as compared to gasoline (30–33 MJ/l) needs more amount of ethanol for the same energy outputs [54].

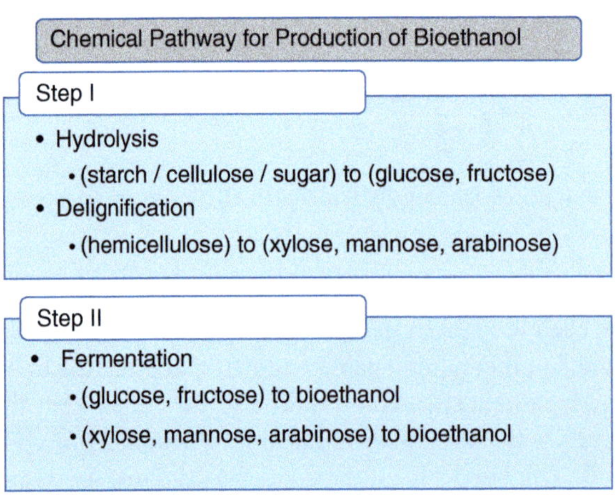

Figure 2.7 The chemical reaction pathway for the production of bioethanol.

However, a higher octane number of ethanol still improves the thermal efficiency and power output, making it sustainable for use.

2.2.1 Feedstock Used for First-generation Bioethanol Production

Depending upon the type of crops produced and the geographic location of the country, different kinds of feedstock can be used in bioethanol production. Cereal grains are the most important crop used in this process. It contains approximately 75% w/w starch, thus proving that it is a rich source for sugar production [55]. However, sugar cane is also preferred as it can directly undergo the fermentation process and directly produces sugar. Maize and wheat do not prove to be a sustainable solution for long-term use, but the crops such as rapeseed, cassava and sugar beet stand well in between its use.

2.2.1.1 Sugar Cane and Sugar Beet as Feedstock

Sugar cane and sugar beet with a juicy sucrose composition are the most commonly used sources for ethanol production. Sugar cane is a semi-perennial plant belonging to the grass family, grown in tropical and subtropical countries, while sugar beet is only grown in moderate climate countries. As it directly produces sugars, direct fermentation can be performed for avoiding an initial heating step (a saccharification step) that is required for starch-containing feedstock. This makes the production process simpler. Brazil produces almost 80% of its sugar cane, proving to be the largest producer of feedstock, with an average increase rate of 1.3% per year in production. The Brazilian government motivated the addition of 25% ethanol to gasoline (E25) and promoted renewable green sources of fuels [56]. Sugar cane can produce about 6000 l of ethanol/hectare. It became world's largest bioethanol-producing country using sugar cane by a simple process of fermentation. However, need, demand and supply of ethanol will rise to 104 billion litres in 2025, which will require a reduction in production costs to withstand the market value [57, 58]. In addition to the cost consideration, advanced technology can provide better environmental performance and greater productivity per unit of land, facilitating its sustainability for the long term.

Sugar beet is a crop from the Amaranthaceae family, which grows in deep soils with neutral pH, high water retention and good aeration. Northwestern Europe uses wheat and sugar beet, while Central Europe and Spain lead in the use of corn. Sugar beets are composed of about 75% water, 18% sugar and 7% insoluble and soluble materials. New sugar beet varieties are being constantly investigated to reduce bioethanol production costs from sugar beet. In the production method (Figure 2.8), separating sugar beet pulps before fermentation improves the equipment utilisation of fermentation and distillation, saves energy consumption and also decreases the production cost.

Sweet sorghum is also preferred as it needs comparatively less water for growth. Sweet sorghum sugars consist of 85% sucrose, 9% glucose and 6% fructose, out of which only sucrose may readily be converted to white sugar. Storage is the major drawback of sugar containing crops. It cannot be stored for a longer time. As they are rich in water, the sugar content quickly decreases once it is cut from the farmland. The ethanol production process is shown by schematic pathway in Figure 2.9.

In the production industry, the sugar cane juice is heated to 1100 °C, concentrated and then fermented. If any sucrose is produced, meanwhile, it can be removed by the centrifuge

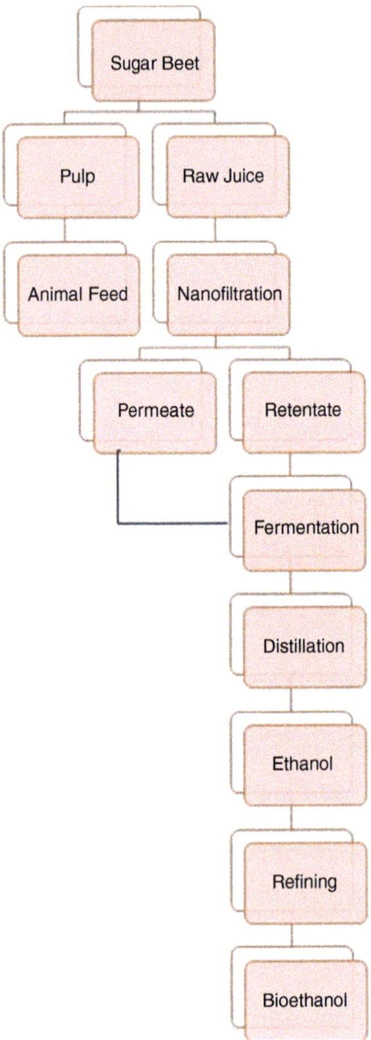

Figure 2.8 Ethanol production from sugar beet.

technique. Then, 1 g of glucose was converted to 0.511 g ethanol. The actual conversion depends on the alcohol content; the higher alcohol content in the feed stock optimises the energy consumption and the ethanol production.

2.2.1.2 Corn as Feedstock

It is a grain plant of the grass family and the genus Zea, originated in Mexico but is widespread in all the continents [59]. Maize crops currently occupy approximately 147 million hectares worldwide [60]. It is used as human as well as animal food with high nutritional value. The growth and crop of this grain is governed by water, temperature and sunlight. Ethanol producers usually separate the corn cob that are left in the fields to improve soil fertility. The grains are transported to the corn mills that allow the storage for 7–12 days and send it to the processing plant to remove debris before milling.

There are two basic production processes of ethanol from corn: dry milling and wet milling. Though the initial treatment of the grain is different in both the processes, but both the processes involve a previous basic step of hydrolysis that breaks down the starch for obtaining glucose syrup, which can be converted into bioethanol by yeasts. Due to the lower capital costs of the dry milling technology, it is more widely used worldwide. In this process, the corn kernel is ground into a powder and mixed well with water and liquefying enzymes (amilase) to break down the starch into simple sugars. pH is well controlled by ammonia addition in the required quantity, which is also used as nutrient for the yeast in the post-fermentation step. The liquid is then cooked to avoid bacterial contamination followed by the saccharification step (starch to glucose conversion) [61, 62]. Co-products of the dry milling process are distiller grains, which are used as animal feed.

In the wet milling, the kernel is soaked in diluted sulphuric acid that separates the germ, gluten, fibre and starch, and the remaining liquor is concentrated and dried, which is sold as feed to the livestock industry. The glucose-rich mash obtained after the saccharification step proceeds to the fermentation process (Figure 2.10).

The USA uses corn feedstock primarily for bioethanol production. More than one-third of USA's corn crop is used to feed livestock, 13% is exported and 40% is used to produce ethanol. The remainder is used in food and beverage production.

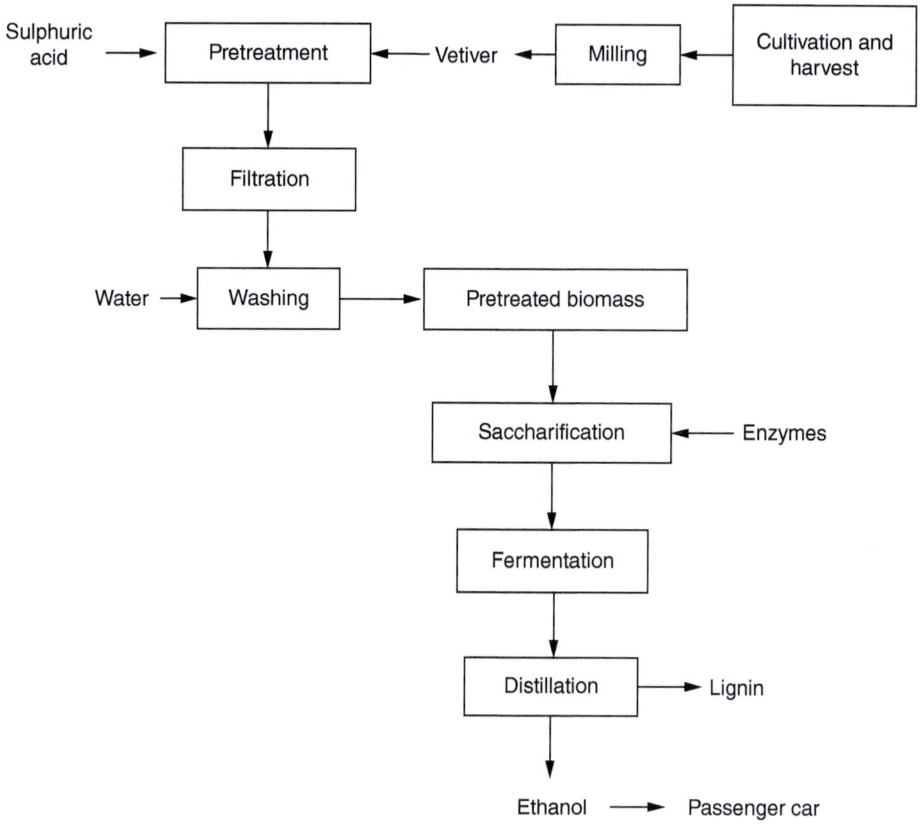

Figure 2.9 Ethanol production from sugar cane.

2.2.1.3 Cassava as Feedstock

Cassava is used by both human and animal as food, in many industrial sectors, (in the form of starch), and recently it is used to produce ethanol [63] (Figure 2.11)

Africa contributes to more than 50% of the global supply of cassava feedstock. Asia promotes the development of cassava crops for industrial and energy purposes. Cassava is still the least performer on the biofuel market. For example, cassava roots, which have a starch content of 30% (w/w), can generate 180 l pure ethanol (96%) per tonne of raw material (500–4000 l per hectare per year) [64, 65].

A generalised flow chart is schematically presented in Figure 2.12.

A significant difference amongst various physical characteristic properties has been still observed between bioethanol and fossil fuels. To improve the fuel quality and make it feasible for use, bioethanol is usually blended with gasoline, as bends of E5–E25 (5–25% ethanol in 95–75% gasoline) require the same diesel engine equipment. It can also be used as additives to increase the oxygen percentage in fuel, which reduces the CO and other aromatic emissions. The two most important liquid biofuels used in transport are bioethanol (substitute for gasoline) and biodiesel (substitute for diesel).

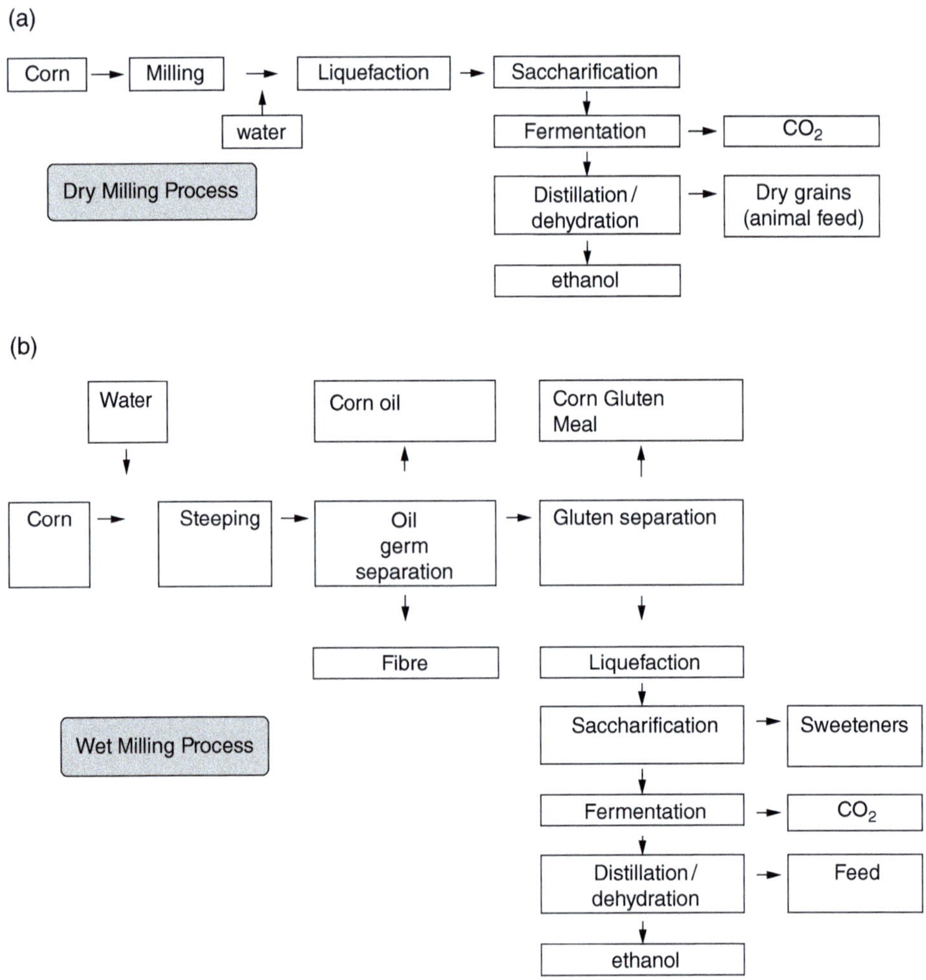

Figure 2.10 Ethanol production from corn (dry and wet milling processes).

2.3 Environmental Aspects of Using Ethanol as Biofuels

In order to achieve success in this field, various factors that hamper and boost the biofuel production should be taken into consideration [66, 67] (Figure 2.13).

Various factors discussed are as follows.

2.3.1 Adaptive Capacity

adaptation of any new technology depends on how smoothly it is diffused and understood to users. Any set-up or improvement in the technology requires new infrastructure and changes in social values. The whole chain involved in the technology should support the

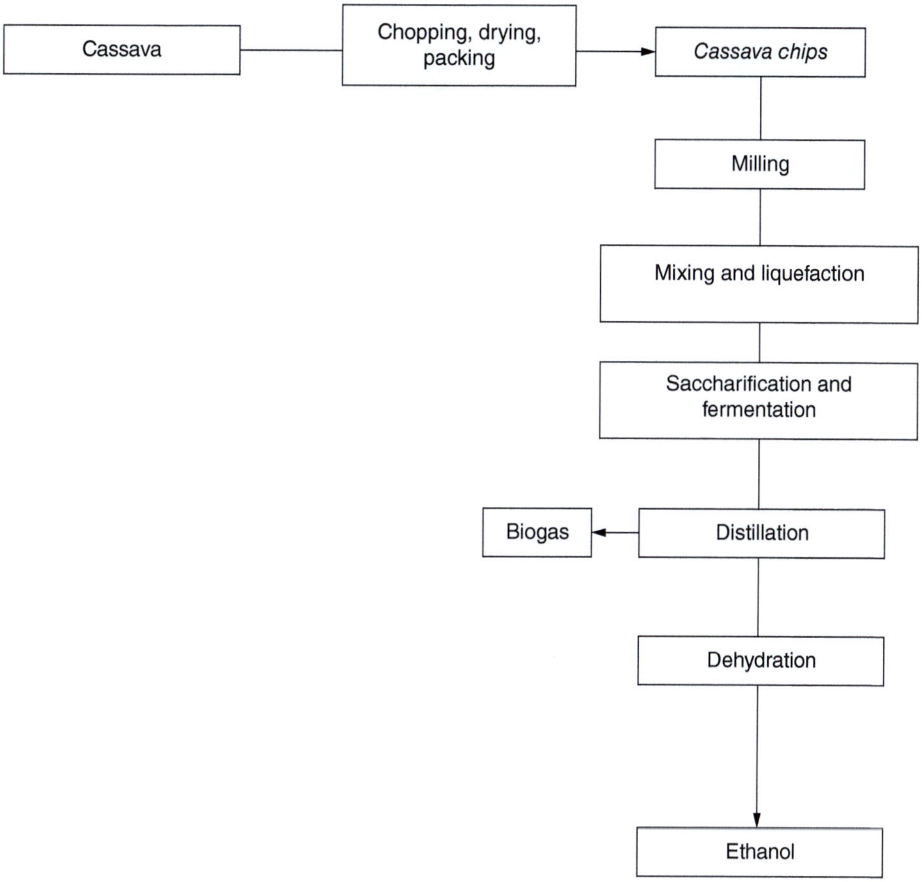

Figure 2.11 Ethanol production using cassava.

development by creating awareness and relevant knowledge. Related society should be smooth enough to adopt new technologies. It could be achieved by supplying the related materials, seminars and trainings at every stage of development.

2.3.2 Biofuel Technology

One has to take brief account of various technologies being used as per the demand and needs of geographical conditions. Any minute improvement in technology may seek attention and can result in a higher profit side [68].

2.3.3 Geographical Impact

As the technology is crossing national boundaries, the specific analysis needs to be considered. Geographical, social and cultural aspects are also very important in the analysis and implementation of the productivity rate.

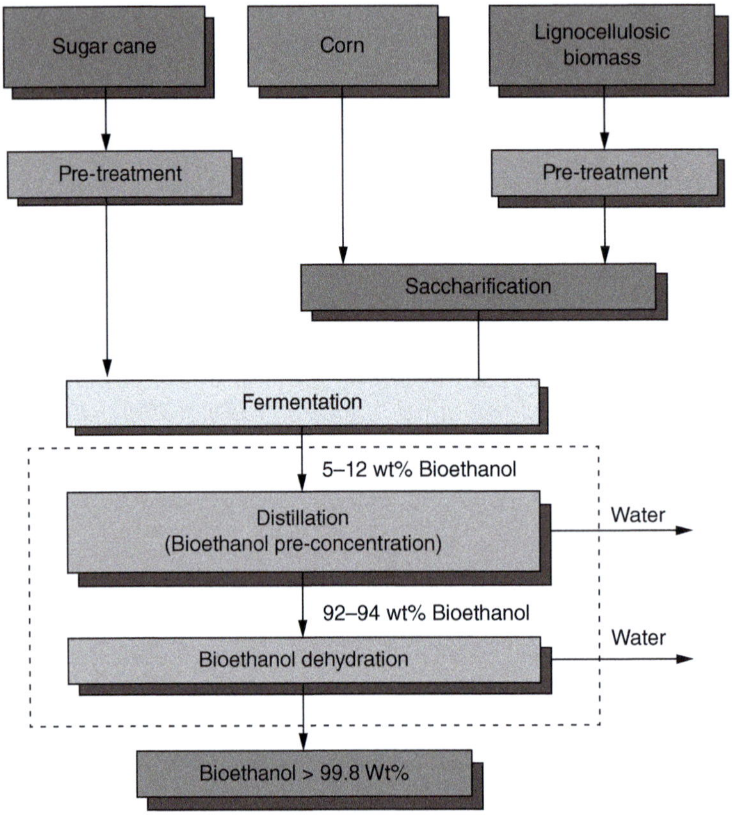

Figure 2.12 Flow chart for first-generation bioethanol.

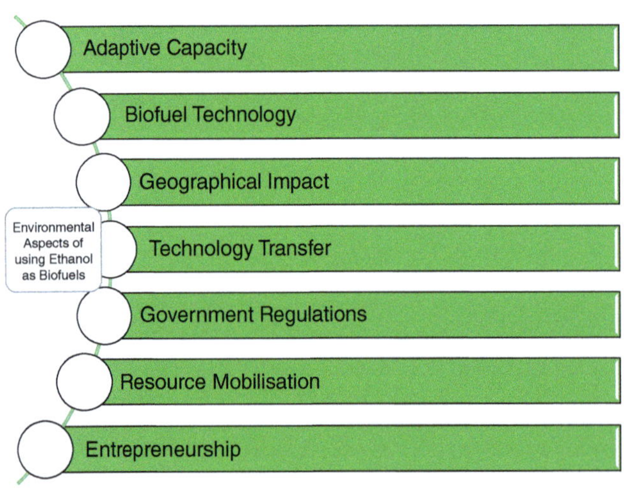

Figure 2.13 Environmental aspects of ethanol as biofuels.

2.3.4 Technology Transfer

Biofuel technology is a newly explored green technology to substitute fossil fuel technology. The innovative technologies should be supported and effectively transferred to the society by regular R&D meetings between different stakeholders. International, national and regional networks should be strong enough to support the sharing of valuable knowledge for further successive development.

2.3.5 Government Regulations

Any national energy policy aims towards the ratio of energy consumption and country's economic growth. Different supports and incentives can be provided to the parties that adapt the new energy technology. Various existing policies framing the regulation may vide after the adoption of new technology. The long-term national vision should be kept in mind, and the policies can be accordingly reframed. By adopting such eco-friendly techniques, any country can achieve various targets such as creation of job opportunities, increasing income levels above the minimum, biofuel feedstock plantation on unused land, reducing the fossil fuel use and thereby decreasing the energy crisis [69, 70].

2.3.6 Resource Mobilisation

Various resources, including natural, human, economic, and social, are the key aspects of innovation systems. Due importance should be given to access, functioning, and mobilising of these resources. Both domestic and foreign investors can be attracted for providing the financial support to the developing technology. MoU, collaborations and partnerships can be signed between international and national companies to strengthen the technology implementation process. Young and literate young force of the country can be educated and trained so that they can also put their share by solving the problem of unemployment. Professionals and researchers with the interest of renewable energy can contribute in reducing climate change and negative environmental impacts.

2.3.7 Entrepreneurship

A very little scope and space is provided for new technology in the market. Especially when the routine technology is in function, it cannot be replaced so easily. The high expectations of governments, focused target, and energy solutions can attract investors and human resources to work around the technology. New companies and small-scale plants can be established expanding the biofuel networking.

Though environmental studies on bioethanol production on a large scale reflect net favourable energy output, they do not contribute to the most sustainable solution for future demand and environmental support [71]. A close analysis and interaction amongst all of the above factors needs to be cross checked by expertise. One should always ensure that there are not too many differences between the new and existing technologies. Otherwise, extra investment, raw material supply and related whole chain may get disturbed. A positive feedback loop produced in this way will surely help in the successful implementation of new technologies.

2.4 Cost Models of Ethanol as Biofuels

The end users are usually sensitive to prices and tend to choose fuels based on the cost/unit energy. This is important as it gives clear-cut vision about the production cost that will influence public willingness to pay for it. However, to obtain the actual cost prediction is very difficult as it highly depends on geographical and climate area, technologies available for conversion, government subsidy plans, existing fuel market and efficiency to meet the supply–demand ratio. With the improvement in ever-growing technology on a commercial scale, the cost is reducing day by day. Thus, the cost will vary as per the region.

Compared to Unites States, Brazil and Europe, the commercialisation of production is less, and bioethanol is produced on a small scale only in European countries, thereby increasing its cost. Brazil is the lowest cost bioethanol producer with large, efficient and commercialised plants for productions that take advantage of sugar cane feedstock.

Gasoline is never used in a pure form. It is often blended with methyl/ethyl tert–butyl ether (MTBE/ETBE). Low blend bioethanol only allows 5–25% addition to gasoline, replacing the oxygen from fossil by some derived oxygenates such as methyl/ethyl tert–butyl ether. Ethanol contains more oxygen than MTBE/ETBE in its molecule. The 2% oxygen demand is fulfilled by 11% MTBE, whereas the same demand is fulfilled by 5.7% ethanol. The lower volume of ethanol satisfies the oxygen demand, so maintaining the same share in gasoline, usually 11% ethanol, is blended in gasoline [72].

High blend bioethanol is also marketed in various countries, allowing 85% – almost 100%, addition. Most of the vehicles in today's market are making use of 10% ethanol. The Brazilian government adopted a programme and made blending of 18–25% ethanol with gasoline mandatory. After 2000, vehicles were used with E100 raising the research interest in bioethanol more, which was reflected in increased production of bioethanol from 17 to 84 billion in only 12 years. The major users of ethanol gradually also became major producers. At that time, 60% bioethanol was used by the United States only, followed by Brazil (21%) and then followed by other countries.

The cost analysis is mainly affected by the following major steps: pre-treatment of the raw of the feedstock, hydrolysis and fermentation of the feedstock for ethanol production, followed by distillation and dehydration to produce concentrated ethanol, transportation to market and its adoptability by consumers. Wu et al. [73] found that the energy consumption in the pre-treatment and dehydration processes was approximately 4.7 and 3.5 $MJ\,l^{-1}$ ethanol, with 80% hydrolysis yield and 92% ethanol yield, respectively.

It was found that the major energy consumption using the 68 atm CO_2 method was 8.6 $MJ\,l^{-1}$ ethanol, while it was 13.2 $MJ\,l^{-1}$ ethanol (reduced by 35%) without adding CO_2 [74]. An economic comparison before and after introducing 68 atm CO_2 is presented in Table 2.2. The energy used was assumed for the electricity consumed, and 10% of the

Table 2.2 Production cost (USD l^{-1} ethanol) estimation for bioethanol production.

Pre-treatment approach	Material	Energy	Other expenditures	Total
190 °C	0.698	0.297	0.099	1.094
190 °C with 68 atm CO_2	0.367	0.192	0.056	0.615

Source: From [73]/with permission of Elsevier.

cost was for the other expenditures resulting in the total cost of bioethanol with 68 atm CO_2 as approximately 0.615 USD l^{-1} ethanol, which was 1.094 USD l^{-1} ethanol (reduced by 44%) without CO_2. The cost of bioethanol production (0.615 USD l^{-1} ethanol) was comparable with other reports, for instance, 0.623 USD l^{-1} ethanol from sugar cane [75] and 0.634 USD m^{-1} ethanol from sugar cane bagasse [76, 77]. It is predicted that optimisation and further research in this technology will improve the processes and will be able to break down the cost in the future.

The factors such as price, availability, preferences for renewable energy, age and income of user play an important role in users' choice between fuels. Amongst these factors, price is a leading factor determining the users' choice. Prices for fuels are dependent on market fluctuations. Considering environmental crisis and climate change issues, bioethanol should always be considered at a cheaper price than fossil fuel to attract users to a greater extent. E5–E25 has already been adopted by a large number of countries due to the ease in its use. It can be used with the same engine infrastructure and is easily available. It needs legal, mechanical, technical, economic and social support to come into the fuel market of any country.

2.5 Sustainability Aspects – Need of Alternative Biofuel

Various issues regarding sustainability aspects of the use of bioethanol have been started since the 1980s. It emphasises various environmental, social and economic issues. Despite some positive values of sustainability, some factors play a role in assigning negative sustainability.

Advantages of first-generation bioethanol are its biodegradability, providing secured energy with reproducible feedstock, social and economic benefits. However, the disadvantages are the competition for agro land, its blending capacity, high carbon footprints, high cost input required for production, increasing food cost and negative impacts on biodiversity. The production of bioethanol has adverse effects on the agricultural sector. It is produced from edible crops that would have reduced the availability of food products, increasing its cost and thus resulting in hunger, poverty and malnutrition. The problems can be severe, especially in developing countries [78].

Besides this, in many countries, forest and woody areas are replaced by biofuel useful plants only. This imbalance adversely affects biodiversity, greenhouse gases (GHGs) and climate change. The new production technology of cellulosic ethanol can tackle some of the issues, also known as second-generation biodiesel. The feedstock for cellulosic ethanol includes non-edible biomass such as harvest residue, switch grass and forestry. However, at the same time, it requires skilled workers and sophisticated technologies for the production, which finally resulted in larger capital cost per unit of production as compared to first-generation bioethanol [79]. On the other hand, the lower cost of feedstock used can curtail those increased cost once the technology is matured enough to commercialise. However, the grass, leftover residue, requires more complex and costly conversion technology to turn it into useful biofuels compared to first-generation bioethanol. The pros and cons of first-generation biofuel are presented in Figure 2.14.

The hampering factors for the use of bioethanol when used in high rotating diesel engine are increase in gas emission, carbon deposits due to incomplete combustions of plant oil, oil ring sticking and gelling of oils. These problems should be overcome for further use of

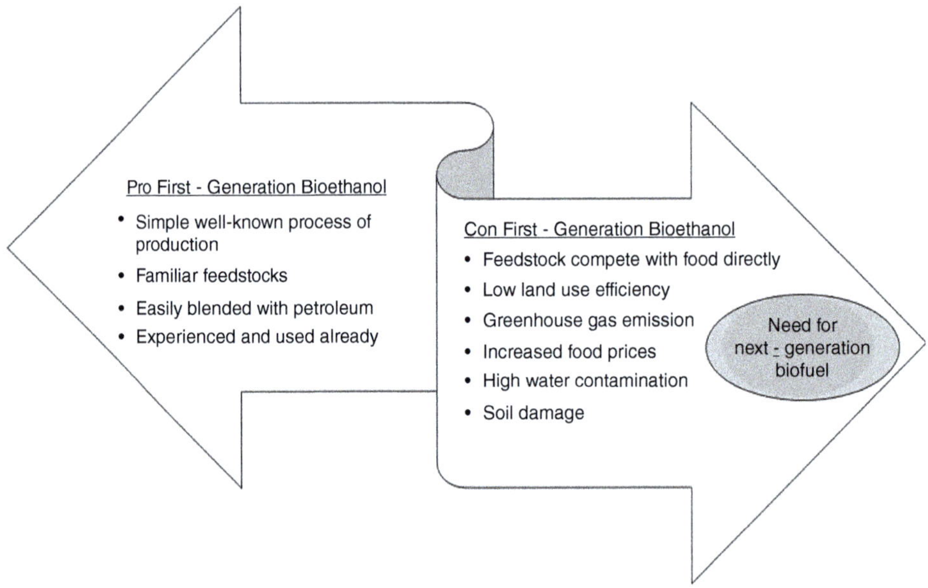

Figure 2.14 Pros and cons of using first-generation bioethanol.

ethanol. Various alternative biofuels and changes in the process technology (pyrolysis, trans-esterification and use of catalyst) may be helpful in overcoming these demerits. Nevertheless, bioethanol has a lower energy density, low vapour pressure and miscible with water, and hence easily incorporating and disturbing environment.

In addition, the countries with limited farmlands face problems in adopting first-generation bioethanol technique. Second-generation bioethanol provides 50% more land use efficiency. Furthermore, the problems of net energy and emission imbalance are minimised in second-generation bioethanol. The use of agro residues can be the solution, though it is still not able to solve the problem completely. Its high capital cost, set-up of large-scale plants, lack of technical knowledge, transportations cost and awareness in society restrict its worldwide use. Catalytic conversion can be a viable solution, which needs further research developments. Thus, when compared with fossil fuels, biofuels proved to be superior alternatives due to their advantages such as environmental approachability, reproducibility, availability in abundance and biodegradable nature, allowing their wide use, sustainability, etc. Biofuel will surely meet our increasing energy demand while minimising the negative impacts on the environment.

As first-generation biofuels (1G) are produced from edible crop, they face severe criticism for problems of food security and GHG emissions as they divert land from producing crops used for food as well as proper land forest uses. As a result, expectations turn to a rapid development of future-generation biofuels produced from various non-edible biomass that do not directly compete with food crops and, furthermore, which are expected to be more efficient in transforming biomass into bioenergy. Agricultural research and techniques should practise more for improvement in yields, high level of productivity protecting environment and, at the same time, also sustaining green biofuel methods.

Due to the high investment in feedstock, production costs of bioethanol, and uncertainties about the future-predicted higher generation bioethanol process, an integration of first and next-generation plants is to be considered for safer side. It will prove its potential in terms of food crop availability, prices and environmental sustainability where first-generation bioethanol lags behind. Thus, an integrated processing plant for higher generation biofuel will surely contribute to the energy market.

In the global market, emerging applications of ethanol play a key role in creating different potential opportunities (Figure 2.15). During the pandemic situation, the outbreak of corona virus disease also increased the usage of ethanol-based sanitizers, which again supported the global market demand of ethanol.

It is also recommended as a disinfectant. A huge demand of ethanol is now encouraging countries for supporting raw material and in turn ethanol production communities.

Top companies focusing on ethanol production are as follows:

- Flint Hill Resources LP
- Braskem
- Andersons Ethanol Group
- Archer Daniels Midland Company
- Cargill Corporation
- Aventine Renewable Energy
- HPCL Biofuels Limited
- Butamax Advanced Biofuels LLC

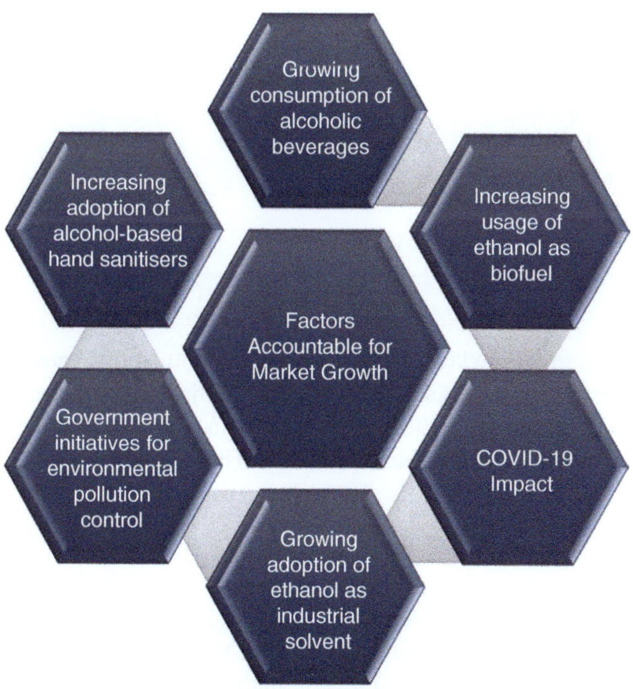

Figure 2.15 Factors accountable for the ethanol market growth.

- Advanced Bioenergy LLC
- British Petroleum

Various research studies are available, which give qualitative and quantitative insights into the ethanol market and growth trend of the potential global market.

Various market segments covered are as follows:

Source Type

- Sugar and molasses based
- Grain based
- Second generation

Purity Type

- Denatured
- Undenatured

Application Type

- Industrial solvents
- Fuel and fuel additives
- Beverages
- Disinfectant
- Personal care
- Others

Geography Type

- North America – USA and Canada
- Europe – Germany, France, United Kingdom and rest of Europe
- Asia Pacific – China, Japan and India
- Southeast Asia and rest of Asia Pacific
- Latin America – Brazil and rest of Latin America
- Middle East and Africa (MEA) – GCC, North Africa, South Africa, rest of Middle East and Africa

The ethanol market sheet shows that the largest market is available in North America, while the fastest growing market is provided by Asia Pacific countries [80]. The global ethanol market size surpassed USD 93.7 billion in 2020 and is expected to reach around USD 155.6 billion by 2030 with a growth rate of 5.2% from 2021 to 2030 (Figure 2.16).

2.6 Summary

At the beginning of the twentieth century, first-generation bioethanol contributed to the majority of the biofuel at the global production level predominantly based on sugar- and starch-containing crops. However, fears over the long-term use of first-generation

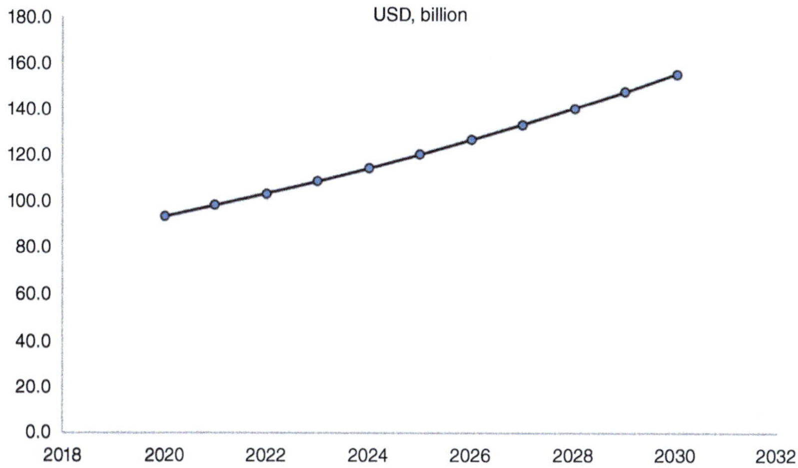

Figure 2.16 Ethanol market size.

bioethanol were still alarming, amongst which the food versus feed was highlighted. In addition, a vast amount of agro land used, impact on water resources and soil contamination with the residue are the other long-term disadvantages that should be handled properly before commercialising the production for the long term. Future research is required to minimise the problems imposed and to also make it cost-effective by modernising the technologies used. Rather than experiencing the new techniques with error or changing the existing engine, some of the methods of first generation can be wisely used to step up the future-generation biofuels. The use of lignocellulosic feedstock will be the answer for 'food versus feed' debut. Thus, integrated biorefineries can be a long-term sustainable solution for future-generation biofuels.

References

1 Hossain, N., Hasan, M.H., Mahlia, T.M.I. et al. (2020). Feasibility of microalgae as feedstock for alternative fuel in Malaysia: a review. *Energ. Strat. Rev.* 32: https://doi.org/10.1016/j.esr.2020.100536.

2 Hassan, S.N., Sani, Y.M., Abdul Aziz, A.R. et al. (2015). Biogasoline: an out-of-the-box solution to the food-for-fuel and land-use competitions. *Energy Convers. Manag.* 89: 349–367. https://doi.org/10.1016/j.enconman.2014.09.050.

3 Kolagar, M., Hosseini, S.M.H., Felegari, R., and Fattahi, P. (2020). Policy-making for renewable energy sources in search of sustainable development: a hybrid DEA-FBWM approach. *Environ. Syst. Decis.* 40 (4): 485–509. https://doi.org/10.1007/s10669-019-09747-x.

4 Pinho, A.R., Cabral, J.A.R., and Leite, L.F. (2007). Ethanol: a green raw material for the petrochemical industry. *ACS Natl. Meet. B. Abstr.* 234: 1155.

5 Stattman, S.L., Gupta, A., Partzsch, L., and Oosterveer, P. (2018). Toward sustainable biofuels in the european union? Lessons from a decade of hybrid biofuel governance. *Sustainability* 10 (11): 1–17. https://doi.org/10.3390/su10114111.

6 Das, A., Ghosh, P., Paul, T. et al. (2016). Production of bioethanol as useful biofuel through the bioconversion of water hyacinth (Eichhornia crassipes). *3 Biotech.* 6 (1): 1–9. https://doi.org/10.1007/s13205-016-0385-y.

7 Klinthong, W., Yang, Y.H., Huang, C.H., and Tan, C.S. (2015). A Review: microalgae and their applications in CO_2 capture and renewable energy. *Aerosol Air Qual. Res.* 15 (2): 712–742. https://doi.org/10.4209/aaqr.2014.11.0299.

8 Searle, S. and Malins, C. (2015). A reassessment of global bioenergy potential in 2050. *GCB Bioenergy* 7: 328–336. https://doi.org/10.1111/gcbb.12141.

9 Redford, D.A. (1993). The need for a marketing strategy for alberta bitumen. *J. Can. Pet. Technol.* 32 (03): https://doi.org/10.2118/93-03-01.

10 Lynch, M.C. and Schneider, S.A. (1984). The oil price revolution. *J. Policy Anal. Manage.* 38 (2): 341–343. https://doi.org/10.2307/3323873.

11 Ki-moon, B. (2008). Kyoto protocol reference manual. *United Nations Framew. Conv. Clim. Chang.* 130: https://doi.org/10.5213/jkcs.1998.2.2.62.

12 Thrän, D. and Ponitka, J. (2016). Government policy on delivering biofuels for the aviation sector. *Biofuels Aviat.* 295–313. https://doi.org/10.1016/b978-0-12-804568-8.00013-5.

13 Putrasari, Y., Praptijanto, A., Santoso, W.B., and Lim, O. (2016). Resources, policy, and research activities of biofuel in Indonesia: a review. *Energy Rep.* 2: 237–245. https://doi.org/10.1016/j.egyr.2016.08.005.

14 Bertrand, E., Vandenberghe, L.P.S., Soccol, C.R. et al. (2016). First generation bioethanol. *Green Energy Technol.* 175–212. https://doi.org/10.1007/978-3-319-30205-8_8.

15 Vijayalakshmi, S., Anand, M., and Ranjitha, J. (2019). Microalgae-based biofuel production using low-cost nanobiocatalysts. Microalgae Cultiv. *Biofuels Prod.* 251–263. https://doi.org/10.1016/B978-0-12-817536-1.00016-3.

16 Bórawski, P., Bełdycka-Bórawska, A., Szymańska, E.J. et al. (2019). Development of renewable energy sources market and biofuels in The European Union. *J. Clean. Prod.* 228: 467–484. https://doi.org/10.1016/j.jclepro.2019.04.242.

17 Potumarthi, R. and Baadhe, R.R. (2013). Issues in algal biofuels for fuelling the future. *Appl. Microb. Eng.* 418–435. https://doi.org/10.1201/b15250-18.

18 DeCicco, J.M., Liu, D.Y., Heo, J. et al. (2016). Carbon balance effects of U.S. biofuel production and use. *Clim. Chang.* 138 (3): 667–680. https://doi.org/10.1007/s10584-016-1764-4.

19 Ben Jebli, M. and Ben, Y.S. (2019). Combustible renewables and waste consumption, agriculture, CO_2 emissions and economic growth in Brazil. *Carbon Manag.* 10 (3): 309–321. https://doi.org/10.1080/17583004.2019.1605482.

20 Kumar, V. and Kant, P. (2013). Biodiesel: beneficial for environment and human health. *Pet. Coal.* 55 (3): 207–215.

21 van Niekerk, A., Drew, B., Larsen, N., and Kay, P. (2020). Impact of low NOx strategies on holistic emission reduction from a CI engine over transient conditions. *Int. J. Engine Res.* 22 (11): 3286–3299. https://doi.org/10.1177/1468087420973887.

22 Lapuerta, M., Hernández, J.J., Rodríguez-Fernández, J., and Calle-Asensio, A. (2020). Vehicle emissions from a glycerol-derived biofuel under cold and warm conditions. *Energy Fuel* 34 (5): 6020–6029. https://doi.org/10.1021/acs.energyfuels.0c00652.

23 Suhud Shote, A. (2019). Biofuel: an environmental friendly fuel. *Anaerob. Dig.* 185–194. https://doi.org/10.5772/intechopen.82856.

24 Mahmudul, H.M., Hagos, F.Y., Mamat, R. et al. (2017). Production, characterization and performance of biodiesel as an alternative fuel in diesel engines – a review. *Renew. Sust. Energ. Rev.* 72: 497–509. https://doi.org/10.1016/j.rser.2017.01.001.

25 Renner, M., Sweeney, S., and Kubit, J. (2008). Green jobs: working for people and the environment. *Worldwatch Pap.* 177.

26 Ali, M. and Akbar, N. (2020). Biofuel is a renewable environment friendly alternate energy source for the future. *Model Earth Syst. Environ.* 6 (1): 557–565. https://doi.org/10.1007/s40808-019-00702-y.

27 Rarbach, M. (2012). Cellulosic ethanol: potential, technology and development status. *DGMK Tagungsbericht* 75–76.

28 de Gorter, H. and Just, D.R. (2010). The welfare economics of biofuel tax credits and mandates. *Handb. Bioenergy Econ. Policy* 33: 347–364. https://doi.org/10.1007/978-1-4419-0369-3_20.

29 Dallemand, J.F., Wani, S.P., Leip, A. et al. (2010). Sustainability of biofuels for transport. *Ann. Arid Zone.*

30 Nogueira, L.A.H., Souza, G.M., Cortez, L.A.B., and de Brito Cruz, C.H. (2020). Biofuels for transport. *Futur. Energy Improv. Sustain. Clean Options Our Planet* 173–197. https://doi.org/10.1016/B978-0-08-102886-5.00009-8.

31 Gracia, A., Barreiro-Hurlé, J., and Pérez, L.P. (2020). Assessing the benefits of sustainability certification of biofuels: how much are consumers willing to pay? *New Medit.* 19 (2): https://doi.org/10.30682/nm2002a.

32 Ashokkumar, V., Salim, M.R., Salam, Z. et al. (2017). Production of liquid biofuels (biodiesel and bioethanol) from brown marine macroalgae Padina tetrastromatica. *Energy Convers. Manag.* 135: 351–361. https://doi.org/10.1016/j.enconman.2016.12.054.

33 Painter, D.S. (2014). Oil and geopolitics: the oil crises of the 1970s and the Cold War. *Hist. Soc. Res.* 39 (4): 186–208. https://doi.org/10.12759/hsr.39.2014.4.

34 Owusu, P.A. and Asumadu-Sarkodie, S. (2016). A review of renewable energy sources, sustainability issues and climate change mitigation. *Cogent Eng.* 3 (1): https://doi.org/10.1080/23311916.2016.1167990.

35 Mills, G. (1987). Alcohols as components of transportation fuels. *Annu. Rev. Energy Environ.* 12: 47–80. https://doi.org/10.1146/annurev.energy.12.1.47.

36 Sandy Thomas, C.E. (2009). Transportation options in a carbon-constrained world: hybrids, plug-in hybrids, biofuels, fuel cell electric vehicles, and battery electric vehicles. *Int. J. Hydrog. Energy* 34 (23): 9279–9296. https://doi.org/10.1016/j.ijhydene.2009.09.058.

37 Permeswaran, P. (2010). Chemurgy: using science innovatively to save American agriculture from overproduction. *Hist. Teach.* 44 (1): 95–125.

38 Frattini, S. (2008). Demand is increasing for renewable resourced resins. *JEC Compos. Mag.* 45 (38): 32–33.

39 Chel, A. and Kaushik, G. (2011). Renewable energy for sustainable agriculture. *Agron. Sustain. Dev.* 31: 91–118. https://doi.org/10.1051/agro/2010029.

40 Erickson, B. and Carr, M.T. (2009). Bio-ethanol development in the USA. *Biofuels* 39–54. https://doi.org/10.1002/9780470754108.ch3.

41 Broin, J. (2017). A conversation with Jeff Broin. *Ind. Biotechnol.* 13 (4): https://doi.org/10.1089/ind.2017.29092.jbr.

42 Qu, Y., Zhu, M., Liu, K. et al. (2006). Studies on cellulosic ethanol production for sustainable supply of liquid fuel in China. *Biotechnol. J.* 1 (11): 1235–1240. https://doi.org/10.1002/biot.200600067.

43 Kumar, L. and Bharadvaja, N. (2020). A review on microalgae biofuel and biorefinery: challenges and way forward. *Energy Sources* 107: 2411–2502. https://doi.org/10.1080/15567036.2020.1836084.

44 Andrade, M.C., de Gorgulho Silva, C.O., de Souza Moreira, L.R., and Ferreira Filho, E.X. (2021). Crop residues: applications of lignocellulosic biomass in the context of a biorefinery. *Front Energy* https://doi.org/10.1007/s11708-021-0730-7.

45 Al-Mulali, U. (2015). The impact of biofuel energy consumption on GDP growth, CO_2 emission, agricultural crop prices, and agricultural production. *Int. J. Green Energ.* 12 (11): 1100–1106. https://doi.org/10.1080/15435075.2014.892878.

46 Dutta, A. (2020). Are global ethanol markets a 'one great pool'? *Biomass Bioenergy* 132. https://doi.org/10.1016/j.biombioe.2019.105436.

47 Yamamoto, H. (2017). An analysis of supply and demand strategy of bioethanol using an agent-based global energy model. *J. Japan Inst. Energy* 96 (7): 239–249. https://doi.org/10.3775/jie.96.239.

48 De Simio, L., Gambino, M., and Iannaccone, S. (2012). Effect of ethanol content on thermal efficiency of a spark-lgnition light-duty engine. *ISRN Renew Energy* 2012: 8. https://doi.org/10.5402/2012/219703.

49 Rasool, U. and Hemalatha, S. (2016). A review on bioenergy and biofuels: sources and their production. *Brazilian J. Biol. Sci.* 3 (5): 3–22. https://doi.org/10.21472/bjbs.030501.

50 Liu, R. and Shen, F. (2008). Impacts of main factors on bioethanol fermentation from stalk juice of sweet sorghum by immobilized Saccharomyces cerevisiae (CICC 1308). *Bioresour. Technol.* 99 (4): 847–854. https://doi.org/10.1016/j.biortech.2007.01.009.

51 Sambusiti, C., Monlau, F., and Barakat, A. (2016). Bioethanol fermentation as alternative valorization route of agricultural digestate according to a biorefinery approach. *Bioresour. Technol.* 212: 289–295. https://doi.org/10.1016/j.biortech.2016.04.056.

52 Manochio, C., Andrade, B.R., Rodriguez, R.P., and Moraes, B.S. (2017). Ethanol from biomass: a comparative overview. *Renew. Sust. Energ. Rev.* 80: 743–755. https://doi.org/10.1016/j.rser.2017.05.063.

53 Kiss, A.A. and Suszwalak, D.J.P.C. (2012). Enhanced bioethanol dehydration by extractive and azeotropic distillation in dividing-wall columns. *Sep. Purif. Technol.* 86: 70–78. https://doi.org/10.1016/j.seppur.2011.10.022.

54 Costa, R.C. and Sodré, J.R. (2010). Hydrous ethanol vs. gasoline-ethanol blend: engine performance and emissions. *Fuel* 89 (2): 287–293. https://doi.org/10.1016/j.fuel.2009.06.017.

55 Vlachos, C.E., Mariolis, N.A., and Skaracis, G.N. (2015). A comparison of sweet sorghum and maize as first-generation bioethanol feedstocks in Greece. *J. Agric. Sci.* 153 (5): 853–861. https://doi.org/10.1017/S0021859614000446.

56 Amorim, H.V., Lopes, M.L., De Castro Oliveira, J.V. et al. (2011). Scientific challenges of bioethanol production in Brazil. *Appl. Microbiol. Biotechnol.* 91 (5): 1267–1275. https://doi.org/10.1007/s00253-011-3437-6.

57 Magaña, C., Núñez-Sánchez, N., Fernández-Cabanás, V.M. et al. (2011). Direct prediction of bioethanol yield in sugar beet pulp using near infrared spectroscopy. *Bioresour. Technol.* 102 (20): 9542–9549. https://doi.org/10.1016/j.biortech.2011.07.045.

58 Bayrakdar Ates, E., Barutcu, B., and Karaosmanoglu, F. (2017). Forecasting on first generation bioethanol production: a comparison of selected linear and non-linear models. *J. Renew. Sustain. Energy* 9 (4): 043101. https://doi.org/10.1063/1.4997058.

59 Buckler, E.S. IV and Stevens, N.M. (2016). Maize origins, domestication, and selection. *Darwin's Harvest* 67–90. https://doi.org/10.7312/motl13316-005.

60 Zhuang, Q., Qin, Z., and Chen, M. (2013). Biofuel, land and water: maize, switchgrass or Miscanthus? *Environ. Res. Lett.* 8. https://doi.org/10.1088/1748-9326/8/1/015020.

61 Di Lena, G., Ondrejíčková, P., del Pulgar, J.S. et al. (2020). Towards a valorization of corn bioethanol side streams: chemical characterization of post fermentation corn oil and thin stillage. *Molecules* 25 (15): 3549. https://doi.org/10.3390/molecules25153549.

62 McKinley, B.A., Casto, A.L., Rooney, W.L., and Mullet, J.E. (2018). Developmental dynamics of stem starch accumulation in Sorghum bicolor. *Plant Direct.* 2 (8): https://doi.org/10.1002/pld3.74.

63 Gallegos, R.K.B., Suministrado, D.C., Elauria, J.C., and Elauria, M.M. (2014). Energy analysis of cassava bioethanol production in the philippines. *J. Japan Inst. Energy* 93 (3): 301–309. https://doi.org/10.3775/jie.93.301.

64 Vang Rasmussen, L., Rasmussen, K., Birch-Thomsen, T. et al. (2012). The effect of cassava-based bioethanol production on above-ground carbon stocks: a case study from Southern Mali. *Energy Policy* 41: 575–583. https://doi.org/10.1016/j.enpol.2011.11.019.

65 Pradyawong, S., Juneja, A., Bilal Sadiq, M. et al. (2018). Comparison of cassava starch with corn as a feedstock for bioethanol production. *Energies* 11 (12): 3476. https://doi.org/10.3390/en11123476.

66 Pulyaeva, V.N., Kharitonova, N.A., and Kharitonova, E.N. (2020). Advantages and disadvantages of the production and using of liquid biofuels. *IOP Conf. Ser. Mater. Sci. Eng.* 976 (1): 012031. https://doi.org/10.1088/1757-899X/976/1/012031.

67 Papong, S., Rewlay-ngoen, C., Itsubo, N., and Malakul, P. (2017). Environmental life cycle assessment and social impacts of bioethanol production in Thailand. *J. Clean. Prod.* 157: 254–266. https://doi.org/10.1016/j.jclepro.2017.04.122.

68 Galletti, A.M.R. and Antonetti, C. (2011). Biomass pre-treatment: separation of cellulose, hemicellulose and lignin. Existing technologies and perspectives. In: *Biorefinery: From Biomass to Chemicals and Fuels* (ed. M. Aresta, A. Dibenedetto and F. Dumeignil), 101–122. Berlin, Boston: De Gruyter.

69 Lechón, Y., de la Rúa, C., Rodríguez, I., and Caldés, N. (2019). Socioeconomic implications of biofuels deployment through an input-output approach. A case study in Uruguay. *Renew. Sust. Energ. Rev.* 104: 178–191. https://doi.org/10.1016/j.rser.2019.01.029.

70 Mohanty, S.K. and Swain, M.R. (2019). Bioethanol production from corn and wheat: food, fuel, and future. *Bioethanol Prod. from Food Crop.* 45–59. https://doi.org/10.1016/b978-0-12-813766-6.00003-5.

71 Guo, M. and Song, W. (2019). The growing U.S. bioeconomy: drivers, development and constraints. *New Biotechnol.* 49: 48–57. https://doi.org/10.1016/j.nbt.2018.08.005.

72 González, U., Schifter, I., Díaz, L. et al. (2018). Assessment of the use of ethanol instead of MTBE as an oxygenated compound in Mexican regular gasoline: combustion behavior and emissions. *Environ. Monit. Assess.* 190 (12): 1–15. https://doi.org/10.1007/s10661-018-7083-7.

73 Wu, Y., Ge, S., Xia, C. et al. (2020). Using low carbon footprint high-pressure carbon dioxide in bioconversion of aspen branch waste for sustainable bioethanol production. *Bioresour. Technol.* 313: 123675. https://doi.org/10.1016/j.biortech.2020.123675.

74 Piccolo, C. and Bezzo, F. (2009). A techno-economic comparison between two technologies for bioethanol production from lignocellulose. *Biomass Bioenergy* 33: 478–491. https://doi.org/10.1016/j.biombioe.2008.08.008.

75 Pratto, B., dos Santos-Rocha, M.S.R., Longati, A.A. et al. (2020). Experimental optimization and techno-economic analysis of bioethanol production by simultaneous saccharification and fermentation process using sugarcane straw. *Bioresour. Technol.* 297: 122494. https://doi.org/10.1016/j.biortech.2019.122494.

76 Unrean, P. and Khajeeram, S. (2016). Optimization and techno-economic assessment of high-solid fed-batch saccharification and ethanol fermentation by Scheffersomyces stipitis and Saccharomyces cerevisiae consortium. *Renew. Energy* 99: 1062–1072. https://doi.org/10.1016/j.renene.2016.08.019.

77 Khajeeram, S. and Unrean, P. (2017). Techno-economic assessment of high-solid simultaneous saccharification and fermentation and economic impacts of yeast consortium and on-site enzyme production technologies. *Energy* 122: 194–203. https://doi.org/10.1016/j.energy.2017.01.090.

78 Mat Aron, N.S., Khoo, K.S., Chew, K.W. et al. (2020). Sustainability of the four generations of biofuels – a review. *Int. J. Energy Res.* 44 (12): 9266–9282. https://doi.org/10.1002/er.5557.

79 Nanda, S., Rana, R., Sarangi, P.K. et al. (2018). A broad introduction to first-, second-, and third-generation biofuels. *Recent Adv. Biofuels Bioenergy Util.* 1–25. https://doi.org/10.1007/978-981-13-1307-3_1.

80 Chen, L., Debnath, D., Zhong, J. et al. (2021). The economic and environmental costs and benefits of the renewable fuel standard. *Environ. Res. Lett.* 16 (3): 034021. https://doi.org/10.1088/1748-9326/abd7af.

3

Advanced Biofuels – Alternatives to Biofuels

3.1 Introduction

Human beings living on the Earth are in constant interaction with nature. The life of human beings is dependent on the natural resources such as air, water, food and various forms of energies. The protection of the environment depends on the behaviour of human beings. Humans have changed the environment as per their needs, such as for agricultural purpose cutting and cleaning the land and constructing the dams for water storage and approaching industrialisation [1, 2]. Factories, power plants and transportation are dependent upon various forms of energies to obtain the desired output such as production, generation, communication, and many more, which totally depend on the sources of energy. Sources of renewable energy such as biofuels, solar energy and wind energy produced by the natural resources such as plants, solar rays, air, water and ocean provide more benefits over the conventional form of energy such as fossil fuels [3]. The root cause of the importance and popularity of renewable energy sources is shown in Figure 3.1 [4].

Biofuels are fuels produced by processing the biomass, such as crops, garbage, wood, landfill gas, alcohol fuel, etc. [5–8]. The general biofuels are bioethanol and biodiesel. Bioethanol is a biofuel obtained through a fermentation process with sugar or starch crops, such as corn [9], sugar cane [10] or sweet sorghum [11], which are used as feedstock as shown in Figure 3.2. It is generally used in the transportation as a fuel in the engines as it is a cleaner energy source and safer to the environment, but its performance characteristic on the engine is low resulting in the reduction of mileage [12]. Furthermore, it is acidic and absorbs water particles that may corrode and damage the engine, which reduces the life of engines [13, 14]. Such transportation vehicles that run on bioethanol charge more as compared to other vehicles that run on fossil fuels.

Bioethanol – the first-generation biofuel – is processed to overcome the following problems:

1) Bioethanol is processed from the food crops, and to produce a large amount of feedstock, a large amount of land is needed to grow the crops, which in turn could greatly impact the biodiversity of our environment [15].

Sustainability in Biofuel Production Technology, First Edition. Pratibha S. Agrawal, Pramod N. Belkhode, and Samuel Lalthazuala Rokhum.

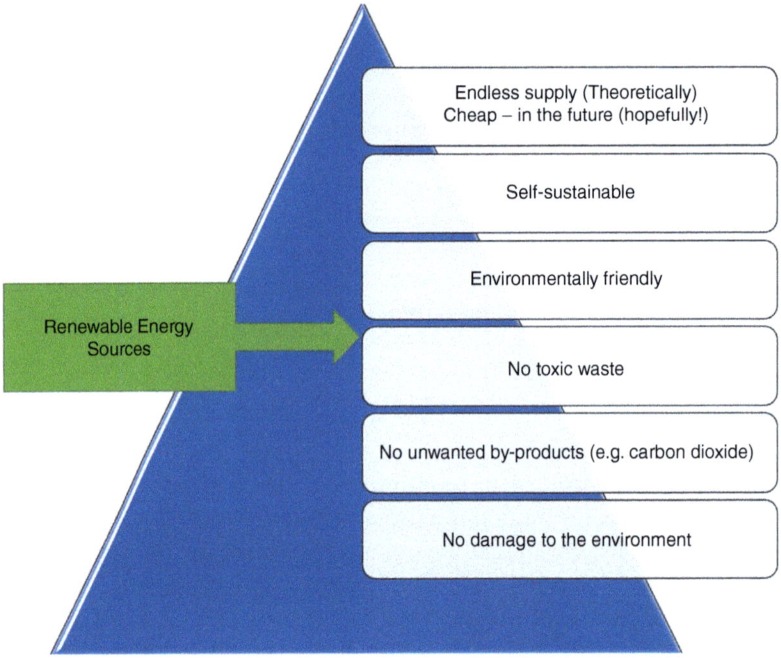

Figure 3.1 Importance and popularity of renewable energy sources.

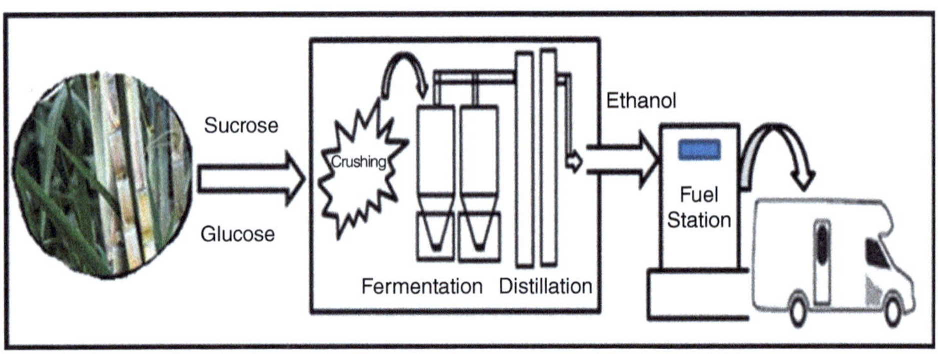

Figure 3.2 Processing of ethanol from sugar cane.

2) Traditional biofuels like ethanol reduce greenhouse gas (GHG) emission by almost 45–48% compared with gasoline, while advance biofuels can effectively reduce these emissions to meet environment emission standard level [2 EESI] [16].
3) Ethanol is difficult for cold starting as it does not burn quickly [17, 18].

These problems can be overcome by further improvement in the biofuels by using ethanol-blended fuels [19], which can reduce the emission of greenhouse gases. In addition, processing the available feedstock of non-food crops from cellulose, hemicellulose or lignin to biomass [20] can be the solution for the debate – 'food versus feed'. Second- and third-generation biofuels are obtained from agro waste materials used as feedstock [21–23].

3.2 Biofuels Deserve Another Look

Researchers have been working on a variety of alternatives to the conventional gasoline fuel as well as first-generation biofuel used in internal combustion engines [24, 25]. Analysis of the biofuels is performed considering various aspects to provide a new look to biofuels.

3.2.1 Economic Model of Biofuels

Biofuel production and consumption depend on the use of feedstock, processing techniques and release of greenhouse gases during the whole practice. Bioethanol produced from first-generation method uses feedstock of food crops that are being used by human beings and animals and are thus directly or indirectly related to food production. Utilising these feedstock for biodiesel may result in their shortage in the market for human consumption, reflecting a higher food cost that is undesirable [26]. Production of biofuels from non-edible crops is more desirable, which will not change the agricultural practice and biodiversity [27]. The limited access to feedstock impacts the biodiesel market, which finally results in less production capacity, a simple chain imposed on the import restriction. Biofuel processing techniques release greenhouse gases that will change the carbon points [28]. Nitrous oxide released during the processing of fertiliser is one of the greenhouse gases that should be prevented [29, 30]. Furthermore, the production of biofuels from the crop such as corn increases the water pollution from nutrients, pesticides and sediments [31]. Economic analysis of the biofuels results in higher crop prices that impact the retail food, thus it is concluded that biofuels should be processed with advance practices with the non-edible crops and waste biomass only without disturbing the harmony of nature. The processing of biofuels with new techniques such as blending of biofuels, uses of non-food crops, uses of algae and uses of waste materials, which is the role model, finally resulted in an advance method of production of biofuels, named as second-, third- and fourth-generation biofuels with the best agreement to the economic model [32].

Development in the bioenergy market and the implementation of the related policies produced larger interconnectivity between energy and market. Thus, bioenergy can be successfully fitted into the bioeconomy model. The supply push and demand pull are analysed time to time and are well reflected in various policies [33, 34]. These policies support the economic growth and environmental emissions. Almost all countries have adopted the use of biofuel blends for one or another purpose. Reduction in GHG emissions, ample renewable sources, increased employment and the future demand are the key focus of any developing industrialised country. As per the new developed policy situation, the global demand for bioenergy was 1277 Mtoe (10% of the total global primary energy use) in 2010, which is expected to increase to 1881 Mtoe in 2035 [35]. The non-traditional biomass was 526 Mtoe (3.3% of the total global primary energy use) in 2010, which is expected to increase to 1200 Mtoe in 2035 (Figure 3.3).

In other words, all these policies promote the shift from the non-renewable-based economy model to the biomass-based renewable energy models [36]. The sustainability of the model depends on the conversion efficiency, blending mandates, tax credits and costs.

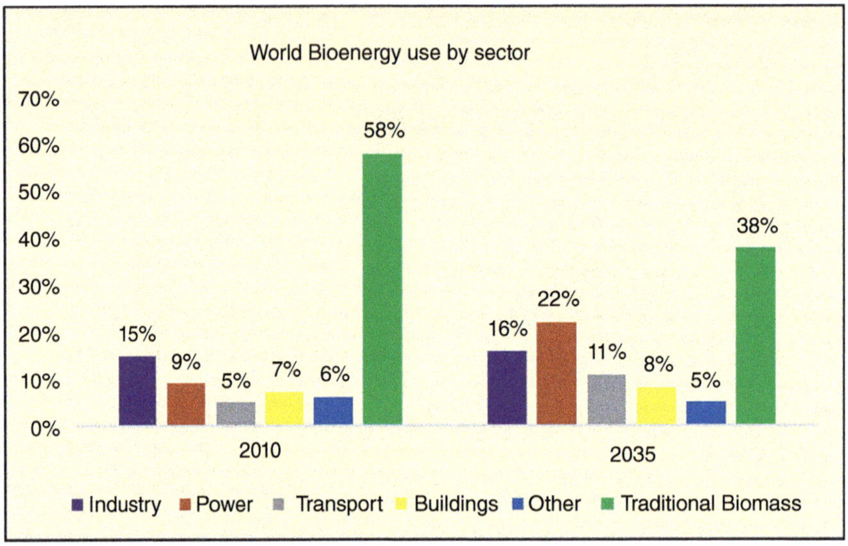

Figure 3.3 Use of bioenergy in different sectors.

3.2.2 Advanced Biofuels

Though biomass has been traditional sources of energy, the advanced development aims to use engineering technique to make the fuels useful for transportation and energy. Sugar-cane- and corn-based biofuel markets were the most competitive but still struggling hard due to higher feedstock cost and low demand. More efficient and advanced technologies are needed to make the dream of bioeconomy realistic. New policy of biofuels recommended to use non-food crops, used cooking oil, residues obtained from animal waste and algae as the feedstock as the raw materials for the production biofuels [37–39] were implemented. Policy advises to develop a new mechanism for processing the waste cooking oil to increase the supply for biodiesel production, which eliminates the reusing in the food flow. Advanced technology will benefit the society as the production of biofuels plants is set up to the commercial scale, which covers cellulosic and renewable fuels [40, 41]. The advanced technology claims to convert wood and agricultural waste into fuels and process municipal corporation waste [42] and microalgae [43, 44] into advanced biofuels. Advanced biofuels can be used in the vehicle engines without any changes in the engines and distribution systems.

Results of the utilisation of advanced biofuels will create new employment opportunities, clean and healthier environment, reduce greenhouse gas emission and prevent the diversion of waste cooking oil into the food chain by fixing the maximum permissible limit of total polar compounds in edible oil [45, 46]. The new biofuel policy also encourages the generation of feedstock by using the wastelands, and for this local communities are encouraged to plant non-edible oilseed – crops such as Karanja (*Pongamia pinnata*) [47], Neem (*Melia azadirachta*) [48], Jatropha (*Euphorbiaceae*) [49], Champa (*Callophylum innophylum*) and Lakshmi Taru (*Simarouba glauca*) for augmenting indigenous feedstock supply for biodiesel production [50, 51]. Farmers are encouraged to grow a variety of different biomasses as a second crop due to rain-fed conditions. Implementation of these policies will enhance the production of advanced biofuels, which will maintain the supply chain mechanism, feedstock for processing and economy mechanism. It will help in the development of biofuels against the disadvantages listed in the conventional biofuels. With the development of advanced biofuels, various schemes, such as subsidies and grants, are being facilitated compared to first-generation biofuels.

The path of conversion of biomass to advanced biofuels and bioproducts depends upon the selection of feedstock, the process adopted to convert feedstock into biofuels and environmental issues. The feedstock used is non-food crops known as energy crops or sustainable crops, which have a positive impact on the environment. The developed crops are converted into biofuels and bioproducts with appropriate technology, fermentation or separation process, where the biomass is broken down into lignin or waste with a synthesis process is converted into biofuels and bioproducts. Bioproducts, such as valuable chemicals, materials, such as plastics, chemicals and fertiliser, and biofuels such as ethanol, biodiesel, green diesel and methane, are processed from the biomass. The flow diagram represents the end product obtained from the biomass, as shown in Figure 3.4.

Some biofuels, such as bioethanol, biodiesel, biogas, biofuel pellets and bioproduct, are discussed. The solid biofuels in the form of biofuel pellet are used as the source of energy for producing the heat in the boiler to convert water into superheated steam, which is used as the working substance in turbines to obtain mechanical energy from thermal energy to generate electricity.

3.2.2.1 Bioethanol

Bioethanol obtained from starch and sugar crops is produced by fermentation, which is a clear colourless liquid, biodegradable, has low toxicity, causes low pollution and produces carbon dioxide when burned [52, 53]. It reduces polluting emission by blending ethanol with gasoline.

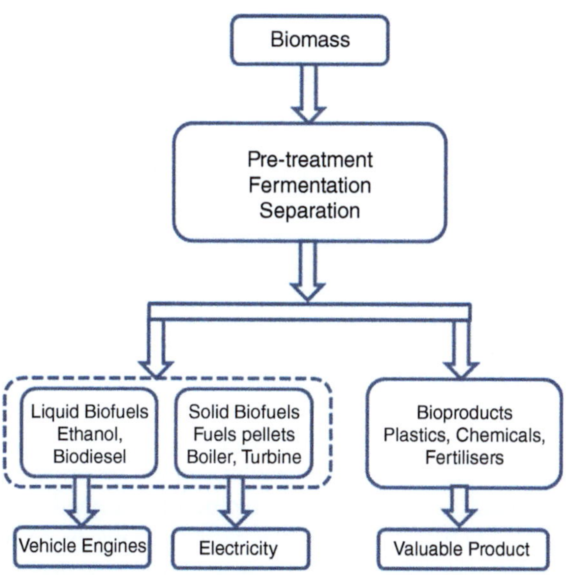

Figure 3.4 End product obtained from biomass.

3.2.2.2 Biodiesel

Biodiesel is a renewable source of energy obtained from vegetable, plant and animal oil with the biochemical and/or chemical process 'transesterification' to make biodiesel that is used in diesel engine [54–56].

3.2.2.3 Biogas

Biogas is produced by processing the organic matter such as plant and animal waste in the absence of oxygen, also known as anaerobic treatment consisting of methane and carbon dioxide [57]. For example, 40 l of biogas is produced with 1 kg of cow manure. Biogas burns without smoke and is not harmful [57].

3.2.2.4 Biofuel Pellets

Biomass pellets are produced from the residues of wood such as sawdust, by-product of wood processing. Biomass palletisation is a standard method for the production of high-density, solid fuel for power plants and other purposes [58].

3.2.2.5 Bioproduct

Bioproducts are produced from renewable energy sources and need a low amount of energy for their production. The most common bioproducts are bioplastics from plant oil and sugars. Chemicals, resins, paints and lubricants are used as the industrial bioproduct. Antibodies and vaccines as the pharmaceutical products are produced from the medicinal plants. Biocosmetic products are products such as soaps, body creams, etc. The most common materials used for bioproducts are soya beans, corn, sunflowers, grass, algae and many others.

It is a time to provide a new look to biofuels by using advanced biofuels, which will recover the losses occurred due to the environmental pollutions by using the fossil fuels, minimise the greenhouse gases and clean the environment.

3.3 Global Production, Need and Demand

Increasing prices and reduction of the resources of the fossil fuels are becoming a serious issue causing environmental pollution. To overcome this issue, renewable energy plays a significant role in facing the challenges of future energy demand. Growing population and advancement in the technology energy requirement are expected to grow around 5–10% annually based on the literature [59, 60]. Fuel requirement for the transportation would be satisfied by the biofuels by approximately 27% in 2050 [61, 62]. The international trade market grows quickly as per the demand. Biofuels can meet the demand as most of the developing countries participate in the farming sectors as the source of income against the dependency upon the land currently used for agricultural purpose of food production. The global biomass energy market is diverse, versatile and variable depending on the geographical area, crops available and fuel-type demand. Production of the biofuels is accepted to be uniform across as the feedstock required to obtain the biofuels is easily available based on the agricultural activities. Requirement of environmentally friendly fuel supply is expected

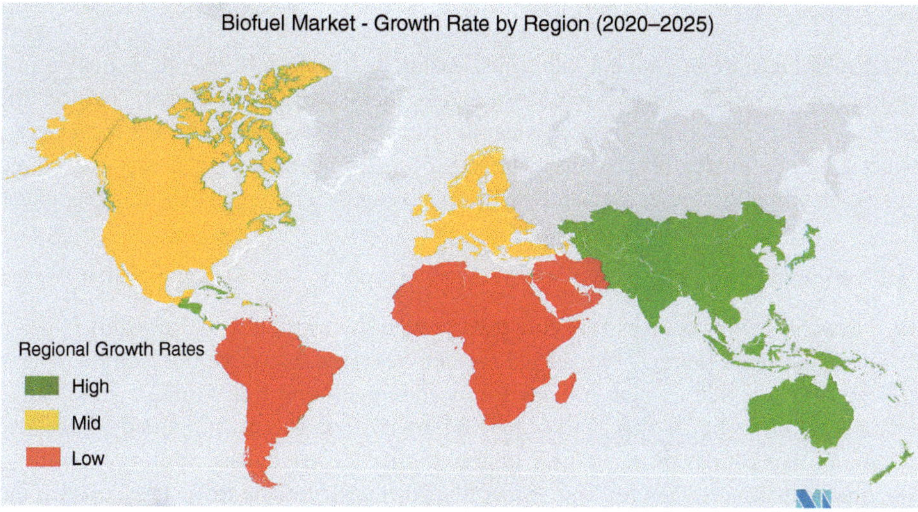

Figure 3.5 Biofuel market. *Source:* Mordor Intelligence.

to explore the need for biofuels. Blending of biofuels [63] in the automobile sector and the government support are expected to enhance the biofuel production to meet the future demand. Major contribution in the renewable energy source is made by the biofuels in the global energy demand. The biofuel market growth rate with region during 2020–2025 as per Mordor Intelligence is shown in Figure 3.5. The most important sector is the transportation, where the demand of biofuels will rise to 20% by 2040; similarly, in the aviation sector, the demand is expected to rise to 20% by 2040 [64].

Global market is expected to be developed due to the rising demand of biofuels. Advancement in biofuels is responsible for the rise in global market as an energy crop obtained from non-food crops and reduces the global consumption of fossil fuels and pollution. Biopolicies are designed in terms of feedstock production considering the future demand in context of current agricultural production. Advanced biofuels are produced by using non-food crops, residues obtained from industrial waste, agricultural waste and municipal waste as a feedstock.

The important highlights that contribute to the growth of global advance biofuel market are as follows.

3.3.1 Environmental Factor

Advanced biofuels are inexpensive and help to increase the bioenergy with a decrease in the dependency of global fossil fuel market. Biofuels can be produced from locally available feedstock that can easily be processed under any environmental condition [65]. This reduces the dependency of fossil fuel market and promotes the agricultural activity with a financial gain in the waste.

3.3.2 Clean Fuel

Advance biofuels are in demand due to increasing prices of fuels, energy security and a continuous rise in the emission level. Alternative fuel – advanced biofuels – reduces the carbon emission and its production depending upon the availability as the local region results in a clean source of energy [66]. The step towards a clean source of energy is explored with the blending of biofuels as the fuel in the diesel engine without changes in the engines.

3.3.3 Biofuel Policies

The renewable energy sources are being used around the world due to the positive impact of advanced biofuel market. Biofuel policies are designed to reduce the technology gap between the conventional biofuels and advanced biofuels and transfer the technology to enhance the biofuel economy market [67]. Many companies are committed to produce 1.7 billion gallons of advanced biofuels, doubling current capacity and working carefully to overcome industrial challenges. This shows that advanced biofuels are on track to meet the targeted emission reduction for clean fuels.

3.3.4 National Biofuel Policy 2018

Policy support for the bioenergy market is provided by all countries. Biofuel policies mainly consider blend mandates, sustainability, tax exemptions and R and D grants. All the stages in the biofuel productions are well supported by the government policies. It covers conversion, production, distribution, consumers' choice, biomass plants and public investments.

The main highlights of the biofuel policy are as follows [68, 69]:

1) Encourage the production of biofuels from the local feedstock, managing natural resources in the future and using them as the fuel in the transportation sector.
2) Utilisation of biofuels in the form of green fuels, energy security, low emission, proper utilisation of land, employment generation, enhancing economic growth, etc., which is expected through this policy.
3) Blending of biofuels, around 20% blend of ethanol in the case of petrol and 5% blend of biodiesel in the case of diesel, to minimise the emission [70–72].
4) Reducing dependence on the non-renewable resources.
5) Adapting to and mitigating climate change [73].
6) Improving trade balance in the developing countries.
7) Ensuring food security [74].
8) In addition, government highlights the steps such as improvement in the production of ethanol and biodiesel by increasing domestic production, set-up of biorefineries, growth in the feedstock for biofuels, technological improvement for biofuel production and facilitating the blending of biofuels with conventional fuels.

The policy objectives will have many direct and indirect effects on the potential development of the bioenergy market. Therefore, policies should be carefully assessed for the sustainability. It will consider all environmental benefits, trade balance and economic growth considering food security as the primary object.

3.4 Feedstock for Advanced Biofuels

Long back in 1925, Henry Ford put the concept that biofuels can be derived almost from everything such as fruits, shrubs, apples, weeds, saw dusts and vegetables through the process of fermentation. According to him, enough alcohol can be obtained by cultivating the potatoes field, which would be used as biofuels to drive the machinery [75]. Henry Ford's statement highlights the production of advanced biofuels from different feedstock. Advanced biofuels, also termed as second- or third-generation biofuels, are obtained from lignocelluloses feedstock such as residues collected from agriculture such as leaves, stalks of crops, fibrous materials remained after crushing sugar cane, non-food crops such as grasses, algae or treated waste obtained from industrial feedstock such as saw dust, pulp and paper waste, or treated waste obtained from human beings [76]. Biofuels obtained from these feedstock reduce greenhouse gases and are therefore named as sustainable feedstock [77]. In other words, these crops can be named as energy crops that result in the production of biofuels with respect to utility of land, marginal land used for production and harvesting.

The sustainable feedstock includes the following:

1) Crops grown on marginal land [78] that is related to the growth of non-food crops.
2) Waste and residues obtained from agricultural, forestry, food, municipal waste, used cooking oil, animal fats and other organic waste and residues [79].
3) Other feedstock related to aquatic plant, algae [80] and microbial biomass [81, 82].

The production of biofuels depends on the supply of feedstock from various supply sectors such as agricultural, forestry, biowastes and non-food crops. The cost of advanced biofuels is higher than that of the conventional biofuels and fossil fuels due to the production cost of the advanced biofuels. The technologist is working in continuous development towards the appropriate technology and the selection of relevant feedstock for advance biofuels to supply biofuels at an affordable rate.

Municipal wastes are being increasingly considered as potential feedback for the biofuel production [7, 83, 84]. Waste is processed to convert into the refuse-derived fuels that can be used as feedstock, resulting in the reduction of landfill. Treatment of used cooking oil and tallow is used as feedstock to obtain biofuels that are blended with diesel. Processing of saw dust and wood residues is used to convert them into wood pellets. Only factor is that the cost depends upon the costing involved in collection, transportation and storage of feedstock to the processing units.

The waste generated by human beings, which is of no use to the society, is treated as municipal solid waste that is treated in the treatment plant to convert it into different forms such as fuel pellets, residual sand waste, and recycled plastics. Figure 3.6 shows the classification of solid waste. The utilisation of municipal solid waste to convert it into the useful energy is explained with the help of an example of power generation using the solid waste. Solid waste can be used as an energy source [85]. The estimation of power generation from the solid waste is carried out as shown in the given example.

Suppose dry solid waste in the form of solid pellets is 6000 kg, which when burns produces the heat in the range of 800–1000 kcal/kg, considering the average value of heat production is 1000 kcal/kg.

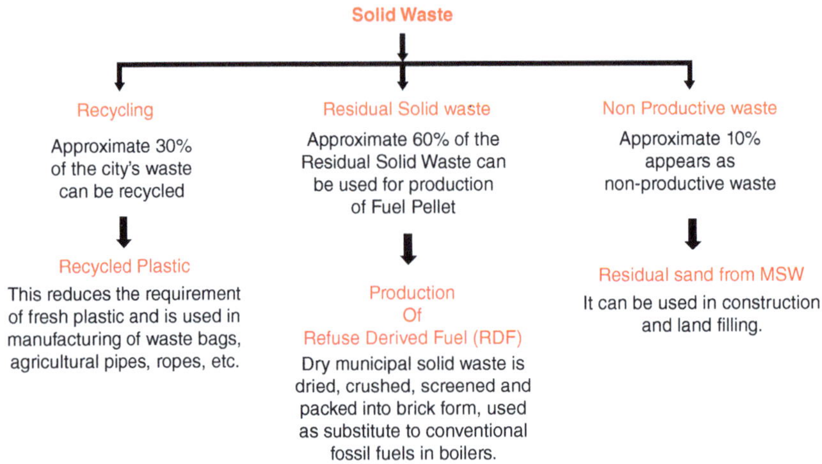

Figure 3.6 Different types of solid waste.

The amount of energy recovered, considering 1 kcal = 1.16×10^{-3} kWh, is 6960 kWh. Considering that the conversion efficiency is 30%,

$$\text{net power generation} = 0.30 \times 6960 = 2088\,\text{kWh}.$$

Power generated from the solid waste can help to utilise the energy to operate any electrical appliances.

The target of achieving the sustainable development scenario depends upon the significant role of advance biofuels in terms of emission reduction, global economy, coal demand and efficiency gain to determine the future path [86]. In addition, the key component to achieve the desired target depends upon the availability of feedstock at national and regional levels and its processing technology.

3.5 Advanced Biofuels for Different Applications

Advanced biofuels satisfy the most important requirement of alternative fuel in transportation, taking care of emission, efficiency and cost involved. Advanced biofuels are used as alternative fuels from the engine to the household appliances. Almost 25% of the energy is used in the transportation across the globe. The renewable energy in other forms, such as solar energy, wind energy and other sources of energy, cannot be used for transportation. Only alternative left for the transportation industry is to depend on biofuels as the most efficient source. Advanced biofuels are used in the engines with a proper blending mechanism to achieve desirable characteristics, efficiency and emissions [87, 88].

Some uses of the advanced biofuels as per the feedstock are as follows:

1) Waste vegetable/cooking oils and residues using hydrotreated vegetable oil techniques are converted into renewable diesel [89, 90], aviation fuels and chemicals.

2) Dry straw feedstock and hard wood are converted into ethanol using fermentation technology, which is used as the transportation fuel, and residue obtained is converted into solid fuels as the wood pellets used in power station [91] as a fuel in boilers.
3) Hydrotreatment of crude tall oil is used as fuel in the transportation sector such as fuel in the engines, machinery and rail road equipment [92, 93].
4) Agro residue with anaerobic digestion produces biocompressed natural gas [94] used as transportation fuels.
5) Food waste, municipal solid waste and crop residue are treated with anaerobic digestion treatment [95] to obtain biofuels used in transportation and power generation units.
6) Rice straw and cotton straw as raw materials with fermentation technology are used to produce ethanol [96], silica and inorganic mineral fertiliser.
7) Algae, wet organic biomass, biowaste such as food waste, sludge, agricultural crop, residue, etc., with the hydrothermal liquefaction treatment are converted into biofuels [96] used in the transportation sector.
8) Domestic feedstock such as wood pellets saw dust, etc., is a process to obtain biomethane using gasification biomass techniques to produce biofuels [97] used in vehicles.
9) Lipid mixture of vegetable oils and residue lipids is processed using a lipid hydrogenation process to produce hydrotreated vegetable oil that can be used as biodiesel in transportation fuel.
10) Hydrolysis process of wood waste and domestic waste is used to convert them into alcohols followed by chemical synthesis to obtain biofuels suitable for aviation, road transportation, and heavy-duty machinery.

3.6 Commercial Development

Biofuels play an important role in the global market as they are achieving global dimension with the largest potential for the commercial development. The biofuel trade depends on the agricultural productive capacity that further depends upon the financial assets, technology and infrastructure. In addition, it also depends upon the environmental condition such as fertility of soil, water availability, seasonal variation and climatic conditions of the particular region where feedstock is grown [98]. Farmers' decisions regarding a particular crop based on the environmental factors decide the availability of feedstock and bioeconomy. The developing countries have a larger scope to grow the feedstock due to the favourable climatic condition and the availability of human labour at a lower cost [99]. The overall development of biofuels depends upon the agricultural productive capacity, human activity, technology and investment in this sector.

Technology can enhance the ecosystem with the modification of the natural productive capacity with the help of innovative practices, advancement in the agricultural machinery, suitable fertiliser and chemical to protect the crops from damaging, which results in the production of energy feedstock on a large scale with low cost. The supply chain of the biofuel's development depends upon the feedstock, which includes the energy crops, residues obtained from agriculture, forest and animal waste.

Advanced biofuel generation is responsible for the commercial development, particularly of developing countries (Figure 3.7). A new type of biofuel is produced from wood, grass or

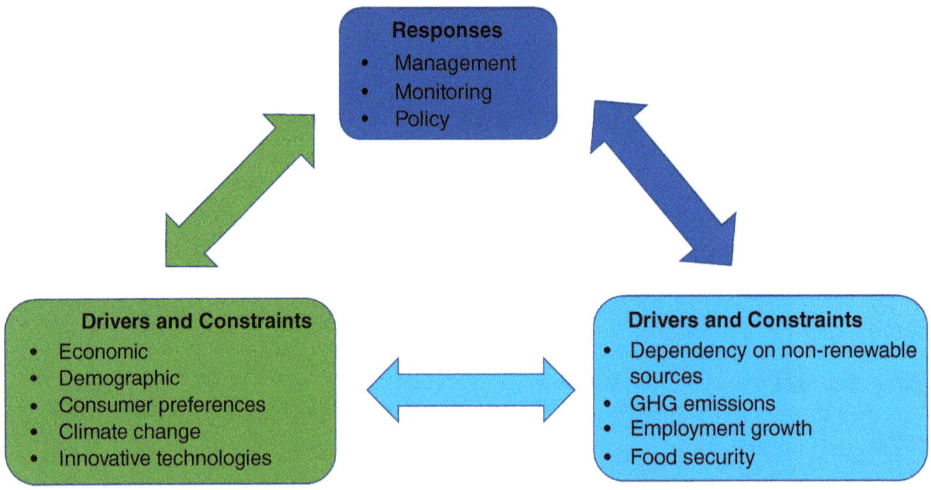

Figure 3.7 Commercial analysis of the bioeconomy growth.

the inedible parts of plants which allow the market to explore with recent policy development across the globe. The United States has the largest installed capacity for second-generation biofuel facilities followed by China, Canada, the European Union and Brazil [100].

The commercial development is explored from the second generation due to the limitation of the first-generation biofuels. International trade governs that the second-generation fuels expected to maintain balance in the market with the acceptance of different methods of producing the biofuels such as indirect use of land, food price effects and greenhouse gas emission.

The following factors are expected to enhance the commercial development of the biofuel industry:

1) Supply the bioenergy at the national level instead of utilising it at the local level [101].
2) Facilitate the technology transfer between domestic organisation and international level to avoid the technology gap between the generation and feedstock.
3) Develop the bioeconomy sector in the industrial development, which will facilitate in achieving the desirable market both at domestic and international levels [102].
4) Promote technology development amongst different regions in order to enhance the advance biofuels with suitable feedstock [103].

All the policies play an important role in cutting down the prices and innovating the newer technologies. There are mainly two types of policies:

Market pull policies – These policies focus on creating demand in market for the technologies.

Technology push policies – These policies focus on production technologies for fostering innovation.

An integrated approach covering all emerging needs and market supply is essential to consider. It should assess the impact of technology, demand and policies on sustainability. A large amount of data collection, developing a mathematical model and a close monitoring of the system should be integrated into the policies.

3.7 Aviation Fuel and Green Diesel

Aviation biofuels reduce the environmental pollution impact due to lower CO_2 emission by almost 20–95% as compared to the conventional fuel used in the aviation [104]. Again, reduction in the emission depends upon the selection of biomass for the production of aviation biofuels. In 2008, the flight with blended fuel with 50% biofuel was tested in 2011, and further blended biofuels are allowed in commercial flights. Plant sources such as Jatropha, tallows, residues of waste oil, algae, palm oil, solid biomass with a pyrolysis process and waste fermentation are used as aviation fuel to operate the gas turbine of the aircraft. Aviation biofuels possess the important characteristics, such as colourless in appearances, a low freezing point and fuel availability [105–107]. Jet A and jet A-1 are the most common types of aviation biofuels used in the commercial flights. Jet fuel consists of different hydrocarbons as per the international specification. Kerosene-type jet fuel that consists of jet A and jet A-1 is usually selected as a lower cost fuel with a higher flash point and less flammability and is considered safe to handle and transport for the aircraft engines. Jet A and jet A-1 aviation fuel is preferred in the United States of America and remaining part of the world [72, 108]. Naphtha kerosene fuel is jet B biofuels and is used to enhance cold weather performance as it is lighter in composition and more dangerous to handle because it has limited use and is generally preferred in the cold climates [109].

Blended biofuel with the desired composition of 30% kerosene and 70% gasoline with low freezing point and flash point is preferred for aircraft. Aviation industries play a vital role in the global economy, which serves 57 million jobs and 2.2 trillion USD in global GDP. Technology improvement has already started with the rise in the aviation demand to reduce the emission of greenhouse gases as the aviation industries are facing significant challenges in improving environmental sustainability. Identification of a suitable technology for the aviation industries was carried out with different programmes at national and international levels by scheduling the workshop for technology identification, different feedstock and aviation biofuels. The important role of stakeholders is to produce cost-competitive biofuels.

In aviation industries, the availability of feedstock such as lignocellulose feedstock reduces the production cost with increased market supply [110]. Used cooking oil, animal fats and vegetable oil are mostly preferred for the aviation biofuels as the feedstock to reduce emission and reduce the environmental impact of the aviation industry improvement in fuel efficiency.

3.8 Conclusion

The concept behind the advanced biofuels is to grow energy crops on the unused land, which is not used for agriculture, with less efforts in irrigation and fertiliser. The energy crops or sustainable crops or environmental feedstock obtained are processed to extract the biofuel with a proper technology, which can be used in a suitable application. The advanced biofuels reduce the emission and are more environmentally friendly. The range of feedstock used for the production of biofuels varies from lignocelluloses to municipal solid wastes. The development of the biofuels is expected to be consistent and depends upon the selection and availability of feedstock, production technology and environmental condition. Advanced biofuels are an alternative to fossil fuels and conventional biofuels as they have the potential to generate a number of benefits and reduce greenhouses gases, which results in the economic model as compared to other conventional fuels [111]. The development of the advanced biofuels is expected by technology transfer, which will result in the reduction of the technology gap between the feedstock and generation and explore the economy sector at the domestic and international levels. A combination of global and local systems ('glocal') can be developed by carefully considering all the areas.

References

1 Yatim, P., Mamat, M.N., Mohamad-Zailani, S.H., and Ramlee, S. (2016). Energy policy shifts towards sustainable energy future for Malaysia. *Clean Techn. Environ. Policy* 18 (6): 1685–1695. https://doi.org/10.1007/s10098-016-1151-x.

2 Wysokiński, M., Domagała, J., Gromada, A. et al. (2020). Economic and energy efficiency of agriculture. *Agric. Econ. (Czech Republic)* 66 (8): 355–364. https://doi.org/10.17221/170/2020-AGRICECON.

3 Morris, A., Owley, J., and Capello, E. (2013). Green siting for green energy. *J. Energy Environ. Law* 5: 17.

4 Yumurtaci, R. (2013). Role of energy management in hybrid renewable energy systems: case study-based analysis considering varying seasonal conditions. *Turk. J. Electr. Eng. Comput. Sci.* 21 (4): 1077–1091. https://doi.org/10.3906/elk-1112-85.

5 Sindhu, R., Binod, P., Pandey, A. et al. (2019). Biofuel production from biomass. *Curr. Dev. Biotechnol. Bioeng.* 79–92. https://doi.org/10.1016/b978-0-444-64083-3.00005-1.

6 Sindhu, R., Binod, P., Pandey, A. et al. (2019). Biofuel production from biomass: toward sustainable development. *Curr. Dev. Biotechnol. Bioeng. Waste Treat. Process. Energy Gener.* 79–92. https://doi.org/10.1016/B978-0-444-64083-3.00005-1.

7 Van, L.M., Chumpi, A., Belur, K. et al. (2022). Bioresource technology reports valorisation of food waste to sustainable energy and other value-added products : a review. *Bioresour. Technol. Rep.* 17: 100945. https://doi.org/10.1016/j.biteb.2022.100945.

8 Laskar, I.B., Gupta, R., Chatterjee, S. et al. (2020). Taming waste: waste mangifera indica peel as a sustainable catalyst for biodiesel production at room temperature. *Renew. Energy* 161: 207–220. https://doi.org/10.1016/j.renene.2020.07.061.

9 Mojović, L., Nikolić, S., Rakin, M., and Vukasinović, M. (2006). Production of bioethanol from corn meal hydrolyzates. *Fuel* 85 (12-13): 1750–1755. https://doi.org/10.1016/j.fuel.2006.01.018.

10 Cardona, C.A., Quintero, J.A., and Paz, I.C. (2010). Production of bioethanol from sugarcane bagasse: status and perspectives. *Bioresour. Technol.* 101 (13): 4754–4766. https://doi.org/10.1016/j.biortech.2009.10.097.

11 Goshadrou, A., Karimi, K., and Taherzadeh, M.J. (2011). Bioethanol production from sweet sorghum bagasse by Mucor hiemalis. *Ind. Crop. Prod.* 34 (1): 1219–1225. https://doi.org/10.1016/j.indcrop.2011.04.018.

12 Senatore, A., Dalena, F., Sola, A. et al. (2019). First-generation feedstock for bioenergy production. *Second Third Gener. Feed. Evol. Biofuels* 35–57. https://doi.org/10.1016/B978-0-12-815162-4.00002-1.

13 Encinas-Oropesa, A., Simms, N.J., Nicholls, J.R. et al. (2014). Effect of biofuel-derived contaminants on coated industrial gas turbines blade materials. *Mater. Corros.* 65 (2): 206–216. https://doi.org/10.1002/maco.201307070.

14 Alves-Fortunato, M., Labaume, J., Cologon, P., and Barré, L. (2018). Biofuel surrogate oxidation: insoluble deposits formation studied by small-angle X-ray scattering and small angle neutron scattering. *Energy Fuel* 32 (9): 9559–9567. https://doi.org/10.1021/acs.energyfuels.8b02055.

15 Tsvei, Y.P., Prysiazhniuk, O.I., Horash, O.S. et al. (2020). Effect of crop rotation and fertilization of sugar beet on the formation of maximum bioethanol yield. *Plant. Arch.* 20 (S1): 268–274.

16 Niphadkar, S., Bagade, P., and Ahmed, S. (2018). Bioethanol production: insight into past, present and future perspectives. *Biofuels* 9 (2): 229–238. https://doi.org/10.1080/17597269.2017.1334338.

17 Iranmanesh, M. (2011). Combustion analysis and study on smoke and NOX emissions of diesel and biodiesel fueled CI engines with ethanol blends. *Am. Soc. Mech. Eng. Intern. Combust. Engine Div. ICE* 44427: 1069–1077. https://doi.org/10.1115/ICEF2011-60011.

18 Antolini, E. and Gonzalez, E.R. (2010). Alkaline direct alcohol fuel cells. *J. Power Sources* 195 (11): 3431–3450. https://doi.org/10.1016/j.jpowsour.2009.11.145.

19 Paul, A., Bose, P.K., Panua, R.S., and Banerjee, R. (2013). An experimental investigation of performance-emission trade off of a CI engine fueled by diesel-compressed natural gas (CNG) combination and diesel-ethanol blends with CNG enrichment. *Energy* 55: 787–802. https://doi.org/10.1016/j.energy.2013.04.002.

20 Pattanaik, L., Pattnaik, F., Saxena, D.K., and Naik, S.N. (2019). Biofuels from agricultural wastes. *Second Third Gener. Feed. Evol. Biofuels* 103–142. https://doi.org/10.1016/B978-0-12-815162-4.00005-7.

21 Mahjoub, N., Sahebi, H., Mazdeh, M., and Teymouri, A. (2020). Optimal design of the second and third generation biofuel supply network by a multi-objective model. *J. Clean. Prod.* 256: 120355. https://doi.org/10.1016/j.jclepro.2020.120355.

22 Boopathi, D., Thiyagarajan, S., Edwin Geo, V., and Madhankumar, S. (2020). Effect of the second generation and third generation biofuel blend on performance, emission and combustion characteristics of CI engine. *Int. J. Ambient. Energy* 41 (7): 767–774. https://doi.org/10.1080/01430750.2018.1492439.

23 Lee, R.A. and Lavoie, J.M. (2013). From first- to third-generation biofuels: challenges of producing a commodity from a biomass of increasing complexity. *Anim. Front.* 3: 6–11. https://doi.org/10.2527/af.2013-0010.

24 Damartzis, T. and Zabaniotou, A. (2011). Thermochemical conversion of biomass to second generation biofuels through integrated process design-A review. *Renew. Sust. Energ. Rev.* 15 (1): 366–378. https://doi.org/10.1016/j.rser.2010.08.003.

25 Berndes, G., Hansson, J., Egeskog, A., and Johnsson, F. (2010). Strategies for 2nd generation biofuels in EU – Co-firing to stimulate feedstock supply development and process integration to improve energy efficiency and economic competitiveness. *Biomass Bioenergy* 34 (2): 227–236. https://doi.org/10.1016/j.biombioe.2009.07.007.

26 Leng, R.A. (2008). Decline in available world resources; implications for livestock production systems in Asia. *Livest. Res. Rural. Dev.* 20 (1): 2008.

27 Scurlock, J.M.O., Dayton, D.C., and Hames, B. (2000). Bamboo: an overlooked biomass resource? *Biomass Bioenergy* 19 (4): 229–244. https://doi.org/10.1016/S0961-9534(00)00038-6.

28 Sharma, V., Bhat, B., Gupta, M. et al. (2018). Role of systematic biology in biorefining of lignocellulosic residues for biofuels and chemicals production. *Sustain. Biotechnol. Enzym. Resour. Renew. Energy* 5–55. https://doi.org/10.1007/978-3-319-95480-6_2.

29 Hui, Z., Huaxiao, Y., Mengmeng, Z., and Song, Q. (2010). Pyrolysis characteristics and kinetics of macroalgae biomass using thermogravimetric analyzer. *World Acad. Sci. Eng. Technol.* 433–440: 1297–1301. https://doi.org/10.4028/www.scientific.net/amr.433-440.1297.

30 Kandasamy, M. and Senthilkumar, D. (2020). Combustion and emission behaviour of honge biofuel in a thermal barrier coated diesel engine suitable for agriculture. *Lect. Notes Mech. Eng.* 121–127. https://doi.org/10.1007/978-981-15-1124-0_10.

31 Nikkhah, A., El Haj, A.M., Rosentrater, K.A. et al. (2020). Comparative review of three approaches to biofuel production from energy crops as feedstock in a developing country. *Bioresour. Technol. Rep.* 10: https://doi.org/10.1016/j.biteb.2020.100412.

32 Saladini, F., Patrizi, N., Pulselli, F.M. et al. (2016). Guidelines for emergy evaluation of first, second and third generation biofuels. *Renew. Sust. Energ. Rev.* 66: 221–227. https://doi.org/10.1016/j.rser.2016.07.073.

33 Saravanan, A.P., Pugazhendhi, A., and Mathimani, T. (2020). A comprehensive assessment of biofuel policies in the BRICS nations: implementation, blending target and gaps. *Fuel* 272: 117635. https://doi.org/10.1016/j.fuel.2020.117635.

34 Fonseca, J.M., Teleken, J.G., de Cinque, A.V. et al. (2018). Microalgae biodiesel production in China : a preliminary economic analysis. *Renew. Sust. Energ. Rev.* 104: 296–306.

35 van Meijl, H., Smeets, E., and Zilberman, D. (2015). Bioenergy economics and policies. *Bioenergy Sustain. Bridg. Gaps.* 72: 683–708.

36 Acharya, R.N. and Perez-Pena, R. (2020). Role of comparative advantage in biofuel policy adoption in Latin America. *Sustain* 12 (4): 1411. https://doi.org/10.3390/su12041411.

37 Shankar, B., Thaiprasert, N., Gheewala, S., and Smith, R. (2017). Policies for healthy and sustainable edible oil consumption: a stakeholder analysis for Thailand. *Public Health Nutr.* 20 (6): 1126–1134. https://doi.org/10.1017/S1368980016003037.

38 Bereczky, A. (2017). Effect of the use of waste vegetable oil based biodiesel on the landscape in diesel engines. *Therm. Sci.* 21(1 Part B): 567–579. https://doi.org/10.2298/TSCI150630280B.

39 Favretto, N., Stringer, L.C., and Dougill, A.J. (2012). Policy and institutional frameworks for the promotion of sustainable biofuels in Mali. *Sustain. Res. Inst. Pap.* 103: 1–35.

40 Seiple, T.E., Skaggs, R.L., Fillmore, L., and Coleman, A.M. (2020). Municipal wastewater sludge as a renewable, cost-effective feedstock for transportation biofuels using hydrothermal liquefaction. *J. Environ. Manag.* 270: https://doi.org/10.1016/j.jenvman.2020.110852.

41 Sims, R.E.H., Mabee, W., Saddler, J.N., and Taylor, M. (2010). An overview of second generation biofuel technologies. *Bioresour. Technol.* 101 (6): 1570–1580. https://doi.org/10.1016/j.biortech.2009.11.046.

42 Mahmoodi, P., Karimi, K., and Taherzadeh, M.J. (2018). Efficient conversion of municipal solid waste to biofuel by simultaneous dilute-acid hydrolysis of starch and pretreatment of lignocelluloses. *Energy Convers. Manag.* 166: 569–578. https://doi.org/10.1016/j.enconman.2018.04.067.

43 Andersson, V., Broberg Viklund, S., Hackl, R. et al. (2014). Algae-based biofuel production as part of an industrial cluster. *Biomass Bioenergy* 71: 113–124. https://doi.org/10.1016/j.biombioe.2014.10.019.

44 Chhandama, M.V.L., Satyan, K.B., Changmai, B. et al. (2021). Microalgae as a feedstock for the production of biodiesel: a review. *Bioresour. Technol. Rep.* 15: 100771. https://doi.org/10.1016/j.biteb.2021.100771.

45 Karp, A. and Richter, G.M. (2011). Meeting the challenge of food and energy security. *J. Exp. Bot.* 62 (10): 3263–3271. https://doi.org/10.1093/jxb/err099.

46 Levidow, L. and Papaioannou, T. (2014). UK biofuel policy: envisaging sustainable biofuels, shaping institutions and futures. *Environ. Plan. A* 46 (2): 280–298. https://doi.org/10.1068/a45711.

47 Mondal, S., Pandey, J.K., and Kumar, S. (2019). Solar thermal pyrolysis of karanja seeds for a sustainable approach for liquid biofuel utilization. *Nat. Environ. Pollut. Technol.* 18 (3): 941–948.

48 Chhabra, M., Saini, B.S., and Dwivedi, G. (2021). Impact assessment of biofuel from waste neem oil. *Energy Sources, Part A* 43 (24): 3381–3392. https://doi.org/10.1080/15567036.2019.1623946.

49 Basili, M. and Fontini, F. (2012). Biofuel from Jatropha curcas: environmental sustainability and option value. *Ecol. Econ.* 78: 1–8. https://doi.org/10.1016/j.ecolecon.2012.03.010.

50 Shikha, K. and Rita, C.Y. (2012). Biodiesel production from non edible-oils: a review. *J. Chem. Pharm. Res.* 4 (9): 4219–4230. https://doi.org/10.32438/wpe.0602149.

51 Singh, A.P. and Anbumani, K. (2011). Use of esterified mustard, karanja and neem oils as a substitute for diesel fuel in C.I. engine. *Int. J. Appl. Eng. Res.* 6 (8): 1061–1073.

52 Susmozas, A., Martín-Sampedro, R., Ibarra, D. et al. (2020). Process strategies for the transition of 1G to advanced bioethanol production. *Processes* 8 (10): 1310. https://doi.org/10.3390/pr8101310.

53 Ayodele, B.V., Alsaffar, M.A., and Mustapa, S.I. (2020). An overview of integration opportunities for sustainable bioethanol production from first- and second-generation sugar-based feedstocks. *J. Clean. Prod.* 245: 118857. https://doi.org/10.1016/j.jclepro.2019.118857.

54 Singh, D., Sharma, D., Soni, S.L. et al. (2020). A review on feedstocks, production processes, and yield for different generations of biodiesel. *Fuel* 262: 116553. https://doi.org/10.1016/j.fuel.2019.116553.

55 Knothe, G. and Razon, L.F. (2017). Biodiesel fuels. *Prog. Energy Combust. Sci.* 58: 36–59. https://doi.org/10.1016/j.pecs.2016.08.001.

56 Changmai, B., Wheatley, A.E.H., Rano, R. et al. (2021). A magnetically separable acid-functionalized nanocatalyst for biodiesel production. *Fuel* 305: 121576. https://doi.org/10.1016/j.fuel.2021.121576.

57 Gustafsson, M. and Anderberg, S. (2021). Dimensions and characteristics of biogas policies – Modelling the European policy landscape. *Renew. Sust. Energ. Rev.* 135: 110200. https://doi.org/10.1016/j.rser.2020.110200.

58 Artiukhina, E. and Grammelis, P. (2016). Modeling of biofuel pellets torrefaction in a realistic geometry. *Therm. Sci.* 20 (4): 1223–1231. https://doi.org/10.2298/TSCI151130156A.

59 Mohammed, Y.S., Mustafa, M.W., Bashir, N., and Ibrahem, I.S. (2017). Existing and recommended renewable and sustainable energy development in Nigeria based on autonomous energy and microgrid technologies. *Renew. Sust. Energ. Rev.* 75 (C): 820–838. https://doi.org/10.1016/j.rser.2016.11.062.

60 Pimentel, D., Marklein, A., Toth, M. et al. (2008). Biofuel impacts on world food supply: use of fossil fuel, land and water resources. *Energies* 1: 4178. https://doi.org/10.3390/en1010041.

61 Bertsch, J. and Müller, V. (2015). Bioenergetic constraints for conversion of syngas to biofuels in acetogenic bacteria. *Biotechnol. Biofuels* 8 (1): 1–12. https://doi.org/10.1186/s13068-015-0393-x.

62 Kröger, M. and Müller-Langer, F. (2012). Review on possible algal-biofuel production processes. *Biofuels* 3: 333–349. https://doi.org/10.4155/bfs.12.14.

63 Lapuerta, M., Rodríguez-Fernández, J., Agudelo, J.R., and Boehman, A.L. (2013). Blending scenarios for soybean oil derived biofuels with conventional diesel. *Biomass Bioenergy* 49: 74–85. https://doi.org/10.1016/j.biombioe.2012.12.002.

64 Deane, P., Gallachóir, B.Ó., and Shea, R.O. (2017). Biofuels for aviation. *Eur. Energy Transit. Insights Policy Mak.* 79–88. https://doi.org/10.1016/b978-0-12-809806-6.00012-2.

65 Usmani, R.A. (2020). Advanced biofuels: review of effects on environment and socioeconomic development. *Eur. J. Sustain. Dev. Res.* 4 (4): em0129. https://doi.org/10.29333/ejosdr/8245.

66 Bergthorson, J.M. and Thomson, M.J. (2015). A review of the combustion and emissions properties of advanced transportation biofuels and their impact on existing and future engines. *Renew. Sust. Energ. Rev.* 42: 1393–1417. https://doi.org/10.1016/j.rser.2014.10.034.

67 Ribeiro, L.A., Pereira da Silva, P., Ribeiro, L., and Dotti, F.L. (2017). Modelling the impacts of policies on advanced biofuel feedstocks diffusion. *J. Clean. Prod.* 142: 2471–2479. https://doi.org/10.1016/j.jclepro.2016.11.027.

68 Su, Y., Zhang, P., and Su, Y. (2015). An overview of biofuels policies and industrialization in the major biofuel producing countries. *Renew. Sust. Energ. Rev.* 50 (C): 991–1003. https://doi.org/10.1016/j.rser.2015.04.032.

69 Das, S. (2020). The national policy of biofuels of India – a perspective. *Energy Policy* 143: 111595. https://doi.org/10.1016/j.enpol.2020.111595.

70 Dagaut, P., Bedjanian, Y., Dayma, G. et al. (2019). Emission of carbonyl and polyaromatic hydrocarbon pollutants from the combustion of liquid fuels: impact of biofuel blending. *J. Eng. Gas Turbines Power* 141 (3): 031028. https://doi.org/10.1115/1.4040712.

71 Manivannan, A., Prabu, R., and Kumar, K.M. (2021). Investigation on influence of blending Jatropha biofuel with diesel to improve fuel quality. *Aust. J. Mech. Eng.* 19 (1): 49–56. https://doi.org/10.1080/14484846.2019.1567029.

72 Kumar, M., Karmakar, S., Kumar, S., and Basu, S. (2021). Experimental investigation on spray characteristics of Jet A-1 and alternative aviation fuels. *Int. J. Spray. Combust. Dyn.* 13 (1-2): 17568277211010140. https://doi.org/10.1177/17568277211010140.

73 Özbuğday, F.C. and Erbas, B.C. (2015). How effective are energy efficiency and renewable energy in curbing CO2 emissions in the long run? A heterogeneous panel data analysis. *Energy* 82: 734–745. https://doi.org/10.1016/j.energy.2015.01.084.

74 Shiferaw, B., Prasanna, B.M., Hellin, J., and Bänziger, M. (2011). Crops that feed the world 6. Past successes and future challenges to the role played by maize in global food security. *Food Secur.* 3 (3): 307–327. https://doi.org/10.1007/s12571-011-0140-5.

75 Limae Silva, L.F., Gonçalves, W.M., Maluf, W.R. et al. (2019). Energy and budget balances for sweet potato-based ethanol production. *Pesqui. Agropecu. Bras.* 54: https://doi.org/10.1590/S1678-3921.PAB2019.V54.26521.

76 Menon, V. and Rao, M. (2012). Trends in bioconversion of lignocellulose: biofuels, platform chemicals & biorefinery concept. *Prog. Energy Combust. Sci.* 38 (4): 522–550. https://doi.org/10.1016/j.pecs.2012.02.002.

77 Kularathne, I.W., Gunathilake, C.A., Rathneweera, A.C. et al. (2019). The effect of use of biofuels on environmental pollution-A review. *Int. J. Renew Energy Res.* 9 (3): 1355–1367.

78 Mehmood, M.A., Ibrahim, M., Rashid, U. et al. (2017). Biomass production for bioenergy using marginal lands. *Sustain. Prod. Consum.* 9: 3–21. https://doi.org/10.1016/j.spc.2016.08.003.

79 Agrawal, T., Jadhav, S.K., and Quraishi, A. (2019). Bioethanol production from an agrowaste, deoiled rice bran by saccharomyces cerevisiae MTCC 4780 via optimization of fermentation parameters. *Environ. Asia* 12 (1): 20–24. https://doi.org/10.14456/ea.2019.3.

80 Mathimani, T. and Pugazhendhi, A. (2019). Utilization of algae for biofuel, bio-products and bio-remediation. *Biocatal. Agric. Biotechnol.* 17: 326–330. https://doi.org/10.1016/j.bcab.2018.12.007.

81 Adeniyi, O.M., Azimov, U., and Burluka, A. (2018). Algae biofuel: current status and future applications. *Renew. Sust. Energ. Rev.* 90: 316–335. https://doi.org/10.1016/j.rser.2018.03.067.

82 Adenle, A.A., Haslam, G.E., and Lee, L. (2013). Global assessment of research and development for algae biofuel production and its potential role for sustainable development in developing countries. *Energy Policy* 61: 182–195. https://doi.org/10.1016/j.enpol.2013.05.088.

83 Thompson, V.S., Ray, A.E., Hoover, A. et al. (2020). Assessment of municipal solid waste for valorization into biofuels. *Environ. Prog. Sustain. Energy* 39 (4): e13290. https://doi.org/10.1002/ep.13290.

84 Gutiérrez Ortiz, F.J., Kruse, A., Ramos, F., and Ollero, P. (2019). Integral energy valorization of municipal solid waste reject fraction to biofuels. *Energy Convers. Manag.* 180: 1167–1184. https://doi.org/10.1016/j.enconman.2018.10.085.

85 Pieta, I.S., Epling, W.S., Kazmierczuk, A. et al. (2018). Waste into fuel – catalyst and process development for MSW valorisation. *Catalysts* 8 (3): 113. https://doi.org/10.3390/catal8030113.

86 Gebreslassie, M.G., Gebreyesus, H.B., Gebretsadik, M.T. et al. (2020). Characterization of municipal solid waste's potential for power generation at Mekelle City as a waste minimisation strategy. *Int. J. Sustain. Eng.* 13 (1): 68–75. https://doi.org/10.1080/19397038.2019.1645757.

87 Hill, J. (2007). Environmental costs and benefits of transportation biofuel production from food- and lignocellulose-based energy crops. A review. *Agron. Sustain. Dev.* 27 (1): 1–12. https://doi.org/10.1051/agro:2007006.

88 Hill, J., Nelson, E., Tilman, D. et al. (2006). Environmental, economic, and energetic costs and benefits of biodiesel and ethanol biofuels. *Proc. Natl. Acad. Sci. U. S. A.* 103 (30): 11206–11210. https://doi.org/10.1073/pnas.0604600103.

89 Skaggs, R.L., Coleman, A.M., Seiple, T.E., and Milbrandt, A.R. (2018). Waste-to-energy biofuel production potential for selected feedstocks in the conterminous United States. *Renew. Sust. Energ. Rev.* 82: 2640–2651. https://doi.org/10.1016/j.rser.2017.09.107.

90 Changmai, B., Rano, R., Vanlalveni, C., and Rokhum, L. (2021). A novel Citrus sinensis peel ash coated magnetic nanoparticles as an easily recoverable solid catalyst for biodiesel production. *Fuel* 286: 119447. https://doi.org/10.1016/j.fuel.2020.119447.

91 Whittaker, C. (2018). Production of wood pellets from waste wood. *Greenh. Gas Balanc. Bioenergy Syst.* 181–191. https://doi.org/10.1016/B978-0-08-101036-5.00012-4.

92 Coll, R., Udas, S., and Jacoby, W.A. (2001). Conversion of the rosin acid fraction of crude tall oil into fuels and chemicals. *Energy Fuel* 15 (5): 1166–1172. https://doi.org/10.1021/ef010018a.

93 Oasmaa, A., McKeough, P., Kuoppala, E., and Kyllönen, H. (1997). Liquid-phase thermal treatment of tall oil soap into hydrocarbon fuels. *Dev. Thermochem. Biomass Convers.* 696–710. https://doi.org/10.1007/978-94-009-1559-6_55.

94 Malode, S.J., Prabhu, K.K., Mascarenhas, R.J. et al. (2021). Recent advances and viability in biofuel production. *Energy. Convers. Manag. X* 10: 100070. https://doi.org/10.1016/j.ecmx.2020.100070.

95 Opatokun, S.A., Lopez-Sabiron, A.M., Ferreira, G., and Strezov, V. (2017). Life cycle analysis of energy production from food waste through anaerobic digestion, pyrolysis and integrated energy system. *Sustain.* 9 (10): 1804. https://doi.org/10.3390/su9101804.

96 Mendoza, T.C. and Mendoza, B.C. (2016). A review of sustainability challenges of biomass for energy : focus in the philippines. *J. Agric. Technol.* 12: 281–310.

97 Madadian, E., Amiri, L., and Lefsrud, M. (2020). Thermodynamic analysis of wood pellet gasification in a downdraft reactor for advanced biofuel production. *Waste and Biomass Valorization* 11 (7): 3665–3676. https://doi.org/10.1007/s12649-019-00663-4.

98 Girard, P. and Fallot, A. (2006). Review of existing and emerging technologies for the production of biofuels in developing countries. *Energy Sustain. Dev.* 10 (2): 92–108. https://doi.org/10.1016/S0973-0826(08)60535-9.

99 Holmatov, B., Schyns, J.F., Krol, M.S. et al. (2021). Can crop residues provide fuel for future transport? Limited global residue bioethanol potentials and large associated land, water and carbon footprints. *Renew Sustain. Energy Rev.* 149 (C): https://doi.org/10.1016/j.rser.2021.111417.

100 UNCTAD (2016). Second generation biofuel markets: satet of play, trade and developing country perspectives. *United Nations Conf. Trade. Dev.* 18–38.

101 Efroymson, R.A., Dale, V.H., and Langholtz, M.H. (2017). Socioeconomic indicators for sustainable design and commercial development of algal biofuel systems. *GCB Bioenergy* 9 (6): 1005–1023. https://doi.org/10.1111/gcbb.12359.

102 Cardona-Alzate, C.A., Serna-Loaiza, S., and Ortiz-Sanchez, M. (2020). Sustainable biorefineries: what was learned from the design, analysis and implementation. *J. Sustain. Dev. Energy, Water Environ. Syst.* 8 (1): 88–117. https://doi.org/10.13044/j.sdewes.d7.0268.

103 Haug, M. (2007). Renewable energy in future energy supply: a renaissance in waiting. *Q. J. Int. Agric.* 46 (4): 305–324.

104 Mohsin, R., Kumar, T., Majid, Z.A. et al. (2017). Assessment of biofuels in aviation industry for environmental sustainability. *Chem. Eng. Trans.* 56: 1189–1119. https://doi.org/10.3303/CET1756199.

105 O'Connell, A., Kousoulidou, M., Lonza, L., and Weindorf, W. (2019). Considerations on GHG emissions and energy balances of promising aviation biofuel pathways. *Renew. Sust. Energ. Rev.* 101: 504–515. https://doi.org/10.1016/j.rser.2018.11.033.

106 Wang, Z., Osseweijer, P., and Posada, J.A. (2020). Human health impacts of aviation biofuel production: exploring the application of different life cycle impact assessment (LCIA) methods for biofuel supply chains. *Processes* 8 (2): 158. https://doi.org/10.3390/pr8020158.

107 Prussi, M., O'Connell, A., and Lonza, L. (2019). Analysis of current aviation biofuel technical production potential in EU28. *Biomass Bioenergy* 130: 105371. https://doi.org/10.1016/j.biombioe.2019.105371.

108 Chuck, C.J. and Donnelly, J. (2014). The compatibility of potential bioderived fuels with Jet A-1 aviation kerosene. *Appl. Energy* 118: 83–91. https://doi.org/10.1016/j.apenergy.2013.12.019.

109 ASTM (2015). Standard specification for jet B wide-cut aviation turbine fuel 1. *Astm* 05 (03): 10. https://doi.org/10.1520/D6615-15AR19.

110 McCollum, C.J., Ramsey, S.M., Bergtold, J.S., and Andrango, G. (2021). Estimating the supply of oilseed acreage for sustainable aviation fuel production: taking account of farmers' willingness to adopt. *Energy. Sustain. Soc.* 11 (1): 1–22. https://doi.org/10.1186/s13705-021-00308-2.

111 Barelli, L., Bidini, G., Pelosi, D., and Sisani, E. (2021). Enzymatic biofuel cells: a review on flow designs. *Energies* 14 (4): 910. https://doi.org/10.3390/en14040910.

4

Biofuel Production Technologies – An Overview

4.1 Introduction

Due to industrialisation and increased energy demands, the biofuels proved to be absolutely necessary for satisfying further thirst. Environmental regulations also supported biofuels. Biofuels can prove to be an effective solution for the ever-growing energy need, both environmentally and from an efficiency point of view. It can successfully replace traditionally used highly polluting fuels.

Advanced biofuels overcome the drawback of the conventional biofuels with respect to emission, feedstock and technology, which enhanced the opportunities and practices in the biofuel market. The most important concept regarding the advance biofuels is the step towards the best from the waste. The two important factors that play a vital role are the feedstock and technology. The selection of suitable feedstock for the biofuel production and suitable technology to process the feedstock has drawn the attention of the global market.

Feedstock such as residues of agricultural waste [1, 2], municipal waste [3], non-food crops [4], microbes [5], and biological plants is selected depending upon the types of biofuels to be produced. Similarly, the selection of a suitable technology for processing the feedstock available at domestic and regional levels to overcome the energy crisis is to be explored. Continuous efforts are being made to explore the biofuel technology on a commercial scale to meet the future demands. The biofuel production technology is expected to bring the new reforms in the biofuel industries in terms of commercial plants, future demands and optimisation costs [6, 7]. The current methods of biofuels, especially biodiesel, are to be optimised for the rapid and extensive use of technology at the industrial level to achieve the biodiesel market as well as sustainability.

Thermal cracking, also known as pyrolysis, microemulsion and transesterification are the primary approaches to produce the biodiesel [8–10]. Amongst these approaches, the transesterification reaction is the general process for biodiesel production, which mostly includes catalytic and non-catalytic processes [11–13]. The advanced biofuel technologies, such as steam gasification [14, 15], ultrasonic [16, 17], microwave intensification [18], catalytic cracking [19–21], and hydro deoxygenate of vegetable oils for biogasoline and green diesel, are different methods for the biofuel production such as biogas, biogasoline, biodiesel and green diesel.

Sustainability in Biofuel Production Technology, First Edition. Pratibha S. Agrawal, Pramod N. Belkhode, and Samuel Lalthazuala Rokhum.

Biofuel innovation mainly considers genetic modifications in plant feedstock and chemo, bio and enzymatic catalysts, changing to dual or flexible fuel engine mode vehicles and searching for alternative energy sources for different geographical landscapes. Renowned companies, R and D sector as well as start-ups are more concerned in reducing the annual petrol consumption and replacing it with biofuel to lower the economic burden. It will gradually save tonnes of CO_2 per year.

4.2 Industry Challenges Associated with Biofuels

The most commonly faced industry challenges are as follows:

1) **Wastewater treatment from biodiesel industry**: Water is potentially a major required factor in any industry. Treatment of water from biodiesel production industry itself poses another challenge to the industries. To treat the water, make the process costlier, and to leave the treatment itself are banned due to environmental danger.
2) **High production cost**: Recent biofuel production estimates show that the biofuels are two to three times more expensive than traditional petroleum fuels when compared on an energy equivalent basis.
3) **Insufficient production of biofuel compounds**: Biofuels are still not produced in sufficient quantities to rely on. Lot of efforts are still needed to make the process viable.
4) **Inconsistent feedstock availability**: Inconsistent industrial-scale production is a major issue due to the lack of feedstock availability. Land crisis for food crops versus feedstock crops is still a major problem. Thus, it leads to high harvesting cost and energy consumption for the production of biofuel.
5) **Conversion technology and integrated biorefinery approach**: Conversion of biomass into value-added products with maximum energy and material recovery is time consuming. Integrated technologies have to be adopted for overcoming all the issues and making the production beneficial in all aspects.

Once the hurdles are rightly guessed, the challenges and solutions can be identified in a more systematic approachable way.

4.3 Edible Vegetable to Non-edible/Low-cost Raw Materials for Biodiesel Production

Varieties of plant species are used for the production of alternative fuels, particularly biodiesel, which is used in the transportation sector to successfully explore the global market. The oil extracted from different plant species is in the form of edible oil or non-edible oil. Biofuels in their first generation were using the food crops as the feedstock, and edible oil was used as the substitution for the conventional food crops for the biodiesel production, which creates imbalance in the food supply chain [22, 23]. This substitution results in a higher cost of edible oil such as palm oil (Obromacao); coconut oil (*Cocos nucifus*); Sunflower oil (*Helianthus annus*); Gingelly oil (*Sesamum indicum*) and soya bean oil (*Glycine max*) [24].

Around 70% of edible oil is imported from other countries mostly from Indonesia and Malaysia. The majority of biodiesel was preferred with the processing of soya bean and canola oil, which created a depletion in the food supply. The production of biodiesel from the processing of non-food crops or edible oil is the solution to overcome the depletion of food supply and maintain the balance with respect to the food supply chain [25] and bioeconomic and global demands. This leads to the development of second-generation biofuels in which non-food crops and non-edible oil benefit in a number of ways, such as sustainable energy, feedstock availability at the local level, involvement of rural people in the production of biofuels, creating job opportunities for the rural people and providing most important solution to the environmental issues such as greenhouse gas emission [26–28].

Different species of non-edible oil are karanja (*Pongamia pinnata* [L]), polanga (*Calophyllum inophyllum*) [29], neem oil (*Azadiracta indica*) [30], jatropha oil (*Jatropha curcas*) [31], rubber (*Hevea brasiliensis*) [32], mahua (*Madhuca indica*) [33], castor oil (*Ricinus communis*) [34, 35] and almond oil (*Prunus dulcis*) [28, 36], which are the renewable sources of biofuels used for the production of biodiesel. The content of methyl ester from karanja oil [37] is around 98% under the optimum condition as compared to the oil content from caster, neem and polanga oil. The properties of the non-edible are shown in Tables 4.1 and 4.2.

The important highlights regarding the production of biodiesel from edible vegetable oil to non-edible vegetable oil are discussed below.

4.3.1 Advantages of Non-edible Oil

The important characteristics of non-edible oil, such as high combustion efficiency, lower sulphur, renewable source of energy, liquid form, biodegradable and easy availability at local and regional levels, attract attention to the use of non-edible oil as the biodiesel fuels.

Table 4.1 Properties of vegetable oil.

Vegetable oil	Kinematic viscosity at 38 °C (Cst)	Cetane no.	Heating value (MJ/kg)	Cloud point (°C)	Pour point (°C)	Flash point (°C)	Density (kg/l)
Diesel	3.06	50	43.8	–	−16	76	0.855
Corn	34.9	37.6	39.5	−1.1	−40.0	277	0.9095
Cottonseed	33.5	41.8	39.5	1.7	−15.0	234	0.9148
Crambe	53.6	44.6	40.5	10.0	−12.2	274	0.9048
Linseed	27.2	34.6	39.3	1.7	−15.0	241	0.9236
Peanut	39.6	41.8	39.8	12.8	−6.7	271	0.9026
Rapeseed	37.0	37.6	39.7	−3.9	−31.7	246	0.9115
Safflower	31.3	41.3	39.5	18.3	−6.7	260	0.9144
Sesame	35.5	40.2	39.3	−3.9	−9.4	260	0.9133
Soyabean	32.6	37.9	39.6	−3.9	−12.2	254	0.9138
Sunflower	33.9	37.1	39.6	7.2	−15.0	274	0.9161
Palm	39.6	42.0	–	31.0	–	267	0.9180

Source: Adapted from Jaichandar and Annamalai [38].

Table 4.2 Biodiesel properties.

Property	Unit	Jatropha	Soyabean	Palm	Waste cooking oil	Biodiesel standards: ASTM D 6751-6702	Biodiesel standards: DIN EN 14214	Diesel fuel
Density at 20 °C	kg/m^3	880	885	880	884	870–900	875–900	850
Viscosity at 40 °C	Mm^2/s	2.37	4.5	5.7	4.5	1.9–6.0	3.5–5.0	2.60
Cloud point	°C	–	1	13	1	–	–	4
Flash point	°C	135	178	164	180	130	120	68
Pour point	°C	2	−7	12	−5.0	−15 to 10	−15 to 10	−20
Water	%	0.025	–	–	0.4	0.03	0.05	0.02
Sulphur	PPM	–	–	–	–	50	50	500
Carbon residue	wt%	0.20	–	–	0.3	–	0.3	0.17
Cetane number	–	61	45	62	57.2	48–60	49	49
Calorific value	MJ/kg	39.2	33.5	33.5	32.9	–	–	42

Source: Adapted from Alnuami et al. [39].

The shorting of the edible oil was discussed in various aspects that are overcome by the non-edible oil or waste edible oil as follows.

1) Biodiesel produced from the processing of the vegetable oil or animal fats through the transesterification process is blended with diesel in a proper proportion to be used in diesel engine [40, 41]. This leads to an increase in the cost of vegetable oil, which again depends upon the production of feedstock. The production of biodiesel from edible oil is currently not economically feasible due to the cost of raw materials.
2) Use of vegetable oil, i.e. edible oil, as a biofuel without any process results in incomplete combustion because the high viscosity of vegetable oil may clog the engine. Generally, fuels with low viscosity are preferred so that they can easily be sprayed for effective combustion [42].
3) Reverse impact may occur due to the utilisation of vegetable oil at a large scale as biodiesel, which may lead to the depletion of food supply that results in an economic imbalance.
4) The important characteristics, such as high viscosity, lower volatility and unburned hydrocarbon chain, are not favourable for use as vegetable oil such as direct fuel as biodiesel. Therefore, vegetable oil is expected to be processed to enhance the characteristics that are suitable for the biodiesel.
5) Non-edible oil produced from the energy crop or sustainable crop, such as karanja, jatropha and mahua, is grown on the marginal land with high oil content, which is favourable to grow under any climatic condition and is environmentally friendly as compared to the edible oil.

4.3.2 Oil Extraction Technologies

Methods commonly adopted to extract the oil from the non-edible oil seed are as follows [43]:

1) **Physical method** [44]:
   An oil expeller is used to extract the oil from seed, and then the resultant oil and seed waste are separated. Furthermore, the extraction process and physicochemical and fatty acid percentages are observed by the analysis of the resultant oil.
2) **Bligh and Dyer method** [45]:
   Grinded seed samples are added to the glass tube and broken with a glass rod with the addition of methanol, chloroform and water to precipitate. The upper layer contains the interfering agent, such as lipid species, which are removed, and the layer that contains the membrane lipids and neural lipids is collected and transferred to a new tube.
3) **Fulch et al. method**:
   Seed samples are homogenised and the mixture is agitated in the orbital shaker at room temperature; homogenate is filtered and washed with water. Furthermore, the mixture is separated into two phases through the centrifuged, and each phase is analysed.
4) **Chemical Solvent Extraction method**:
   Oil is extracted from the seed through repeated washing or percolation with chemicals such as hexane or ether under reflux in a special extraction chamber.

4.3.3 Biodiesel Standards and Characterisation of Non-edible Biodiesel

Non-food crops that are grown on unproductive land contain toxic substance that cannot be used for human beings and are widely explored for producing biodiesel as they do not create fuel versus food nexus. The problem associated with the production of biofuels from edible oil results in high prices as it is not economical as compared to non-edible oil [46, 47].

Structural features of the fatty acid in non-edible oil influenced the physical and chemical properties of biodiesel. The iodine number indicates the total unsaturation within the mixture of fatty materials. A high value of the iodine number indicates higher unsaturation in fats and oils. Heating of fatty acid that is unsaturated results in glycerides during polymerization, which deteriorates the lubricating properties of the thick sludge in the engine and decreases the performance of the engine.

The liquid form of oil is maintained by the unsaturated fatty acid, but if the concentration of poly-unsaturated fatty acids is beyond a certain limit, it results in polymerisation of glycerides, which will block the engine. The cloud point is increased by the high concentration of saturated fatty acid, which makes them undesirable liquid fuels. Therefore, the performance of the biodiesel depends upon the characteristic of oil that depends on the structural features of fatty acid [48].

Similarly, the acid number plays a significant role, which changes during the blending of biodiesel due to the normal oxidation process over time. The change in the acid number is an indication of fuel degradation and emission [49]. A higher acid value will damage fuel pumps and fuel filters. The acid number indicates the quality of oil and expresses the amount of potassium hydroxide necessary to neutralise full fatty acids contained in 1 g of oil.

4.3.4 Technologies of Biodiesel Production from Non-edible Oil

Biodiesel is produced from the non-edible oil, such as jatropha, karanja, mahua and castor, by using a transesterification method with different catalysts such as sodium hydroxide, potassium hydroxide and sulphuric acid. Alcohol reacts with triglycerides of fatty acid (non-edible oil) in the presence of catalysts in the transesterification process for the production of biodiesel. The two commonly used methods for the production of biodiesel from non-edible oils are base-catalysed transesterification and acid-catalysed transesterification. The base-catalysed transesterification technology is mostly preferred over the acid-catalysed transesterification technology due to a faster reaction rate of the former, and hence it is more efficient.

Considering the advantages of edible oil, the technology involved in the biodiesel production, characterisation and properties and its comparison with the biodiesel and ASTM standards are explored to produce the biodiesel from the non-edible oil.

4.4 Development of Chemical Conversion Technologies

The chemical conversion process is a technology that step towards 'waste to energy' in which energy is converted into a useful product using a chemical reaction. The thermal energy is controlled based on the heat density of chemical changes, which is comparatively more than physical energy due to the chemical reaction. Chemical conversion technologies refer to the transfer of chemical energy due to the chemical reaction in the systems such as battery, fuel cell, internal combustion engine and many more. The energy storage system will play an important role in the future for storing energy due to its instability in supply and demand and due to renewable and decentralised energy and seasonal changes. The chemical reaction has the potential for the storage of energy and the transfer of stored energy to the desired system.

The conversion technology is used to convert solid waste into useful products and fuels [50]. This technology refers to the recovery of energy from waste material such as solid waste recovery system, utilisation of municipal waste to convert into solid fuels [51], utilising organic waste to convert into biofuels [52], which reduces the burden of landfilling and environmental pollution. The conversion technology facilitates the conversion of energy available in organic waste, solid waste to value-added products, such as chemicals, and useful products, such as (i) biodiesel, ethanol and oil as liquid fuels, (ii) power generation, heat and steam from the combustion of biogas, (iii) chemical and consumable products from the oil and (iv) food processing through the activated carbon [53].

The conversion technology facilitates the number of processes to enhance the technology, which are mainly classified as chemical, thermochemical and biochemical. The chemical conversion technology includes the transesterification. The value-added products, such as chemical and fuels, are derived from suitable feedstock using chemical catalytic conversion, and the reactions involved during chemical conversion are hydrolysis, dehydration, condensation, reforming, hydrogenation and oxidation.

Lignocellulosic biomass is processed to derive value-added products and chemicals and advanced biofuels using the chemical conversion technology. Lignocellulosic biomass consists of three primary chemical fractions namely cellulose, a glucose polymer and a hemicelluloses process to achieve value-added products as shown in Figure 4.1 [56].

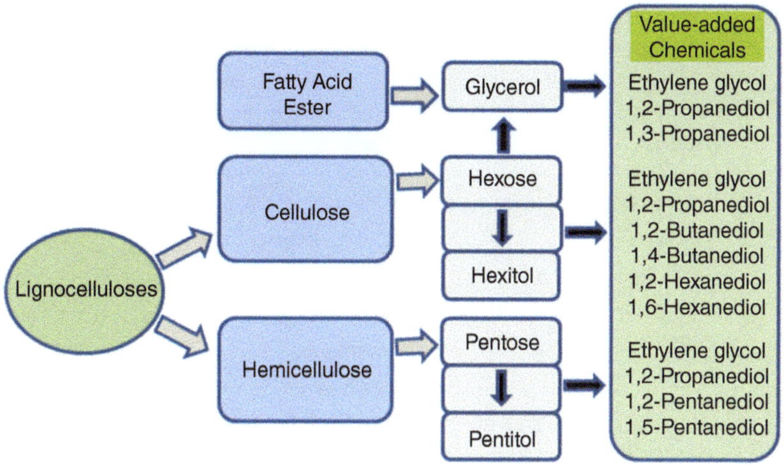

Figure 4.1 Chemical conversion of biomass. *Source:* Adapted from [54, 55].

4.5 Development of Thermochemical Conversion Technologies

The heating or oxidation or both are controlled during the conversion of biomass into a suitable form of energy through thermochemical conversion. Compared to other forms of conversion, it is comfortable with the existing facilities required to form conversion with high productivity. The most common processes of thermochemical in practice for converting biomass into biofuels and chemicals are (i) pyrolysis [57], (ii) liquefaction and (iii) gasification [58]. Various thermochemical processes with the conversion of feedstock into end product are presented in Table 4.3.

Pyrolysis is the main thermochemical process preferred for the conversion of biomass conducted in the absence of oxygen around the temperature range of350–700 °C. The non-availability of oxygen opposes the combustion and thermally decomposes the chemical compounds into the end products such as charcoal, oil and various forms of gases. The flow of conversion of biomass into biofuels is presented in Figure 4.2.

The liquefaction process is the biofuel production from the biomass using a thermochemical conversion process in which biomass is processed in a hot, pressurised water environment for a certain period to convert the solid biopolymeric structure into a liquid component. The long carbon chain molecules in the biomass are broken at high temperature and high pressure, and oxygen is removed in the form of water and carbon dioxide. The chemical reactions during the hydrothermal liquefaction process produce biocrude oil. The properties of end product liquid fuel can be controlled with the operating temperature and pressure in the presence of suitable catalysts and the duration of the chemical reaction [60]. The process in presented in Figure 4.3.

Synthesis gas (syngas) is produced using a gasification process in which solid biomass is converted into fuel gas. The solid biomass contains carbon-based raw material such as coal in the process of producing convenient gaseous fuel using the gasification thermochemical conversion process. This gaseous fuel can be further used in the boiler, engines and gas

Table 4.3 Chemical conversion process.

			End product	
Feedstock	**Process**		**Fuels**	**Co-product**
Dry biomass municipal waste wood and farm residues	Destructive distillation		Charcoal (methanol)	Wood chemical
	Pyrolysis		Chars, oil gases	
	Hydroliquefication		Oils	
	O2 partial combustion	Synthesis gas	Methane	
			Ammonia	
			Methanol	
	Air partial combustion	Low Btu Gas	Heat, steam, electricity	Mineral ash
	Total combustion			

Source: Adapted from Final Report, SERI [59].

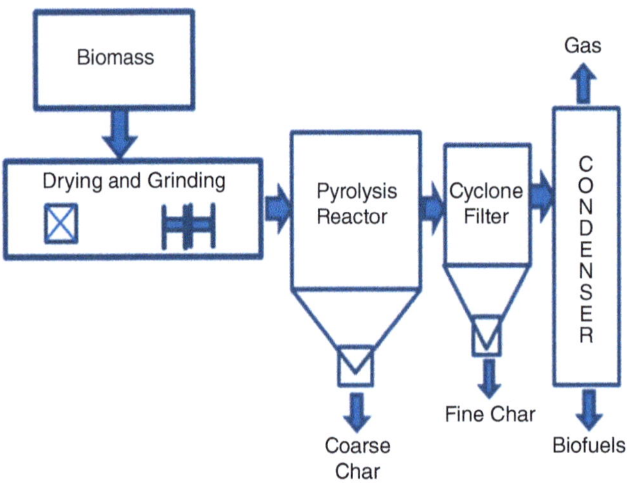

Figure 4.2 Biomass pyrolysis process.

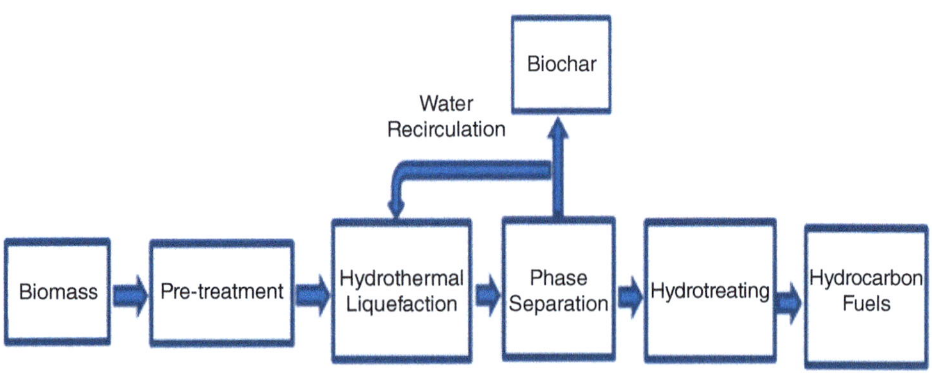

Figure 4.3 Hydrothermal liquefaction process.

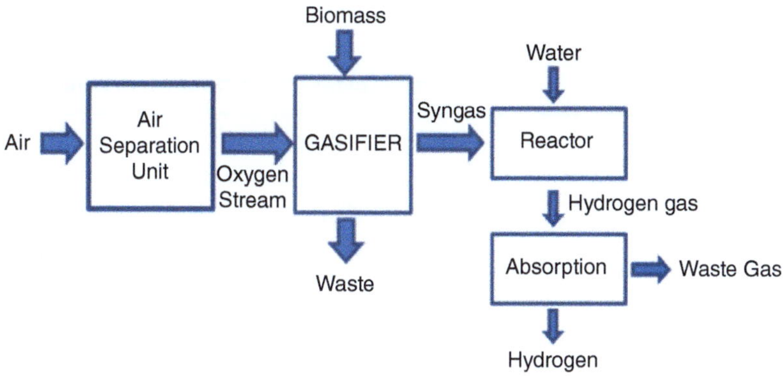

Figure 4.4 Biomass gasification process.

turbine to generate electricity. Hot gas generated by burning the biomass is used as fuel in the boiler to convert water into superheated steam, which is further expanded in the steam turbine to produce electricity. The end product obtained through gasification is the gas with a low calorific value, which is almost in the range of 1000–1250 kcal/(N m^3) with a low heating value around 4–6 MJ/m^3 and 8–14 vol% H_2 content [61, 62]. The gasification thermochemical process is presented in Figure 4.4.

4.6 Development of Biological Conversion Technologies

The biological technology is the branch of biology involved in the use of living system and organisms to develop or produce biochemical, medical or agricultural products. Biofuels are produced by microbes engineered using synthetic biology. Biological technologies help in response to COVID-19 in the identification of the virus and treatment. Biological development plays a vital role in the economy as it is involved in multiple sectors, such as it is applied in the transformation of energy, food, healthcare and value-added products. Molecular biology and biochemical and genetic biology are the branches of the biological technology with the capabilities of delivering value-added products and services with a minimum cost [63, 64].

The conversion of lignocellulosic biomass to ethanol using biological conversion follows (i) the separation of cellulose and hemicellulose of structural polymer lignin from the plant tissue, (ii) the production of free sugars from cellulose and hemicellulose using depolymerisation in which polymers are broken down into nanomeric components and (iii) the production of ethanol from the fermentation of mixed hexose and pentose sugars. Biological conversion of lignocellulosic is possible in the presence of lignin-degrading microorganisms with an ecophysiological environmental bioreactor. Table 4.4 shows different biological conversion processes with various types of feedstock.

4.7 Development of Biochemical Conversion Technologies

Biochemical conversion is a branch in which energy in the form of light or heat is chemically converted into fuels, electricity or stored energy. The fuels from biomass, thermochemical hydrogen production and fuel cell conversion of solar-derived fuels are the separate divisions

Table 4.4 Biological conversion processes.

Feedstock	Biological conversion processes	Biofuels
Sugar cane/sugar beet/corn	Sugar fermentation	Bioethanol
Lignocellulosic	Separate hydrolysis and fermentation	
Algal biomass		
Gasification of different biomass feedstock or municipal solid waste	Syngas fermentation	

of chemical conversion. The energy content of biomass residues grown on the marginal lands is converted into heat, fuels and electricity with the help of different chemical conversion processes, such as anaerobic digestion to methane, microbial digestion to molecular hydrogen and hydrolysis of cellulose to sugar using fermentation to produce alcohols.

The development of the chemical conversion process is shown in Figure 4.5 and Table 4.5.

Organic matter that includes the animal or food waste is broken into pieces in an anaerobic digestion process in the absence of oxygen to produce the biogas and biofertiliser in the digester (Figure 4.6) [63].

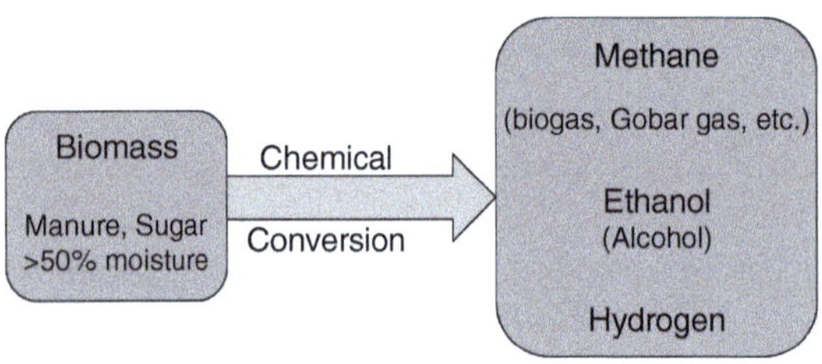

Figure 4.5 Natural flow of biomass through a chemical conversion process.

Table 4.5 Chemical Conversion Process.

		End product	
Feedstock	**Process**	**Fuels**	**Co-product**
Biomass	Anaerobic digestion	Methane	Sludge
Manure	Microbial digestion	Hydrogen	Sludge
Algae	Hydrolysis	Alcohol	Protein
Waste	Sugar fermentation	Ethanol	Sludge
	Enzymatic digestion		
	Aerobic digestion	Compost	

Source: Adapted from Final Report, SERI [59].

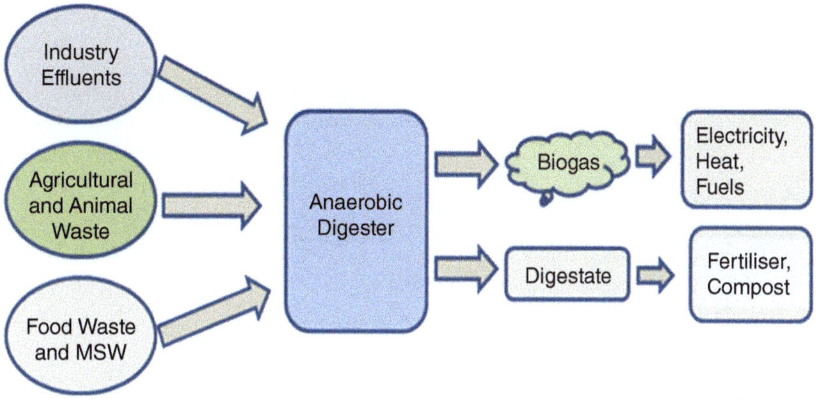

Figure 4.6 Anaerobic digestion process.

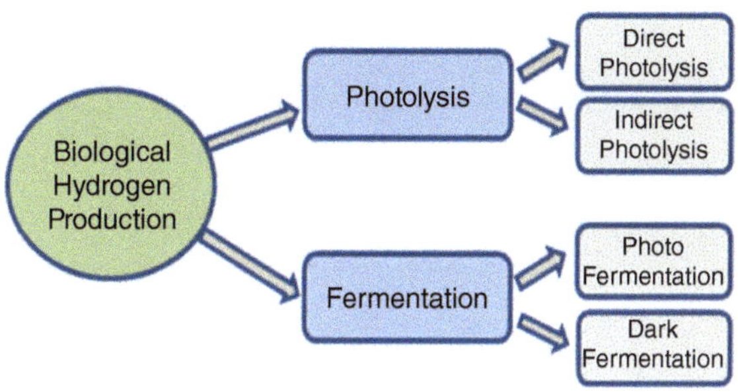

Figure 4.7 Microbial digestion process.

Anaerobic digestion produces biogas as renewable energy from biomass, manure and waste cause less environmental pollution with lower sludge mass generation. It requires more investment for an anaerobic digester and takes much time due to its slow process.

Microorganism is consumed, and biomass is digested to release the hydrogen that takes place in the microbial biomass conversion process. As no light is required in this process, therefore the process is also called the dark fermentation process (Figure 4.7) [65, 66].

The cellulose and hemicellulose present in a feedstock are broken down to form free sugar such as glucose in the hydrolysis process. The free sugar in the form of glucose is fermented to produce ethanol (Figure 4.8).

4.8 Technology Innovation in Biofuel Production

Various biofuel production technologies are being evolved by modifying the process at different production stages. The use of the latest modified equipment has also proven to provide the expected results. Some notable technologies are shown in Figure 4.9.

One of the technologies developed for the biofuel production is visionary fibre technologies developed by the United States, which helps to improve the production efficiency with the

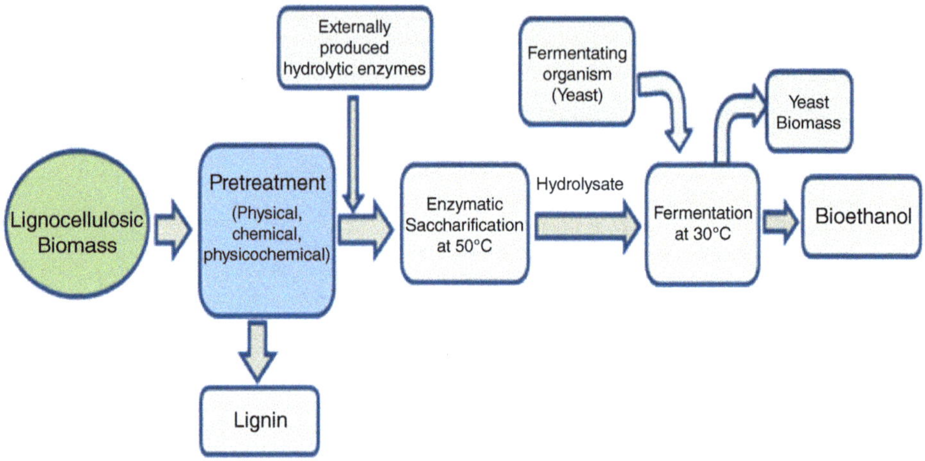

Figure 4.8 Hydrolysis sugar fermentation process.

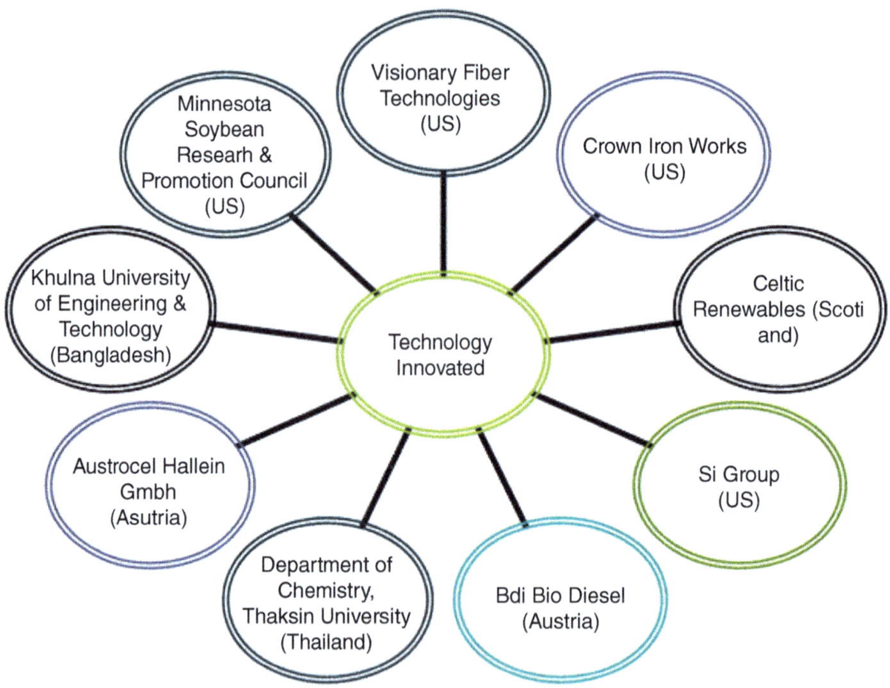

Figure 4.9 Technology developed in biofuel production.

enhancement in the production environment. This includes the modification of the surface area for enhancing the reaction rate, which results in an increase in the efficiency and improvement in the cost analysis. The process cost is improved with the help of suitable chemistry and the separation of reaction product by utilising the fibre reactor technology.

The antioxidant stabiliser technology developed by the Si Group from the United States promotes the fuel stability without the use of any catalyst, process or feedstock in the fuel and is susceptible to oxidation as this antioxidant is developed for fuel stability.

The Celtic Renewable Technology from Scotland is related to bioeconomy developed by the biorefineries in which around 50 000 tonnes of residues are processed each year and promotes the low carbon green bioeconomy.

The Bdi Biodiesel Technology from Austria promotes the production of biodiesel from the multiple feedstock using complete waste oils and fats, which involves a fully continuous process without settling processes.

The Crown Iron Works Technology from the United States promotes the production of biodiesel with a lower operating cost and higher efficiency by using advanced catalyst reduction and economisation.

The Chemistry Department of the Thaksin University from Thailand developed biofuel from the coconut meal and yeast with a cost-efficient process and achieved a high yield through dry processing.

Technology developed by Austrocel Hallein Gmbh from Austria produces high-yield ethanol from wood sugar by fermenting and distilling wood sugar from the viscose pulp mill.

The Minnesota Soybean Research and Promotion Council from the United States produces biodiesel at low cost using liquid plasma and electricity, powered by wind or solar.

Technology developed by Khulna University of Engineering from Bangladesh promotes the production of a healthy feedstock for microalgae cultivation in PBRs using the hybrid anaerobic baffled reactor in which organic and solid are removed and which significantly helps to improve the wastewater at lower cost and enhance the biofuel production.

4.9 Process Integration and Biorefinery

Process integration relates to the combination of different processes into a single piece of equipment or unit to enhance the efficiency of a process plant with the existing infrastructure. In integrated biorefinery, biomass is processes to convert into biofuels and value-added products with high energy content due to the possibility of utilising the input from other processes as well as the existing biofuel industry [67]. Integrated biorefinery has the potential to produce the bioproduct and biofuels with strong, sustainable and environmentally friendly effects at high efficiency. Based on the feedstock, the biorefinery is sugar refinery, vegetable oil refinery, lignocelluloses refinery, and thermochemical and syngas refinery. Feedstock is pre-treated at the initial stage using the decomposition or separation process followed by a conversion process that may be biological, chemical, thermochemical or biochemical processes to convert into biofuels or bioproducts [68, 69]. The maximum conversion efficiency of the biorefineries is expected to be achieved by integrating processes into the industrial cluster with the existing equipment such as boilers, air separation, heat exchanger, reactors, etc. instead of processing at unit level that will increase the positive impact on the economy. Some major advantages of integrated biorefinery are as follows: (i) the utility of present equipment, (ii) the distribution of heat between sources of heat and heat consumers and (iii) the use of the end product obtained from the biorefinery as a feedstock.

An effective utilisation of the waste generated from the feedstock and the combination of the processes to produce the bioproduct is the part of the process integration. Various processes, such as chemical, biochemical, biological or thermochemical, are combined to produce biofuels or bioproducts in order to utilise the resource, minimise the costs and improve the present generation of biofuels. The design of a new plant involves the concept of process integration that increases the effectiveness, efficiency and productivity [70]. The concept of

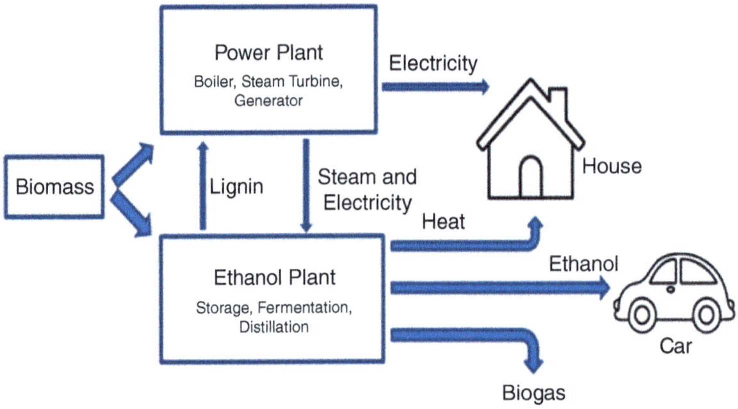

Figure 4.10 Bioenergy from the biomass – integration process.

process integration involves (i) heat integration for optimisation of energy in which the transfer of energy takes place from a cooling or condensation process to a heating or evaporation process using the pinch analysis and (ii) combination of different multiple unit operations such as reaction, separation or heat transfer in one single unit of equipment that is termed as equipment integration. Equipment integration results in the optimisation of design, value engineering, cost reduction and is environmentally friendly, which involves (iii) the integration of product in which the product obtained from one process is considered as the raw materials for another process to form a new product (Figure 4.10) [71].

4.10 Alternatives to Biofuel Production

Some alternative sources for biofuel production can be categorised into three parts as shown in Figure 4.11.

4.11 Technology Survey

During the production of biofuels, various problems/shortcoming are observed and further difficulties that have been overcome by various organisations are discussed.

Agglomeration occurs during the mixing of palm oil because the drier sticks up with the product, which results in a decrease in efficiency of the drier.

The process is improved as the moisture content of palm oil effluents is lowered by the process of evaporation, which is followed by the mixing of palm fruit fibre in the form of bulk material to produce biomass that is used to obtain biofuels or biofertilisers. This environmentally friendly and cost-efficient method is processed by Palmite Processing Engineering.

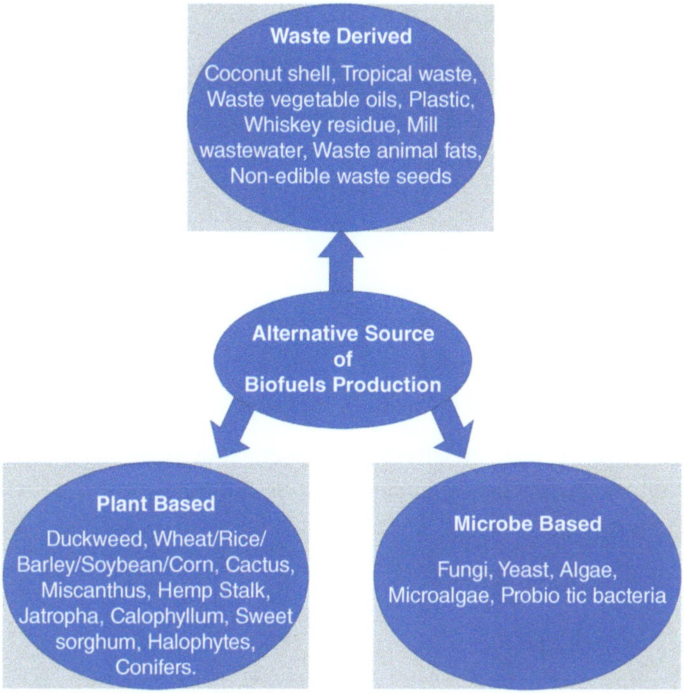

Figure 4.11 Alternative sources for biofuel production.

One of the shortcomings of an unmodified diesel engine is related to efficiency and exhaust emission characterisation.

Engine performance with exhaust emission is improved in biofuel generations. The investigated results of fuel characterisation are enhanced in the third-generation fuel results, highlighting that the use of microalgae biodiesel blends used for diesel engine gives the cleaner fuel.

The oil quality obtained from the fermentation process and the presence of high level free fatty acids affected the production of end products.

Unique properties are achieved by directly adding esterase such as lipase to the fermentation process to facilitate the chemical modification as per Poet Res Inc. of the United States, which increases the utility of corn oil.

Biodiesel production using a domestic process involved a two-step reaction, which is heterogeneous with a long ester exchange reaction time and a low conversion rate with high energy consumption.

The process involved for producing biodiesel using microreaction transesterification using device having simple integral structure results in a short reaction time and a high conversion rate with a high production efficiency and is promoted by Zhejiang Jiaao Environmental Protection Technology Co. Ltd.

4.12 Key Collaborations for Biofuel Production

Various organisations have been collaborated for the enhancement of the biofuel production in view of integrating and updating the technology to promote the production of sustainable biofuels. Some key collaborations for biofuel production are discussed in Table 4.6.

4.13 Market Research on Biofuels

Global biofuel market trends are discussed in Table 4.7.

4.13.1 Technology Commercialisation of Innovation in Biofuel

Innovation in biofuel in technology commercialisation is presented in Figure 4.12, which is based on different countries. Technology commercialisation innovation enhances the economy and production technology.

Table 4.6 Key collaboration for biofuel production.

Key collaborations	Agreement	Work towards
Clariant, Switzerland (1995), has joined forces with Ineratec Germany (1999)	Challenge for a greener future in 2020	Upgrading catalysts were used to support Ineratec's ground breaking gas-to-liquid technology
Sekab E-technology, Swedish (1985) and Praj Industries, India (1983)	Geared towards the production of sustainable biofuels	Coniferous residue conversion into biofuels, green chemicals and bio-based materials
Synthetic Genomics, California based (2005) having partnered with ExxonMobil Texas based (1999)	Producing biofuel from algae	Producing biofuel from algae at an unprecedented scale in 2017

Table 4.7 Global biofuel market trends.

	Compound annual growth rate (CAGR)	USD billion	
		2020	2025
The global biofuel market size	6.9%	135.7	230.5
The global biodiesel market size	7.0%	35.1	49.2
The global bioethanol market size	14.0%	33.7	64.8
The global biobutanol market size	6.9%	7.86	17.78

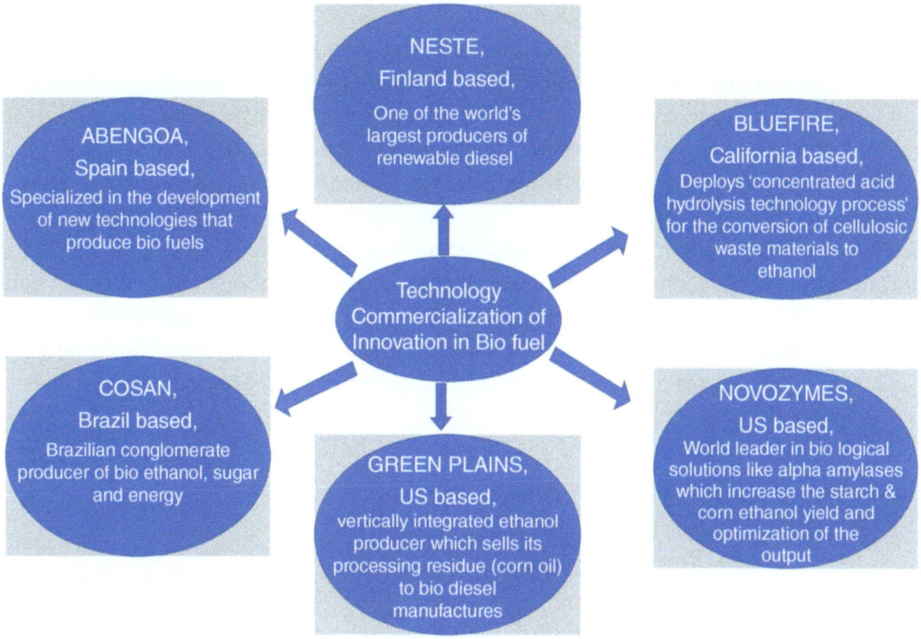

Figure 4.12 Innovation in technology commercialisation.

4.13.2 Start-up Innovation In Biofuel Technology

Biofuel Technology	Country	Year of Technology	Innovation
MANT BIOFUEL	**US**	**2014**	Cost competitive renewable fuel by cheaply farming algae
CELTIC RENEWABLES	SCOTLAND	2011	Produce biofuel from the by-products of the scotch whisky industry
BIOFUEL EVOLUTION	ENGLAND	2016	Convert mixed food and agricultural waste into clean, localised and decentralised bioenergy
INNOLTEK	CANADA	**2010**	Manufacture and market renewable fuels and their biodiesel is made from animal fats and non-edible by-products from food industry

4.14 Future Trends

Recently, all the biofuel industries have pursued the development of the biofuel conversion technologies to meet the global requirement, particularly transportation fuel needs. Furthermore, these industries are growing interest in the enhancement of the advance technology with process integration to optimise large productivity and cost reduction in the production process. New employment in the region, restoration of degraded lands for the production and reduction in greenhouse gas emission are the promising key factors in the production of biofuels. Result of thermochemical conversion technology show an increase in economy due to simple technology in which equipment requirement is the same as that of fossil fuel conversion and the large-scale production capacity is relatively at a lower cost. The trend of biofuels is gaining significant importance globally due to energy demand and the development of conversion systems. It is necessary to promote the conversion technologies in terms of energy production, their limitation, cost, operations and need of conversion systems. Considering the future requirement, direct conversion systems such as storage battery and fuel cell system should be reviewed to fulfil the energy demand. Individual energy conversion system will not be predominating over the other conversion systems instead continuous efforts should be made to conversion processes to meet the overall system requirement in the most effective manner.

4.15 Summary

Biomass conversion technologies produce heat and electricity, which include biochemical, biological, thermochemical and chemical technologies. Conversion technologies follow the combustion, gasification, pyrolysis and anaerobic digestion processes in which biomass feedstock is broken down to produce different types of biofuels. Integrated biorefineries are the new bioindustries equipped with the present process equipment integrated to optimise the process, raw material, and energy to enhance the efficiency. In simple words, the set of different process units should be combined into a single unit to utilise the maximum resources through the concept of heat integration, equipment integration and product integration to fulfil the future demand at a lower cost.

References

1 Johansson, K., Liljequist, K., Ohlander, L., and Aleklett, K. (2010). Agriculture as provider of both food and fuel. *Ambio* 39 (2): 91–99. https://doi.org/10.1007/s13280-010-0017-4.

2 Farmanbordar, S., Karimi, K., and Amiri, H. (2018). Municipal solid waste as a suitable substrate for butanol production as an advanced biofuel. *Energ. Conver. Manage.* 157: 396–408. https://doi.org/10.1016/j.enconman.2017.12.020.

3 Susmozas, A., Martín-Sampedro, R., Ibarra, D. et al. (2020). Process strategies for the transition of 1G to advanced bio ethanol production. *Processes* 8 (10): 1310. https://doi.org/10.3390/pr8101310.

4 Lin, C.Y. and Lu, C. (2021). Development perspectives of promising lignocellulose feedstocks for production of advanced generation biofuels: a review. *Renew. Sustain. Energy Rev.* 136: 110445. https://doi.org/10.1016/j.rser.2020.110445.

5 Kumar, S., Shrestha, P., and Abdul Salam, P. (2013). A review of biofuel policies in the major biofuel producing countries of ASEAN: production, targets, policy drivers and impacts. *Renew. Sustain. Energy Rev.* 26: 822–836. https://doi.org/10.1016/j.rser.2013.06.007.

6 Delvi, M.K., Soudagar, M.E.M., Khan, H. et al. (2019). Biodiesel production utilizing diverse sources, classification of oils and their esters, performance and emission characteristics: a research. *Int. J. Recent Technol. Eng.* 8 (2S3): doi: 10.35940/ijrte.B1183.0782S319.

7 Krishnaprabu, S. (2019). A review on biodiesel production as alternative fuel. *Int. J. Pure Appl. Biosci.* 7 (2): 258–266. https://doi.org/10.18782/2320-7051.7441.

8 Gashaw, A., Getachew, T., and Teshita Mohammed, A. (2015). A review on biodiesel production as alternative fuel soil animal diversity in agroecosystems view project a review on biodiesel production as alternative. *Fuel* 4 (2): 80–85.

9 Ong, Y.K. and Bhatia, S. (2010). The current status and perspectives of biofuel production via catalytic cracking of edible and non-edible oils. *Energy* 35 (1): 111–119. https://doi.org/10.1016/j.energy.2009.09.001.

10 Crouch, L.I., Labourel, A., Walton, P.H. et al. (2016). The contribution of non-catalytic carbohydrate binding modules to the activity of lytic polysaccharide monooxygenases. *J. Biol. Chem.* 291 (14): 7439–7449. https://doi.org/10.1074/jbc.M115.702365.

11 Piazzi, S., Menin, L., Antolini, D. et al. (2021). Potential to retrofit existing small-scale gasifiers through steam gasification of bio mass residues for hydrogen and biofuels production. *Int. J. Hydrogen Energy* 46 (13): 8972–8985. https://doi.org/10.1016/j.ijhydene.2021.01.004.

12 Larsson, A., Kuba, M., Berdugo Vilches, T. et al. (2021). Steam gasification of bio mass – typical gas quality and operational strategies derived from industrial-scale plants. *Fuel Process. Technol.* 212: 106609. https://doi.org/10.1016/j.fuproc.2020.106609.

13 de Medeiros, E.F., Vieira, B.M., de Pereira, C.M.P. et al. (2019). Production of biodiesel using oil obtained from fish processing residue by conventional methods assisted by ultrasonic waves: heating and stirring. *Renew. Energy* 143: 1357–1365. https://doi.org/10.1016/j.renene.2019.05.079.

14 Naveena, B., Armshaw, P., and Tony Pembroke, J. (2015). Ultrasonic intensification as a tool for enhanced microbial biofuel yields. *Biotechnol. Biofuels* 8 (1): 1–13. https://doi.org/10.1186/s13068-015-0321-0.

15 Balaraman, H.B. and Rathnasamy, S.K. (2020). Kinetics and microwave-assisted extractive transesterification studies of high octane methyl esters (HOME) from karanja and chicken lard oil using protic deep eutectic solvent. *Fuel* 268: 117299. https://doi.org/10.1016/j.fuel.2020.117299.

16 Widayat, S.A., Saputro, E.M., Ginting, A.N., and Annisa, H. (2017). Satriadi, biofuel production by catalytic cracking method using Zn/HZSM-5 catalyst. *ARPN J. Eng. Appl. Sci.* 12 (22): 6347–6351.

17 Sirajudin, N., Jusoff, K., Yani, S. et al. (2013). Biofuel production from catalytic cracking of palm oil. *World Appl. Sci. J.* 5 (11): 9. https://doi.org/10.5829/idosi.wasj.2013.26.nrrdsi.26012.

18 Taufiqurrahmi, N. and Bhatia, S. (2011). Catalytic cracking of edible and non-edible oils for the production of biofuels. *Energ. Environ. Sci.* 4 (4): 1087–1112. https://doi.org/10.1039/c0ee00460j.

19 Mohr, A. and Raman, S. (2015). Lessons from first generation biofuels and implications for the sustainability appraisal of second generation biofuels, Effic. *Sustain. Biofuel Prod. Environ. Land-Use Res.* 63: 114–122. https://doi.org/10.1016/j.enpol.2013.08.033.

20 Gasparatos, A., Stromberg, P., and Takeuchi, K. (2013). Sustainability impacts of first-generation biofuels. *Anim. Front.* 3 (2): 12–26. https://doi.org/10.2527/af.2013-0011.

21 Gebremariam, S.N. and Marchetti, J.M. (2018). Economics of biodiesel production: review. *Energ. Conver. Manage.* 168: 74–84. https://doi.org/10.1016/j.enconman.2018.05.002.

22 Barth, S. (2009). Genetic improvement of bio energy crops. *Ann. Bot.* 103 (6): viii–ix. https://doi.org/10.1093/aob/mcp028.

23 Banković-Ilić, I.B., Stamenković, O.S., and Veljković, V.B. (2012). Biodiesel production from non-edible plant oils. *Renew. Sustain. Energy Rev.* 16 (6): 3621–3647. https://doi.org/10.1016/j.rser.2012.03.002.

24 Rani, S., Kaur, S., Shivani, and Kaur, H. (2020). Production of biodiesel from non-edible oil (WCO). *Int. Res. J. Sci. Technol.* 1 (3): 268–271. https://doi.org/10.46378/irjst.2020.010313.

25 Fadhil, A.B., Al-Tikrity, E.T.B., and Albadree, M.A. (2017). Biodiesel production from mixed non-edible oils, castor seed oil and waste fish oil. *Fuel* 210: 721–728. https://doi.org/10.1016/j.fuel.2017.09.009.

26 Sahoo, P.K. and Das, L.M. (2009). Combustion analysis of Jatropha, Karanja and Polanga based biodiesel as fuel in a diesel engine. *Fuel* 88 (6): 994–999. https://doi.org/10.1016/j.fuel.2008.11.012.

27 Banik, S., Rouf, M., Rabeya, T. et al. (2018). Production of biodiesel from neem seed oil. *Bangladesh J. Sci. Ind. Res.* 53 (3): 211–218. https://doi.org/10.3329/bjsir.v53i3.38268.

28 Widayat, W. and Suherman, S. (2012). Biodiesel production from rubber seed oil via esterification process. *Int. J. Renew. Energy Dev.* 1 (2): 57. https://doi.org/10.14710/ijred.1.2.57-60.

29 Ghadge, S.V. and Raheman, H. (2005). Biodiesel production from mahua (Madhuca indica) oil having high free fatty acids. *Biomass Bioenergy* 28 (6): 601–605. https://doi.org/10.1016/j.bio mbio e.2004.11.009.

30 Osorio-González, C.S., Gómez-Falcon, N., Sandoval-Salas, F. et al. (2020). Production of biodiesel from castor oil: a review. *Energies* 13 (10): 2467. https://doi.org/10.3390/en13102467.

31 Sreenivas, V.P. (2011). Development of biodiesel from Castor oil. *Int. J. Energy Sci.* 1 (3): 192–197.

32 Fadhil, A.B., Al-Tikrity, E.T.B., and Ibraheem, K.K. (2019). Transesterification of bitter almond oil as a new non-edible feedstock with mixed alcohols system: parameter optimization and analysis of biodiesel. *Waste Biomass Valorization* 10 (6): 1597–1608. https://doi.org/10.1007/s12649-017-0172-y.

33 Uddin, M.R., Ferdous, K., Mondal, S.K. et al. (2017). Preparation of biodiesel from Karanja (Pongamia Pinnata) oil. *J. Chem. Eng.* 29 (1): 24–28. https://doi.org/10.3329/jce.v29i1.33815.

34 Jaichandar, S. and Annamalai, K. (2011). The status of biodiesel as an alternative fuel for diesel engine – an overview. *J. Sustain. Energy Environ.* 2 (2): 71–75.

35 Toivanen, H. and Novotny, M. (2017). The emergence of patent races in lignocellulosic biofuels, 2002–2015. *Renew. Sustain. Energy Rev.* 77 (C): 318–326. https://doi.org/10.1016/j.rser.2017.03.089.

36 Elkelawy, M., din Etaiw, S.E.H., Bastawissi, H.A.E. et al. (2020). Study of diesel-biodiesel blends combustion and emission characteristics in a CI engine by adding nanoparticles of Mn (II) supramolecular complex. *Atmos. Pollut. Res.* 11 (1): 117–128. https://doi.org/10.1016/j.apr.2019.09.021.

37 Kim, J.K., Yim, E.S., Jeon, C.H. et al. (2012). Cold performance of various biodiesel fuel blends at low temperature. *Int. J. Automat. Technol.* 13 (2): https://doi.org/10.1007/s12239-012-0027-2.

38 Perumal Venkatesan, E., Kandhasamy, A., Sivalingam, A. et al. (2019). Performance and emission reduction characteristics of cerium oxide nanoparticle-water emulsion biofuel in diesel engine with modified coated piston. *Environ. Sci. Pollut. Res.* 26 (26): 27362–27371. https://doi.org/10.1007/s11356-019-05773-z.

39 Ortiz Moreno, A., Dorantes, L., Galíndez, J., and Guzmán, R.I. (2003). Effect of different extraction methods on fatty acids, volatile compounds, and physical and chemical properties of avocado (*Persea americana* Mill.) oil. *J. Agric. Food Chem.* 51 (8): 2216–2221. https://doi.org/10.1021/jf0207934.

40 Yin, H., Solval, K.M., Huang, J. et al. (2011). Effects of oil extraction methods on physical and chemical properties of red salmon oils (oncorhynchus nerka), JAOCS. *J. Am. Oil Chem. Soc.* 88 (10): 1641–1648. https://doi.org/10.1007/s11746-011-1824-x.

41 Vasconcelos, B., Teixeira, J.C., Dragone, G., and Teixeira, J.A. (2018). Optimization of lipid extraction from the oleaginous yeasts *Rhodotorula glutinis* and *Lipomyces kononenkoae*. *AMB Express* 8 (1): 1–9. https://doi.org/10.1186/s13568-018-0658-4.

42 Mohammed, M.N., Atabani, A.E., Uguz, G. et al. (2020). Characterization of hemp (*Cannabis sativa* L.) biodiesel blends with euro diesel, butanol and diethyl ether using FT-IR, UV–vis, TGA and DSC techniques. *Waste Biomass Valorization* 11 (3): 1097–1113. https://doi.org/10.1007/s12649-018-0340-8.

43 Kudre, T.G., Bhaskar, N., and Sakhare, P.Z. (2017). Optimization and characterization of biodiesel production from rohu (*Labeo rohita*) processing waste. *Renew. Energy* 113: 1408–1418. https://doi.org/10.1016/j.renene.2017.06.101.

44 Doudin, K.I. (2021). Quantitative and qualitative analysis of biodiesel by NMR spectroscopic methods. *Fuel* 284: 119114. https://doi.org/10.1016/j.fuel.2020.119114.

45 Zahedi, S., Sales, D., García-Morales, J.L., and Solera, R. (2018). Obtaining green energy from dry-thermophilic anaerobic co-digestion of municipal solid waste and biodiesel waste. *Biosyst. Eng.* 170: 108–116. https://doi.org/10.1016/j.bio systemseng.2018.04.005.

46 de Sousa, M.H., da Silva, A.S.F., Correia, R.C. et al. (2021). Valorizing municipal organic waste to produce biodiesel, bio gas, organic fertilizer, and value-added chemicals: an integrated bio refinery approach. *Biomass Convers. Biorefinery* 1–15. https://doi.org/10.1007/s13399-020-01252-5.

47 Surendra, K.C., Olivier, R., Tomberlin, J.K. et al. (2016). Bio conversion of organic wastes into biodiesel and animal feed via insect farming. *Renew. Energy* 98: 197–202. https://doi.org/10.1016/j.renene.2016.03.022.

48 Muradin, M. (2020). Environmental impact assessment of organic waste conversion technology for additives to liquid fuels. *Polityka Energy* 23 (1): 135–150. https://doi.org/10.33223/epj/118731.

49 Wang, H., Yang, B., Zhang, Q., and Zhu, W. (2020). Catalytic routes for the conversion of lignocellulosic bio mass to aviation fuel range hydrocarbons. *Renew. Sustain. Energy Rev.* 120: 109612. https://doi.org/10.1016/j.rser.2019.109612.

50 Murat Sen, S., Henao, C.A., Braden, D.J. et al. (2012). Catalytic conversion of lignocellulosic bio mass to fuels: process development and technoeconomic evaluation. *Chem. Eng. Sci.* 67 (1): 57–67. https://doi.org/10.1016/j.ces.2011.07.022.

51 Tu, W.C. and Hallett, J.P. (2019). Recent advances in the pretreatment of lignocellulosic bio mass. *Curr. Opin. Green Sustain. Chem.* 20: 11–17. https://doi.org/10.1016/j.cogsc.2019.07.004.

52 Jahirul, M.I., Rasul, M.G., Chowdhury, A.A., and Ashwath, N. (2012). Biofuels production through bio mass pyrolysis – a technological review. *Energies* 5 (12): 4952–5001. https://doi.org/10.3390/en5124952.

53 La Villetta, M., Costa, M., and Massarotti, N. (2017). Modelling approaches to bio mass gasification: a review with emphasis on the stoichiometric method. *Renew. Sustain. Energy Rev.* 74: 71–88. https://doi.org/10.1016/j.rser.2017.02.027.

54 Toor, S.S., Rosendahl, L., and Rudolf, A. (2011). Hydrothermal liquefaction of bio mass: a review of subcritical water technologies. *Energy* 189–217. https://doi.org/10.1016/j.energy.2011.03.013.

55 Griffin, D.W. and Schultz, M.A. (2012). Fuel and chemical products from bio mass syngas: a comparison of gas fermentation to thermochemical conversion routes. *Environ. Prog. Sustain. Energy* 31 (2): 219–224. https://doi.org/10.1002/ep.11613.

56 Shamsul, N.S., Kamarudin, S.K., and Rahman, N.A. (2017). Conversion of bio-oil to bio gasoline via pyrolysis and hydrothermal: a review. *Renew. Sustain. Energy Rev.* 80 (C): 538–549. https://doi.org/10.1016/j.rser.2017.05.245.

57 Samiran, N.A., Jaafar, M.N.M., Ng, J.H. et al. (2016). Progress in bio mass gasification technique – with focus on Malaysian palm bio mass for syngas production. *Renew. Sustain. Energy Rev.* 62: 1047–1062. https://doi.org/10.1016/j.rser.2016.04.049.

58 Sánchez, Ó.J. and Cardona, C.A. (2008). Trends in bio technological production of fuel ethanol from different feedstocks. *Bioresour. Technol.* 99 (13): 5270–5295. https://doi.org/10.1016/j.bio rtech.2007.11.013.

59 Dikshit, P.K., Jun, H.B., and Kim, B.S. (2020). Bio logical conversion of lignin and its derivatives to fuels and chemicals. *Korean J. Chem. Eng.* 37 (3): 387–401. https://doi.org/10.1007/s11814-019-0458-9.

60 Zhang, L., Zhang, B., Zhu, X. et al. (2018). Role of bio reactors in microbial bio mass and energy conversion. *Green Energy Technol.* 39–78. https://doi.org/10.1007/978-981-10-7677-0_2.

61 Sun, H., Zhao, W., Mao, X. et al. (2018). High-value bio mass from microalgae production platforms: strategies and progress based on carbon metabolism and energy conversion. *Biotechnol. Biofuels* 11 (1): 1–23. https://doi.org/10.1186/s13068-018-1225-6.

62 Chia, S.R., Chew, K.W., Show, P.L. et al. (2018). Analysis of economic and environmental aspects of microalgae bio refinery for biofuels production: a review. *Biotechnol J.* 13 (6): e1700618. https://doi.org/10.1002/bio t.201700618.

63 Kumar, L. and Bharadvaja, N. (2020). A review on microalgae biofuel and bio refinery: challenges and way forward, energy sources, part a recover. *Util. Environ. Eff.* 1–24. https://doi.org/10.1080/15567036.2020.1836084.

64 Clark, J.H., Luque, R., and Matharu, A.S. (2012). Green chemistry, biofuels, and bio refinery. *Annu. Rev. Chem. Bio Mol. Eng.* 3: 183–207. https://doi.org/10.1146/annurev-chembio eng-062011-081014.

65 Mailaram, S., Kumar, P., Kunamalla, A. et al. (2021). Bio mass, bio refinery, and biofuels. *Sustain. Fuel Technol. Handb.* 51–87. https://doi.org/10.1016/b978-0-12-822989-7.00003-2.

66 Kim, S., Dale, B.E., Jin, M. et al. (2019). Integration in a depot-based decentralized bio refinery system: corn Stover-based cellulosic biofuel. *GCB Bio Energy* 11 (7): 871–882. https://doi.org/10.1111/gcbb.12613.

67 Chia, S.R., Chew, K.W., Show, P.L. et al. (2018). Analysis of Economic and Environmental Aspects of Microalgae Biorefinery for Biofuels Production: A Review. *Biotechnol J.* https://doi.org/10.1002/biot.201700618.

68 Kumar, L. and Bharadvaja, N. (2020). A review on microalgae biofuel and biorefinery: challenges and way forward. *Energy Sources, Part A Recover Util Environ Eff* https://doi.org/10.1080/15567036.2020.1836084.

69 Clark, J.H., Luque, R., and Matharu, A.S. (2012). Green chemistry, biofuels, and biorefinery. *Annu Rev Chem Biomol Eng* https://doi.org/10.1146/annurev-chembioeng-062011-081014.

70 Mailaram, S., Kumar, P., Kunamalla, A. et al. (2021). Biomass, biorefinery, and biofuels. *Sustain. Fuel Technol. Handb.* https://doi.org/10.1016/b978-0-12-822989-7.00003-2.

71 Kim, S., Dale, B.E., Jin, M. et al. (2019). Integration in a depot-based decentralized biorefinery system: Corn stover-based cellulosic biofuel. *GCB Bioenergy* https://doi.org/10.1111/gcbb.12613.

5

Chemically Produced Biofuels

5.1 Introduction

A biofuel is produced from plant-derived biomass. A plant fixes atmospheric CO_2 producing polysaccharides [1]. As biofuels are derived from such plants, atmospheric CO_2 does not increase the maintenance of the harmony of environmental attributes. There are various technologies and processes for the production of various types of biofuels [2], anaerobic digestion (AD) [3, 4], bio-syngas from gasification [5, 6], and bio-oil from pyrolysis [7, 8]. Biodiesel is a mixture of long-chain fatty acid methyl ester (FAME) that is formed from biomass through transesterification of triacylglycerol, with methanol [9, 10]. The biodiesel synthesis can be expressed by various steps, and suitable kinetics can be studied. The reactions can be acid/base/enzyme catalysed. In principle, any triacylglycerol could be the best participant used for biodiesel production. Research is going on for the efficient conversion of TG to FAME. Different technologies are also upgraded to obtain maximum efficiency [11–13].

A wide variety of biofuels can be produced by fermentive [14], biological [14], chemical, and thermal processes [15, 16]. Anaerobic digestion [17, 18] for biogas production has also progressed rapidly in the recent years. Burning biomass for satisfying different human needs is the long-ago method. Based on that, a technique converting biomass into bio-syngas, bio-oil and biochar at elevated temperature in a limited supply of air has also been developed [19, 20]. This chapter addresses different chemical routes of the processes involved, kinetics of transesterification reaction, catalysts used and the worldwide environmental impacts of biofuels. Such green technology will surely answer the global demand for energy.

5.2 Triglycerides – Best Participant as Fuels

Usually, alcoholysis, acidolysis and interesterification are studied in the transesterification head [21, 22]. Transfer of acyl groups between an ester and an alcohol molecule (alcoholysis)/an acid molecule (acidolysis)/another ester molecule (Interesterification)

Sustainability in Biofuel Production Technology, First Edition. Pratibha S. Agrawal, Pramod N. Belkhode, and Samuel Lalthazuala Rokhum.

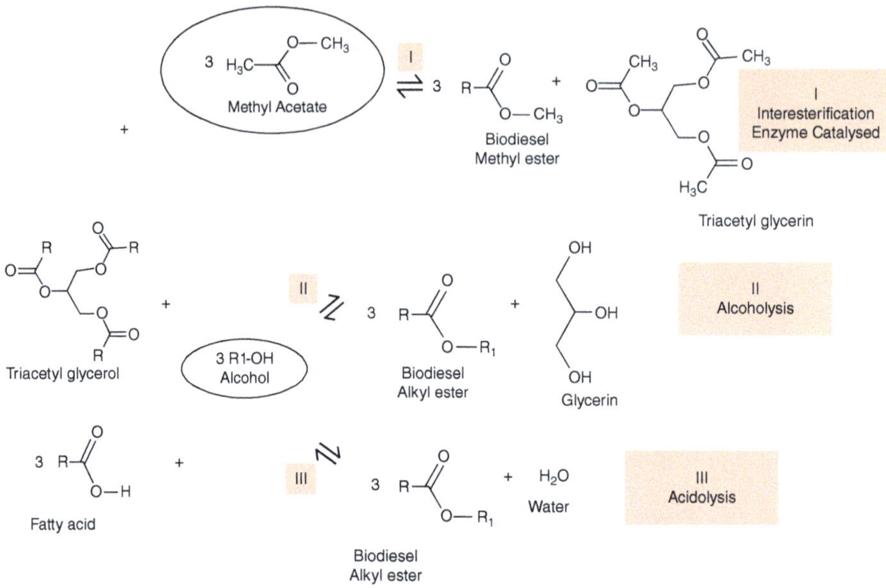

Figure 5.1 Scheme of the transesterification reaction for biodiesel production.

takes place in the commonly used reversible transesterification reactions, for biodiesel production. The most commonly used method is the one in which the oil/fat is reacted with a monohydric preferably short-chain alcohol in the presence of a catalyst. These reactions are illustrated in Figure 5.1,

In Scheme I, the acyl acceptor molecule is methyl acetate and triacetylglycerin, a valuable side product, is obtained. In Scheme II, acyl acceptor molecule is methanol or ethanol (alcoholysis), and glycerine is produced as a side product with a lower commercial value. In Scheme III, direct esterification is carried out. Water needs to be removed continuously to enhance the biodiesel yield. Since reactions are reversible, an excess amount of any of the substrates or removing the side product formed will shift the equilibrium towards biodiesel formation.

5.2.1 Base-catalysed Transesterification Process

Conventionally, the bulk of biodiesel is produced with alkaline catalysts, such as NaOH and KOH, due to their large availability and low price. It contains four-step reactions, in which the first step is the formation of alkoxide, which is followed by three consecutive reversible steps where triglycerides (TG) are converted into diglyceride (DG) and finally to monoglyceride (MG) and glycerine (GL). In each step, short-chain alcohol (A) is used, and esters (E) are produced [23]. Higher chain alcohols have lower polarity, which are difficult to separate from alkyl esters. Thus, for a single molecule of oil, three molecules of alcohol (methanol or ethanol) are required. Alcohol is used in various ratios of 1:1 to 6:1% w/w, out of which 6:1% w/w is used mostly. Such a large amount of alcohol is also essential to develop the yield by shifting the reaction towards the product side. As this is a two-phase reaction (TG is non-polar and methanol is polar), the concentration of oil in methanol is initially low, and a time lag in the ester production will be observed. However, as the reaction proceeds, more oil will be shifted to the methanol phase, thereby leading to an increased rate of conversion. Methanol is more toxic, and ethanol is preferred by industries in the biodiesel production. Propanol or butanol can also be used as these two alcohols enhance the miscibility of alcohol–oil phases. Higher temperature, increasing the stirring rate or adding cosolvent can also increase the mutual solubility of TG and methanol, favouring the product formation.

The reaction pathways predicted for the production of biodiesel are as follows [24, 25]:

$$\mathrm{TG + A \underset{K2}{\overset{K1}{\rightleftharpoons}} DG + E} \tag{5.1}$$

$$\mathrm{DG + A \underset{K4}{\overset{K3}{\rightleftharpoons}} MG + E} \tag{5.2}$$

$$\mathrm{MG + A \underset{K6}{\overset{K5}{\rightleftharpoons}} GL + E} \tag{5.3}$$

The overall reaction can be combined in a single step as follows:

$$\mathrm{TG + 3A \underset{K8}{\overset{K7}{\rightleftharpoons}} GL + 3E} \tag{5.4}$$

Base-catalysed reaction is always preferred over acid-catalysed reaction by the industrialists, as in acid-catalysed reactions, the acid used may corrode the reactors, thereby decreasing the conversion efficiency. As the reaction is base catalysed, sodium hydroxide (NaOH), potassium hydroxide (KOH) or sodium alkoxide is directly used in the reaction along with alcohol. NaOH or KOH are cheaper than alkoxide but are less active [26, 27]. Nevertheless, it is always recommended to use the NaOH or KOH catalyst in the amount 0.5–1% w/w, though reports are available for its use in the minimum amount of 0.005–0.35% w/w. In such reactions, it is always better to produce sodium alkoxide first to obtain better yield.

$$\mathrm{R - CH_2OH + NaOH \rightleftharpoons H2O + R - CH_2ONa}$$

However, sodium methoxide causes the formation of sodium salts as waste by-product. It also required the high quality of oil. When potassium hydroxide is used, the by-products formed can be neutralised using phosphoric acid. This neutralisation process produces potassium phosphate, which is used as fertiliser. Thus, KOH is preferred more over NaOH in the transesterification process [28].

The feedstock used can be any plant part of vegetables, such as sunflower, corn, palm, palm kernel, canola, olive, peanut, soya bean, etc. The list of sources are available for this in the literature. The selection of feedstock depends on the reproducibility, easy accessibility, and hence the comparatively lower market price of the crude feedstock.

The preferred reaction temperature is 60 °C as the methanol boils at 68 °C; however, still depending upon the catalyst used, the reaction condition varies and may occur at temperature between 25 and 120 °C [29, 30].

The limitation of this process occurs at its sensitivity. The reaction is very much dependent on the quality and purity of the reactants. If the amount of water exceeds the sample, soap formation takes place, which leads to emulsification of the product. This will make recovery and purification of glycerol more difficult due to increased viscosity as well as making the process costlier.

$$\mathrm{RCOOR + H_2O \rightleftharpoons RCOOH + ROH}$$

$$\mathrm{RCOOH + NaOH \rightleftharpoons RCOONa\,(soap) + H_2O}$$

The percentage of free fatty acid (FFA) is expected to be 2% w/w in normal waste cooking oil. If that amount exceeds, FFA in the feedstock again reacts with the base to produce soap. The gel produced will increase the viscosity. It will again not only consume the alcohol but emulsification will also lead to the increased processing cost. Water present supports the creation of FFA, which will again neutralise the soap and generating soap again. The total water content and catalyst should be less than the range of 0.1–0.3 wt%, which will keep it anhydrous. A pre-treatment using sulphuric acid is recommended followed by a normal base-catalysed process. Sulphuric acid, a homogeneous catalyst, can catalyse the esterification and transesterification reaction, but at the same time the corrosion of the reactor and the possibility of waste produced in the reaction cannot be ignored. The use of heterogeneous catalyst may solve those problems [31, 32]. The use of potassium carbonate is also recommended as it avoids soap formation. Instead of water, it produces bicarbonate that restricts the hydrolysis of the esters.

$$\mathrm{K_2CO_3 + ROH \rightleftharpoons ROK + KHCO_3}$$

Other heterogeneous and non-ionic base catalysts, such as amines [33], pyridines, guanidinesphosphozenes [34], have also been tested producing promising results.

The kinetic pathway equations for transesterification of triglycerides as given by Darnoko and Cheryan [35] can be reconsidered as follows:

$$\frac{dTG}{dT} = -K1[TG][A] + K2[DG][E] = -K7[TG][A]^3 + K8[G][E]^3 \quad (5.5)$$

$$\frac{dDG}{dT} = K1[TG][A] - K2[DG][E] - K3[DG][A] + K4[MG][E] \quad (5.6)$$

$$\frac{dMG}{dT} = K3[DG][A] - K4[MG][E] - K5[MG][A] + K6[GL][E] \quad (5.7)$$

$$\frac{dGL}{dT} = K5[MG][A] - K6[GL][E] = K7[TG][A]^3 - K8[GL][E]^3 \quad (5.8)$$

$$\begin{aligned}\frac{dE}{dT} = -\frac{dA}{dT} &= K1[TG][A] - K2[DG][E] + K3[DG][A] - K4[MG][E] \\ &\quad + K5[MG][A] - K6[GL][E] + K7[TG][A]^3 - K8[GL][E]^3\end{aligned} \quad (5.9)$$

If R1 = K1/K2, R2 = K2/K3 and R3 = K4/K6, then experimentally it was observed that R3 > R2 > R1 by many researchers [36, 37].

The rate constant of the first forward step (the transesterification of TG) had the largest value, whereas the further steps were proven to be smallest by many researchers. Whichever may be the slowest step, oil, alcohol and catalyst would be the prime components altering the kinetics of transesterification. The reaction is proposed to be a pseudo-second-order reaction for the initial stages, followed by first-order or zero-order kinetics in the later stages of the reaction [38]. More theoretical formulation based on the experimental observation is required to clearly identify the character and importance of each component.

Advantages of the homogeneous alkali-catalysed reaction: It is most frequently preferred due to [39]

- low operating conditions of reaction temperature (60 °C) and pressure (20 Psi),
- good yields of the ester (98%) minimising the side reactions,
- a fast reaction thereby reduces the reaction time and
- conversion to methyl ester directly without any transitional steps.

Disadvantages of homogeneous alkali-catalysed processes: There are some difficulties [40] stumble upon by their use:

- high energy demand,
- at the completion of the reaction, a separation step of catalyst from the product is required,
- soap formation occurred due to the reaction of FFA and water during the course of the reaction,
- soap formation makes recovery of glycerol difficult after the reaction completion,
- treatment of the alkaline wastewater is essential at the end of reaction to avoid its adverse environmental effects and
- the use of acid as catalyst and heterogeneous catalyst systems can address some of these problems.

5.2.2 Acid-catalysed Transesterification Process

The low-cost feedstock usually contains high amounts of FFAs [30, 41]. The most important advantage of the acid-catalysed transesterification reaction is that it can be used when the feedstock contains FFA. This FFA can be converted to fatty acid of methyl esters in an acid-catalysed process [42, 43]. The transesterification reaction can be catalysed by Bronsted acids such as H_3SO_4, HCl, H_3PO_4 and preferably using sulphuric and sulphonic acids as though the reaction is slow (3–48 hours), but it gives a very high yield at temperature above 100 °C [44].

The reaction pathways predicted for the production of biodiesel are as follows:

$$\mathrm{TG + A \underset{K2}{\overset{K1}{\rightleftarrows}} DG + E} \tag{5.10}$$

$$\mathrm{DG + A \underset{K4}{\overset{K3}{\rightleftarrows}} MG + E} \tag{5.11}$$

$$\mathrm{MG + A \underset{K6}{\overset{K5}{\rightleftarrows}} GL + E} \tag{5.12}$$

The overall reaction can be combined in a single step as follows:

$$FFA + A \underset{K8}{\overset{K7}{\rightleftarrows}} W + E \tag{5.13}$$

The alcohol used can be methanol at 65 °C (reaction time 50 hours) or ethanol at 78 °C and butanol at 117° C (reaction time 3–4 hours) preferred in the ratio 30:1. 3 d block metals, lanthanum and salts of amino acids can be used to obtain a reasonable rate of reaction [45, 46]. It can be increased by using larger amounts of catalysts (typically suggested between 1% and 5% by weight). However, if the catalyst exceeds beyond this, then catalyst should be neutralised using CaO (added proportionately to the acid concentration used) before the product is separated. The high amount of acid used increases the high requirement of CaO, which increases the cost and still results in high waste generation [47]. Homogeneous transesterification using liquid acid is not used for its thousand times slower rate than homogeneous base-catalysed reaction.

As stated earlier, acid-catalysed reactions are used specifically for low-quality feedstock, such as waste cooking oil or waste animal fat, which contains a relatively higher amount of FFA in it. However, these catalysts are insignificant for the water content (more than 0.5%) in the feedstock. This is because sulphuric acid has a higher affinity for water. Thus, the acid catalyst preferentially reacts with water, which decreases its catalytic activity. The large quantity of acid catalyst used can also dehydrate alcohol.

The following assumptions are made for the kinetic model [48] used in this process:

a) the reaction must be reversible and homogeneous in nature,
b) the chemical reaction should be occurred in the oil phase,
c) the amount of alcohol (methanol) used should be high, which can be considered to remain constant throughout the process.

Based on the above assumptions, the kinetic reaction equations can be formulated as follows:

$$\frac{dTG}{dT} = -K1[TG][A] + K2[DG][E] \tag{5.14}$$

$$\frac{dDG}{dT} = K1[TG][A] - K2[DG][E] - K3[DG][A] + K4[MG][E] \tag{5.15}$$

$$\frac{dMG}{dT} = K3[DG][A] - K4[MG][E] - K5[MG][A] + K6[GL][E] \tag{5.16}$$

$$\frac{dGL}{dT} = K5[MG][A] - K6[GL][E] \tag{5.17}$$

$$\begin{aligned} \frac{dE}{dT} = -\frac{dA}{dT} = {} & K1[TG][A] - K2[DG][E] + K3[DG][A] - K4[MG][E] \\ & + K5[MG][A] - K6[GL][E] + K7[FFA][A] - K8[W][A] \end{aligned} \tag{5.18}$$

$$\frac{dFFA}{dT} = -K7[FFA][A] + K8[W][A] \tag{5.19}$$

Using these models, the reaction was expected to be pseudo first order precisely in the forward direction due to the presence of the excess amount of alcohol and typically second order in the reverse direction. Higher temperature, increased molar ratio of alcohol to oil and acid catalyst also increase the ester production. The use of heterogeneous catalyst can solve the problems occurred due to homogeneous acid catalysts.

5.2.3 Enzyme-catalysed Transesterification Process

Though acid- and base-catalysed productions are well-set processes for industrial applications, they still require several processes of phase procession, separation and purification, waste water treatment methods, along with the risk of oil oxidation at high temperature. Many researchers have promoted lipase-mediated biodiesel production [49, 50]. As the enzymes are specific in the action, a highly purified product is obtained, which reduces the cost of further purification processes. A minimal amount of waste is generated, and mil conditions also rests oil oxidation.

The kinetic model is developed considering the successive cleavage of fatty acid with methanol and catalysed by the enzyme [51]. The reaction proceeds as follows.

The mechanism follows first hydrolysis and then transesterification of triglyceride. It can be seen that enzyme E can bind with any of the substrate, either triglyceride or with methanol. If E reacts with A, it forms an EA complex, which retards the rate reaction and inhibits the reaction to proceed further. The other possibility is the reaction with TG forming ETG complex [52].

$$E + A \underset{K-m}{\overset{Km}{\rightleftharpoons}} EA \quad (a) \text{ inhibitory step and} \tag{5.20a}$$

$$E + TG \underset{K2}{\overset{K1}{\rightleftharpoons}} ETG \quad (b) \text{ promising step} \tag{5.20b}$$

$$ETG + A \underset{K4}{\overset{K3}{\rightleftharpoons}} EDG + E \tag{5.20c}$$

$$EDG + A \underset{K6}{\overset{K5}{\rightleftharpoons}} EMG + E \tag{5.21}$$

$$EMG + A \underset{K8}{\overset{K7}{\rightleftharpoons}} EGL + E \tag{5.22}$$

$$EGL \underset{K10}{\overset{K9}{\rightleftharpoons}} 3E + GL \tag{5.23}$$

The overall reaction can be combined in a single step as follows:

$$E + TG + 3A \underset{K2}{\overset{K1}{\rightleftharpoons}} ETG \underset{K12}{\overset{K11}{\rightleftharpoons}} GL + 3E \tag{5.24}$$

The mild reaction conditions of pH, temperature and pressure, required for enzymatic-catalysed reaction, permit less energy used and improved product quality because of the negligible thermal oxidation of the oil. In lipase-catalysed reactions, the recovery of glycerine and the purification of biodiesel are easier, making the process economic and environmentally friendly. However, the low stability of enzymes and the high cost of lipase at the commercial level hinders its use [53]. More research is required for reducing the cost of enzyme-catalysed biodiesel production at the industrial level. Lipase production, immobilisation and enzymatic biodiesel production in supercritical fluid are some of the ways for achieving these targets.

The kinetic reaction equations can be formulated as follows:

$$\frac{dETG}{dT} = K1[TG][E] - K2[ETG] - K3[ETG][A] + K4[EDG][E] \quad (5.25)$$

$$\frac{dEDG}{dT} = K3[ETG][A] - K4[DG][E] - K5[EDG][A] - K6[EMG][E] \quad (5.26)$$

$$\frac{dEMG}{dT} = K5[EDG][A] - K6[EMG][E] - K7[EMG][A] + K8[EGL][E] \quad (5.27)$$

$$\frac{dGL}{dT} = K9[EGL] - K10[GL][E] \quad (5.28)$$

The Michaelis–Menten kinetic model can be used to estimate the kinetic constants of the reaction [54].

Lipases are carboxylic ester hydrolases derived from microbial or plants, showing catalytic activity in both aqueous and non-aqueous phases. Plant-based enzymes are readily available, comparatively cheaper and hence preferred in making the process more cost effective. The percentage of water content in the reaction should be in the range of 0.5–1%. The presence of water increases the internal flexibility of the lipase molecule, increasing its activity more, but it can also act as a substrate (especially in hydrolytic reactions) producing many unwanted side reactions, resulting in the decreased productivity. Washing of enzyme preferably with *n*-hexane maintains its stability until many cycles of synthesis. The nature of solvent used and its physiochemical properties affect the catalytic property and stability for the reuse of the enzymes to a greater extent. Hydrophilic solvents usually deactivate enzymes, whereas hydrophobic solvents are well used in enzymatic reactions as they do not disturb the water layer formed on the polar surface [55, 56]. The retarding effect of alcohol and glycerol may cause enzyme deactivation. Glycerol produced as a side product could limit mass transfer and also reaction rate reduction.

The widely used short-chain length methanol due to its high polarity is very much useful, but at the same time, it can cause the inhibition of lipase by deactivating it. This may result in the folding of enzyme chain, immiscibility of two phases and blocking the entrance of TG. Three equivalent moles of methanol are needed in the transesterification reaction, but the presence of more than one mole will deactivate lipase. These problems can be solved by the continuous or successive addition of methanol during the course of the reaction or by choosing ethanol instead of methanol. Ethanol causes less inhibitory action. The stepwise addition can surely increase the reaction rate of biodiesel formation.

Supercritical fluids of carbon dioxide can also be used with enzymes catalysed reactions [57]. It can easily be removed just by decreasing the pressure. Its toxicity is also low when compared to other solvents. As with hexane, non-polar compounds can be dissolved in it, but for polar compounds, the addition of co-solvent will be needed further [58]. In addition, enzymes are not soluble in supercritical fluids of carbon dioxide, so heterogeneous process using solid enzymes are used. A 60--70% conversion efficiency was reported in eight hours at a very low temperature of 45 °C as compared to non-catalytic processes. The hunt for the best enzyme in biodiesel production has never ended. Lipase that can sustain high temperature, organic solvent, pH and able to sustain mechanical stress could make production more

Table 5.1 Advantages and disadvantages of different biodiesel production processes.

Biodiesel production methods	Advantages	Disadvantages
Chemical catalyst –homogeneous	High production yield Low cost Can convert FFA (acid catalyst)	Wastewater treatment is required Product separation and purification step is required Catalyst is not recovered easily Alkali catalyst may lead to saponification
Chemical catalyst –heterogeneous	Relatively fast reaction Comparatively high yield Catalyst can be reused Catalyst can be used in continuous process Can convert FFA (acid catalyst)	High energy is needed Catalyst preparation is tedious Catalyst leaching may occur Alkali catalyst may lead to saponification
Biochemical catalyst – enzymes	Medium production yield Can convert FFA (acid catalyst) Low energy usage High purity of product and by-product Catalyst can be reused Wastewater is not produced	Inhibition of the reaction by alcohol or by-product, high enzyme cost and relatively slow reaction
Non-catalysed –supercritical alcohol	Superfast reaction Higher yield Can convert FFA No catalyst Product purification is easier No waste is produced	High temperature and pressure High cost of the reactor High molar ratio of alcohol to oil

Source: Adapted from [59].

viable. Enzyme stability and activity are increased by the suitable compressed fluid used and the optimal amount of water used in the reaction mixture. Various techniques are under development for commercialising the biocatalysed reactions. Different sources of lipases production, various methods of enzyme pre-treatment, enzyme post-treatment, enzyme immobilisation, different alternatives of solvents and reactor design will contribute to the reduction of biodiesel production cost that will cause minimum environmental hazards. Advantages and disadvantages of different biodiesel production processes are listed in Table 5.1.

5.3 Biogas Using Anaerobic Digestion

World is facing energy crisis at one end, while the waste generated is difficult to manage on the other hand. Industrial waste treatment is also costly and harmful to environment. Its recovery, storage and treatment are always a challenging task for all. One of the brilliant solutions on this problem is to separate organic waste form it and convert it into renewable energy (bio-oil, biogas and biochar) by an anaerobic digestion technique [60]. Thus, it can shift the environmental pollutants into an economical process. The separated waste can be from any industries and thus must be different in properties. Therefore, the waste should be treated carefully and differently, and accordingly techniques can be modified to make it cost effective [61]. It is a much cheaper technique than aerobic waste treatment processes. At the beginning of the twentieth century, Europe promoted this technology by constantly increasing its production [62]. Germany is the leading producer in Europe with more than 1000 biogas plants. Biomass contains carbohydrates, proteins, cellulose and fats, which are useful feedstock in the conversion. Nowadays, co-substrate is also added to increase the organic matter collected from various waste sources. The composition, yield and efficiency of biogas production largely depend on feedstock and co-substrate types [63].

Amongst many renewable energy sources, methanol is created in environment by using various human activities as well as through the decay of organic waste. Though it may not replace the energy option completely, it can be helpful in decreasing the energy cost and hence the energy facility available. Organic matter in the absence of oxygen decays to biogas, a mixture of CO_2, CH_4 and the trace amounts of SOx. CO_2 has a lower calorific value, H_2S is responsible for corrosion in pipes and reactor, but CH_4 has a significant energy value. Many researchers are conducting experiments for conversion of huge amounts of municipal, agricultural and industrial wastes to energy [64–66].

The four main processes involved in anaerobic digestion are hydrolysis, acidogenesis, acetogenesis and methanogenesis. Organic waste matter is first hydrolysed to soluble matter, which is further biodegraded to alcohol and volatile fatty acids using heterogeneous microbial (antigens) in acidogenesis. The compounds carbon dioxide and hydrogen are produced from volatile fatty acids in the acetogenesis phase. Lastly, methanogens microbes convert useful organic compounds into methane during methanogenesis [67]. Any bacteria cannot produce biogas alone, and hence all work together to carry out the process towards completion (Figure 5.2).

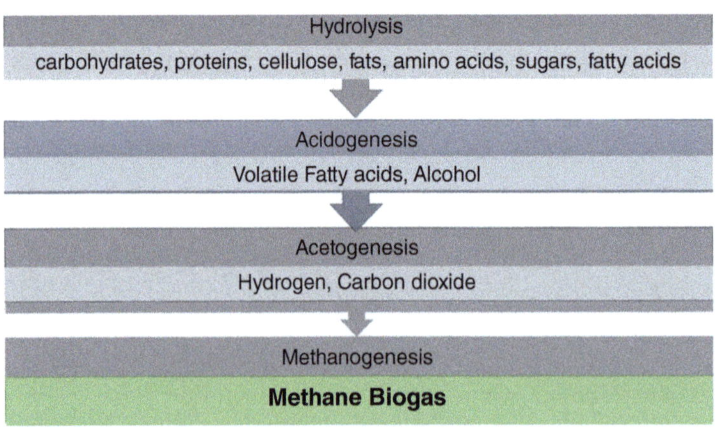

Figure 5.2 Steps involved in methane production.

Step I: Hydrolysis
A chemical bond is broken, and the compound is hydrolysed with water. A hydrogen atom is attached to one end of the compound, and the remaining hydroxyl group of water is attached to the other end. In this step, carbohydrates are hydrolysed to sugars, proteins to amino acids, fats to fatty acids and so on.

Step II: Acidogenesis (fermentation)
In the second step, the fermentative bacteria transform sugars and other organic products obtained in the first hydrolysis step to organic acids, alcohols, hydrogen, carbon dioxide and ammonia.

Step III: Acetogenesis
This is the third step of the AD process where the acetogenic bacteria convert the products obtained from the fermentation step into hydrogen, carbon dioxide and acetic acid by using present carbon of biomass and the dissolved or bound oxygen. Hence, the acid producing bacteria create anaerobic environment for the methanogens.

Step IV: Methanogenesis
Furthermore, methanogenesis is the final step of the AD process where useful products of step III are converted into methane and carbon dioxide, and trace quantities of nitrogen, hydrogen sulphide and other components.

Methane possesses a calorific value of 6 kwh/m^3 that compete to half a litre of diesel oil [68]. The by-products obtained in this process (digestate) are biologically more or less stable and rich in nutrients and thus act as fertilisers, soil conditioner used for crop cultivation and show the capacity of replacing commonly used mineral fertilisers. The composition of biogas so obtained is mostly methane (50–75%), carbon dioxide (25–50%), hydrogen (5–10%), nitrogen (1–2%) and hydrogen sulphide (traces) [69, 70]. This is a well-proven and established technology for treating industrial effluents and solid waste.

Many parameters affect the efficiency of microbes used to keep the process efficiency stable. pH and temperature are the most influencing factors. The AD process usually proceed in a neutral range (pH 6.5–7.6), and at mesophilic (30–45 °C) or thermophilic (45–65 °C) temperature range. Volatile fatty acids are formed as intermediate in the process, which is finally converted into methane gas, but its proper retention time in the reactor is controlled by total alkalinity to keep the pH value stable. The organic waste may contain biodegradable and refractory organic waste in different compositions, which will also be a prominent factor in the AD process efficiency. It has been observed that carbohydrate- and protein-rich organic waste shows a faster production of biogas than fats, but the fat-rich waste shows a higher biogas yield. The efficiency will also be increased by the pre-treatment of feedstock, and other chemical, mechanical, thermal and enzymatic processes can also be used to speed up the process with increased efficiency [71]. From a varied range of reactor outlines, fluidising bed reactors are the most commonly used due to the ease of operation, high stability under pyrolysis conditions and high oil yields.

Biorefineries will employ all conversion technologies, such as fermentation, gasification, and anaerobic digestion. Waste water treatment will also be the ley process in this. Thus, biorefineries will treat various complex biomass and thus will be the promising industries of next generation supporting the industrial revolution.

Some key challenges for future investigation are as follows:

- Understand the limitations scope of improvements of the processes.
- Economic analyses of pyrolysis plants to reduce the production cost.
- Modify the rate of the reaction to make the process more beneficial.
- Mend the quality and consistency of biogas production.
- Development of deoxygenated catalysts for processes to improve the yield.
- Optimise the process by maximising the yield qualitatively and quantitatively.

Thus, the AD process is continuously being improved for waste treatability, process feasibility and efficiency of biogas production. Although the technique is good, still no appreciable product is generated by this process [72]. A detailed cost-effective assessment of the industrial scale plant is required to establish the technology globally.

5.4 Catalytic Biofuel Production

Renewable and sustainable biofuels are of two types:

1) Nonoxygenated (syngas, biogas and solid biofuels)
2) Oxygenated (butanol, ethanol and biodiesel)

Different biofuels involved altered catalytic processes for their production.

5.4.1 Biomass Gasification

Gasification is a process in which solid or liquid carbonaceous biomass reacts with air, oxygen and/or steam to produce syngas or producer gas containing CO, H_2, CO_2, CH_4 and N_2 in various proportions [73]. Producer gas is usually used in the production of electricity and/or heat, while syngas is used to produce H_2 (water-gas shift [WGS] reaction), transportation fuels and alkanes (Fischer–Tropsch synthesis [FTS]), isobutane and isobutene (isosynthesis), ethanol (fermentation) and other chemicals.

The gasification could be

- pyrolysis (the thermal decomposition in the absence of air or steam),
- partial oxidation (using less amount of oxygen than that the required, stoichiometric amount for complete combustion of biomass),
- steam oxidation (the reaction of water with the biomass-derived feedstock to produce CO, CO_2 and H_2),
- The WGS reaction (water and CO react to form H_2 and CO_2) and
- methanation (CO and H_2 react to form CH_4 and H_2O).

5.4.2 Production of Hydrogen

Hydrogen can be produced from biomass by biological and thermal techniques. Biological methods (microbial conversion and fermentation) require less energy, but the yield produced is also lower in biological methods as compared to thermal techniques [74]. Steam methane reforming is a proven technology for hydrogen production involving basically the following four steps [75]:

(a) Feed pre-treatment (desulfurisation), (b) steam reforming, (c) CO shift conversion and (d) hydrogen purification.

Feed pre-treatment (desulfurisation): The catalyst used is a hydrogenator followed by a zinc oxide bed.

Steam reforming (produces CO, CO_2 and H_2): Nickel-based catalysts on a mineral support (alumina, cement or magnesia).

If the feedstock is heavy, then the risk of formation of coke on the catalyst surface increases. In such cases, metals such as potassium, lanthanum, ruthenium and cerium are used to increase steam gasification of solid carbon reducing coke formation. For heavier feedstock, nickel-free catalysts containing strontium, aluminium and calcium oxides are used. For feedstock containing high amounts of sulphur, uranium oxide and chromium oxide are used for the acceptance to sulphur poisoning.

5.4.3 Fischer–Tropsch Synthesis

FTS is used for the production of hydrocarbon such as olefins, paraffin and various oxygenated products. It contains the following four steps: (a) syngas generation, (b) gas purification, (c) FT synthesis and (d) product upgrading. Various factors such as temperature, composition of gas, pressure and catalyst used affect the product efficiency [76]. Metals and metal oxides of group VIII exhibit good catalytic activity [77] in the following order: Ru > Fe > Ni > Co > Rh > Pd > Pt. Poisoning of these catalysts is also possible due to the deposition of carbon, chemical poisoning and coke deposition, making it an inactive oxide. Sulphur compounds if present also poison catalysts such as iron and cobalt. ZnO can also be used for desulfurisation of catalyst [78].

5.4.4 Isosynthesis

In the isosynthesis reaction, isobutene is produced over a thorium- or zirconium-based catalyst under high temperature and pressure conditions (150–1000 atm and 450 °C). Promoters such as Zn, Cr and alkali metals with sulphur tolerance and higher lifetime are used [79]. Some of the catalysts used in biofuel production are given in Table 5.2.

5.4.5 Methanol To Gasoline (MTG Process)

The conversion of methanol to gasoline (MTG) occurs over zeolite catalysts. It is a two-step process. Initially, crude methanol (17% water) is heated and dehydrated partially over alumina at 27 atm and 300 °C to give a mixture of methanol, dimethyl ether and water. It is then added to a reactor containing ZSM-5 zeolite along with syngas at 350–366 °C and 19–23 atm to produce hydrocarbons (44%) and water (56%) [82].

5.4.6 Biofuels Production

The fatty acids and the vegetable oils can be used for the production of biofuel through different approaches. The palladium and platinum catalysts, d block metals such as nickel and iron, and molybdenum are used for biofuel production.

Table 5.2 Catalyst used in biofuel production.

Starting material	Catalysts used	Products formed	Name of reaction
Syngas ($CO + H_2$)	Alkali doped – ZnO/Cr_2O_3 Cu/ZnO $Cu/ZNO/Al_2O_3$ $CuO/CoO/Al_2O_3$ MoS_2	Mixed alcohols	
	Fe Co Ru	Wax Olefins	Fischer–Tropsch synthesis
	ThO_2 ZrO_2	Isobutene	Isosynthesis
	H_2O	H_2	
	$HCo(CO)_4$ $HCo(CO)_3P(Bu_3)$ $Rh(CO)(PPh_3)_3$	Aldehydes Alcohol	Oxosynthesis
	Co Rh	Ethanol	
	Cu/ZnO	Methanol	
Methanol	Ag	Formaldehyde	
	Isobutylene	Methyl tert butyl ester	
	Co	Ethanol	Homologation
	Al_2O_3	Dimethyl ester	
	Zeolites	Olefins Gasolines	MTG process
	$CH_3OH + CO$ Co Rh Ni	Acetic Acid	Carbonylation

Source: Adapted from [80, 81].

In order to overcome various environmental issues and increase the production yield, newer catalytic systems, mixed catalysts and supported catalysts are used for obtaining better results in terms of fuel and energy.

5.5 Nanoparticles Potential in Biofuel Production

The probable applications of nanobiotechnology for the generation of bioenergy that can be maintained have fortified researchers nowadays to explore innovative nanoscaffolds for synthesising nanobiocatalytic systems for biofuel production which are robust [83, 84].

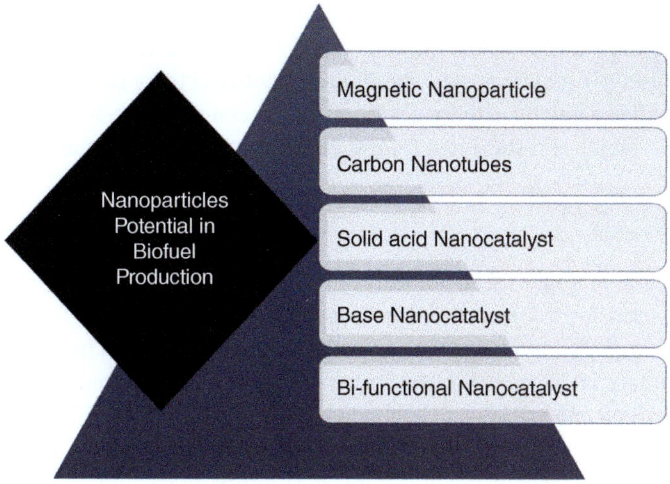

Figure 5.3 Various nanoparticles used in biofuel production.

Nanoparticles have unique characteristics of producing quantum effects because they can confine their electron due to their nanosize. Some elementary characteristic properties and features of nanocatalysts such as high stability and selectivity, activity, ease of separation from the reaction mixture, high surface-to-volume ratio and energy efficiency make it a virtuous material for biofuel production. The type of metal nanoparticles, ability to modify acid–base properties and porous nature are the additional properties that also help in increasing the conversion efficiency of biomass to biofuels. The efficiency of biofuel in terms of production and stability can be improved with the use of nanoparticles. Nanocatalysts from metal oxide, nanozeolites, magnetic nanocatalysts and nanohydrotalcites are the efficient catalysts as they give better yield and have high selectivity. Two or more nanocatalysts when combined together further enhance the efficiency, and simple purification techniques enable fast recovery [85]. Different types of nanoparticles and nanomaterials that are employed in the production of biofuel are shown in Figure 5.3.

5.5.1 Magnetic Nanoparticle

Magnetic nanoparticles (MNPs) are the nanoparticles that consist of two components, often iron or cobalt and nickel, that act as a magnetic material, and functionality is provided by the chemical component. Some of the unique properties of MNPs include the effect of immobilisation, increased ratio of surface to volume and quantum property, which make it stand out for the biofuel production. MNPs have a non-toxic effect as they can be used as a highly convenient catalyst, making immobilised particles easily removable by the application of suitable magnetic fields. MNPs can be coated with other nanomaterials that are catalytically active, making them suitable nanocatalysts, apart from the support of enzyme immobilisation. Photo-oxidation, inductive heating and hydrogenation can be performed by these nanocatalysts by applying high-frequency magnetic fields. [86, 87].

The catalytic performance of a magnetic nanoparticle in biofuel synthesis is higher than that of conventional catalysts, which can be attributed to a strong ionic interaction between the particles provided by magnetic attraction, resulting in high activity and stability. Studies have shown that Fe_3O_4 magnetic-nanoparticle-immobilised lipase showed enhanced tolerance to variations in solution pH and improved stability, and better activity compared to the free enzyme [88]. Effective catalytic activity under optimum conditions was obtained for magnetic nanoparticles, which was indicated from the maximum yield of biofuel.

5.5.2 Carbon Nanotubes

Carbon nanotubes (CNTs) are graphite sheets folded in the cylindrical tubes such as their functional groups are exposed for the reaction. The large surface area of CNTs enables loading enzymes with high capacity and low diffusion resistance. CNTs are used in biofuel production because this material is vital for the immobilisation of biomolecules, 3-D electro-active area, conductivity and porosity, which increase the concentration of enzyme and other redox compounds on its surface. Metal oxides (silica, alumina, zirconia, etc.), mesoporous niobium oxide and covalent organic framework are various supports with CNTs that display high efficiency in biofuel production [89].

The surface attracts the organic reactants and unfavourable reactions caused because the presence of H_2O can be retarded by the hydrophobic property of carbon nanotubes, which helps in regaining and reusing it. The immobilisation efficiency can be increased with the use of an ionic liquid with CNTs through providing an improved dispersion of carbon nanotubes than buffer solution. Recent research has established that enzymes coupled with carbon nanotubes show greater stability and activity.

5.5.3 Solid Acid Nanocatalyst

The idea of a solid acid catalyst comprises the use of solid particles, in which protons can be donated or electrons are accepted in reactions. The acidic canter presents a catalytic function, present mainly on their surface. The advantages of stronger acidic nanocatalysts are porous structures and high surface areas, which lead to higher catalytic activity, high selectivity, long catalyst life, and ease in recovery and recycle. Many conventional liquid acids can be replaced with these nanocatalysts as they have the potential to be efficient and environmentally friendly in biofuel production processes [90].

Some of different types of solid acid catalysts are metal oxides, functionalised silica, acid resins, supported metals, H-form zeolites, heteropoly acids, magnetic acids, immobilised ionic liquids, carbonaceous acids, and hydrotalcite nanoparticles [91]. Solid acid nanocatalyst (SO_4^{-2}/Fe_2O_3) are utilised in biofuel production. The feedstock that can be used as waste cooking oil for biofuel production is allowed with acid nanocatalysts as they catalyse both esterification of FFAs and triglycerides, which are transesterified simultaneously.

5.5.4 Base Nanocatalysts

Base nanocatalysts are the nanocatalyst solids with Brønsted basic and Lewis basic activity centres. In these nanocatalysts, protons can be accepted from reactants, and electrons are supplied to them for the biofuel synthesis process. The unique features of base nanocatalysts includes easy regeneration and have a corrosive nature, which is less, resulting in more environmentally friendly operations that are safer as well as cheaper. Amongst several base nanocatalysts for biofuel production, oxides of calcium, zeolites and hydrotalcites have drawn more attention [92]. Amongst these, calcium oxides have been vastly used due to their basicity that is on the higher side and activity, long catalyst lifetimes, mild reaction conditions and low cost. Boron group-based and waste-based catalysts are the catalysts that are predominantly classified as base nanocatalysts. The solid base catalysts have therefore lately attracted significant attention in biofuel synthesis from biomass.

5.5.5 Bi-functional Nanocatalysts

Acid catalysts are known to tolerant the purity (FFA content) of the feedstock, while the alcoholysis reactions are accelerated by base catalysts. Hence, the idea of a bi-functional nanocatalyst was introduced to improve the biofuel production process in one step. A bi-functional catalyst is the progressive explanation for biofuel production in a one-step reaction [93]. The presence of both acid and base sites in these catalysts promotes the esterification and transesterification at the same time. Two-step reactions being replaced with a one-step reaction, and employing low-cost equipment could also reduce the cost of biofuel production. Mixed oxide nanocatalysts exhibited distinct features that resulted in attaining effective performance in the transesterification reaction. Bi-functional nanocatalysts possessing great catalytic performances could offer an easy and economically favoured way with a promising potential to be employed as an industrial method for biofuel production. An example of a bi-metallic nanocatalyst for the biofuel production includes Au/Ag bimetallic catalyst applied in the transesterification of sunflower oil.

5.6 Production Cost Analysis

The Department of Fuel and Energy is constantly working on biofuel economics and trade to evaluate the potential of biofuels cost-effectively displacing the use of petroleum-based fuels in transport. A certain biorefinery can predict the capital costs and product yields using the best available technologies, for a particular time period. However, the varying feedstock price, geographical conditions and market value over time cannot be estimated. In addition, the technologies used are changing drastically in search of a more viable solution on a commercial scale. Thus, the estimated costs may vary more or less than the real-world projects. In general,

$$\text{biofuel cost of production} = \left(\text{the feedstock value over time} + \text{the calculated conversion cost}\right) - \text{the co} - \text{product cost.}$$

Studying the energetic evaluation and reduction potentials of CO_2 as well as the market prospects of biofuels for transport feedstock production and market cost is the most important factor [94, 95]. The geographical conditions, availability and reproducibility of feedstock in terms of time and cost are the key factors guiding the market value of feedstock. Plant costs also depend on local variables such as labour rates and tax rates. The other factors deciding the cost, for example ethanol plants, may include the engineering

Table 5.3 Yield and energy comparison between two advanced technologies.

Technology	Biofuel yield (litres/dry t)	Energy content (MJ/l)	Energy yields (GJ/t)	Production cost USD/lge		
				2010	2030	2050
Biochemical						
Enzymatic hydrolysis for ethanol production	110–300	21.1	2.3–6.3	0.8	0.55	0.55
Thermochemical						
Syngas to Fischer–Tropsch diesel	75–200	34.4	2.6–6.9	1.00	0.6	0.55

cost, crushing cost, grain cost, fermentation cost, distillation and evaporation plant cost, dehydration cost, drying systems, boilers, oxidisers used, storage and cooling towers, west water treatment plant, water treatment plants, electrical systems used, and miscellaneous along with start-up cost, administration, building, furnishing cost and internal cost.

It is supposed that (i) there will not be any increase in the amount of land until 2050, and the similar amount of land presently used for biofuels remains constant; (ii) no additional policies favouring the production and use of biofuels will be applied; and (iii) there is no tax concern regarding its use for transportation so that the future scenario in 2050 can be analysed as follows.

Based on the observations, it is expected that prices of fossil fuels, feedstock (oil seeds and cereals) and wood-based resources (e.g. wood industry residues, waste wood) will increase by 3%, 2%, and 1%, respectively, per year until 2050 [96].

Compared to the two advanced production routes as shown in Table 5.3, the route that produces ethanol only is biochemical, while the thermochemical method can synthesise a large number of biofuels used in different fields including aviation and marine purposes.

The cost of production of a second-generation plant is based on the type of project. A single plant has to consider every minute aspect, whereas a co-located plant can take advantages of many commonly occurring facility reducing the cost of biofuel production. In a single plant, the construction of its plant, cost of whole chain required for the procurement of feedstock and charges for logistics must be considered, whereas in a co-located plant some infrastructures and operations that are already present can be shared reducing the overall cost of production. The worldwide pandemic situation has had a great impact on the cost of the production process as indicated in Table 5.3. It can be seen that the fuel prices will be negligibly dropped in next coming years due to the losses in production technologies and less transportation.

Within the entire process, 'pre-treatment' costs 30–50% of the total equipment cost. Common pre-treatment steps are (i) physical pre-treatment (particle size is reduced and surface area is increased), (ii) chemical pre-treatment (maintaining pH of the medium, SO_2 and CO_2 explosion) and (iii) biological pre-treatment (microorganisms or fungi are used). However, all pre-treatment methods are not equally sustainable in terms of economy [97].

- A decentralised pre-treatment process increases the durability of biomass for long-term storage.
- Pre-treatment procedures that require large volumes of water make the method more expensive and consequently less profitable.
- Large-scale production is more feasible.
- Methods that require less energy, inexpensive chemicals and moderate temperatures lower processing cost making them more profitable.
- Processes that use supercritical fluids (water and CO_2) require high pressure increasing the cost.

After pre-treatment, enzymatic hydrolysis was responsible for 25–30% of the operational cost data. The biggest restraining factor has been the absence of published cost. Many researchers reported that total costs per gallon compared to grain-based ethanol production would be 44% more for the biochemical process related to 48% higher for the thermochemical process. It would need about 6.8 times and 7.7 times higher initial capital dollars to build the respective plant, respectively. Facts and figures show the unprofitable nature of all three types of production due to the absence of policies and subsidies.

Despite a lot of discussion and argument over the technology choice, the biochemical process always proved to be economically viable because the technology supports environmentally friendly practices and a satisfactory return on capital [98]. The financial analysis must consider a period of 10 years for capital. It may produce conservative results, and the yield and efficiency may be seen to be increasing due to changes in technologies adopted.

5.7 Environmental Footprints of Chemical Processes

The burning and combustion of fuels releases heavy metals and other pollutants such as CO_2, CO, unburned hydrocarbons, NOx, SOx, particulates and lead, affecting the surrounding and finally global air quality, which may finally result in global warming [99]. This will damage the crop production efficiency, affecting agricultural sector that in turn will be harmful to human health. Road transport adversely affects the air quality (Table 5.4).

However, as related to fossil fuels, the harmful impacts after utilising biofuels are significantly lower. Bioenergy producing crops reduce greenhouse gas (GHG) emissions by directly removing CO_2. In addition, these crops produce many protein-rich co-products that can be used as animal feed.

Large amount of water is required for processing the feedstock into fuel, especially ethanol. It is required for biodiesel refining, seeds processing, cooling the system at desired temperature and product washing. A typical US soya bean requires 19 kg water per tonne of oil produced. But in Brazil, sugar cane, for example, requires 3900 l for production. The treatment plants must be flushed properly putting a large quantity of organic substance into local water reservoirs. These air and odour emissions increase with an increase in the processing plant capacity.

The appropriate government policies, regulations, and technologies are supporting in minimising these pollutions significantly. Various precautionary measurements are also used in reducing the emissions, such as installing a new NOx burner reduces NOx

Table 5.4 Environmental impacts of fuel combustion emission.

Combustion product	Impacts
CO_2	Contributes to global warming and climate change
CO	Results from incomplete combustion or burning. In the atmosphere, CO reacts with oxygen to form ozone, a highly reactive molecule that damages plant leaves and human and animal lungs
Benzene	The smallest aromatic hydrocarbon and a highly toxic carcinogen
NO and NOx	Ozone precursors; they also react with atmospheric water and create acid rain
SO_2 and SO_3	Acid rain precursors
Lead	Has been phased out from gasoline in most of the countries, but it is still used as an octane enhancer
Particulate matter	Formed from SOx, NOx and hydrocarbons; particulates contribute to ozone formation and affect visibility and hence global warming

Source: Adapted from [100].

emission, mixing fuels at the pollutant collection centre will reduce volatile organic compounds (VOC) emissions, and the use of renewably generated power technology is helpful in reducing the emissions. Regardless of the precautions or the technologies used, care must be taken to maintain the air, soil and water nutrient levels.

5.7.1 Water Pollution

Gasoline and petroleum diesel spills have harmful effects on marine animals. Pure ethanol and biodiesel fuels are comparatively beneficial to environment. Due to their increased solubility in water, the risk of suffocation decreases, reducing their harmful impact on soil and water. Many researchers have reported that the biodiesel obtained from rapeseed oil can be biodegraded within half of the time that required for petroleum diesels [101]. Blending with biodiesel also increases the biodegradation rate, while reverse is the case when using ethanol. The rapid breakdown of ethanol reduces the oxygen level in soil and water which in turn slows the breakdown of gasoline. This slow breakdown will be responsible for more exposure of pollutants in the environment.

5.7.2 Air Pollution

The increased use of fuels in transportation, marine engines or aerospace finally increases its share in air pollution. The possibility of spill and evaporation during storage, fuelling or transport will be dangerous in all terms. Evaporative emission also depends on the blend level. Usually, E1-E10 has the highest level of volatility, which decreases to a minimum from E10-E40. However, the NOx level increases slightly as the blend level increases. Incomplete combustion of ethanol produces carcinogenic by-products such as acetaldehyde, formaldehyde and peroxyacetyl nitrate. However, the emission of pollutants, such as benzene, 1,3-butadiene, toluene and xylene, that are more hazardous decreased when blended with ethanol.

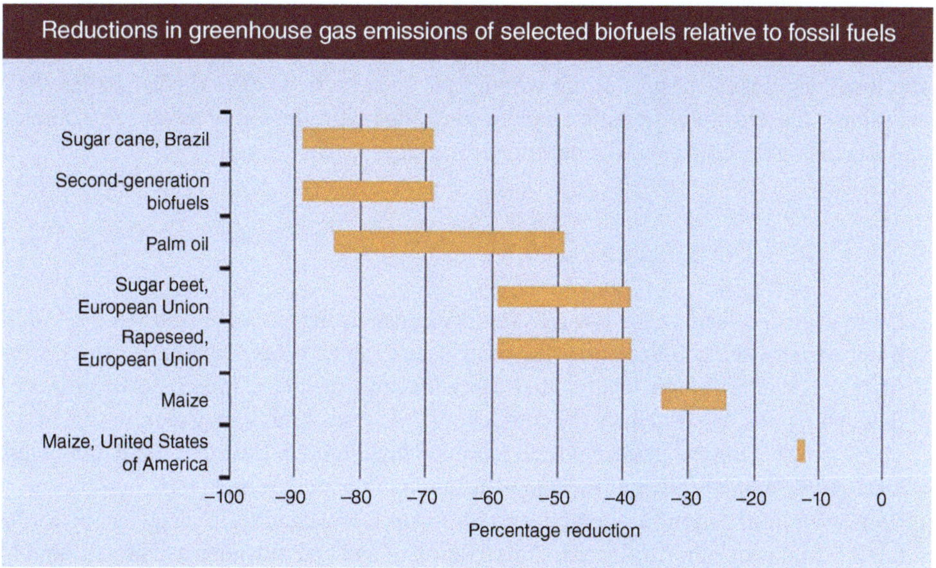

Figure 5.4 Reduction in GHG emissions for various feedstock used.

Ethanol-mixed with gasoline increases fuel oxygen content, making complete burning of hydrocarbons in the fuel and reducing release of CO, sulphur and hydrocarbon [102]. For example, Brazil transport reported the CO emission of 50 g/km in 1980 and that is reduced to 1 g/km in 2000 due to ethanol use. However, the excess addition of ethanol than the recommended one in gasoline can increase NOx emissions. Ethanol blended with diesel can be beneficial to air quality. E15-E20 significantly lowers the harmful emissions as compared to the use of a pure diesel engine.

Biodiesel – either used in pure form or blended form – always results in decreased level emissions of pollutants as compared to diesel. It has also been found that compared to conventional diesel, pure soya bean diesel results in a reduction of particulate matter by 40%, CO by 44%, unburned hydrocarbons by 68%, polycyclic hydrocarbons by 80%, carcinogenic nitrate by 90% and sulphate by 100% (Figure 5.4). It has also been observed that in spite of decreased emission in other pollutants, the NOx level is increased by the use of ethanol and biodiesel, which can increase acid rain resulting in lung damage.

Advantages of biodiesel are as follows [103–105]:

- Renewable energy source
- Less toxic than diesel
- Higher degradation rate
- Lower emission of harmful gases thereby reducing the health risk
- No COx and SOx emissions
- Easily blended with traditionally used fuels
- Engine does not require technical modifications

There are still certain disadvantages that cannot be ignored. It shows higher nitrous oxide (NOx) emissions than diesel fuel [106, 107]. It is used in blends to answer such problems.

Newer vehicles are designed with an effective catalyst system that can reduce those emissions further in ethanol–gasoline blends. NOx emission can also be controlled by using specified blend of biodiesel. Several works have been carried out to further reduce NOx emissions. Air and water pollution can be controlled with the aid of advanced refineries and increasing the efficiencies of natural resources.

5.8 Future Demand and Scope

- Large degraded lands have low soil fertility constraining the plant growth. Pollution, water scarcity and soil depletion may also arise in some areas. Biofuel feedstock crops such as Jatropha and eucalyptus that can withstand such conditions can be grown on these land parts. It will reduce the conflict of food versus feed. It may require some extra efforts for growing the plants on such marginal land, but, on the other hand, it will also increase the employability of the people residing in the nearby area.
- Conservation of agricultural practices with modern tools and technology can be set for farmers and people in rural areas. Conservation of soil and nutrients can be achieved by planting the biodiesel feedstock in rotation. Good agricultural practices can be coupled with the good forestry practices to reduce the cost of feedstock growth.
- Suitable knowledge for the sustainable growth of feedstock can be addressed to the stakeholders for making the process globally. It also includes information and awareness about the natural resources management as per the increasing energy demand and climate changes.
- The life cycle assessment model of biodiesel production will be helpful in analysing various biophysical and chemical factors that can help to lessen the cost of biofuel production.
- The global bioenergy production needs harmony between all techniques and the stakeholders to further stimulate the process of biofuel production. Dedicated research, technology investment and reinforced institutions and infrastructure will be essential to improve the yield of biofuel.
- Government policies must be upgraded as per the time demands. They must be uniform for global level marketing, although there must be some variations as per the climate of the local region.

5.9 Conclusion

Biodiesel is a substituent green fuel to diesel, and bioethanol can replace petrol. Traditional petroleum fuel can be blended with the biodiesel, whereas bioethanol can be used on its own or in a blend. Biodiesel is formed by mixing oil-seed crops with methanol and either sodium hydroxide or potassium hydroxide. The fatty acids and glycerol formation take place with the fatty acids reacting further to produce biodiesel. The reaction can be acid or base or enzyme catalysed. A proper use of various catalysts, nanoparticles, temperatures, pH values and different reaction conditions will increase the productivity of the process.

Many countries are using biodiesel from Jatropha feedstock, which can be cultivated on poor land that is inappropriate for agriculture and stand drought conditions and so does not compete with food production. Brazil, France, the USA, the UK, Argentina, and South Africa are the leading countries that support biodiesel production and use.

Under certain situations, many biofuels may help reduce harmful emissions. The use of persistent feedstock on marginal or degraded lands may offer capacity for globally efficient biofuel production. Good agricultural practices, technological developments and improvements will lead to increased yields. The economic feasibility of biofuels can be transformed with unforeseen variations in energy markets. Advanced technologies should be adopted that convert biomass fibres into synthetic cleaner burning biodiesel. A combination of various generations of biofuels can play a significant role in reducing air pollution and meeting the demand of biofuel on a global level.

References

1 Leschine, S.B. and Canale-Parola, E. (1989). Carbon cycling by cellulose-fermenting nitrogen-fixing bacteria. *Adv. Space Res.* 9 (8): 149–152. https://doi.org/10.1016/0273-1177(89)90039-2.

2 Kheybari, S., Mahdi Rezaie, F., and Rezaei, J. (2021). Measuring the importance of decision-making criteria in biofuel production technology selection. *IEEE Trans. Eng. Manage.* 68 (2): 483–497. https://doi.org/10.1109/TEM.2019.2908037.

3 Wang, J. (2014). Decentralized biogas technology of anaerobic digestion and farm ecosystem: opportunities and challenges. *Front. Energy Res.* 2: 10. https://doi.org/10.3389/fenrg.2014.00010.

4 Kumar, A. and Samadder, S.R. (2020). Performance evaluation of anaerobic digestion technology for energy recovery from organic fraction of municipal solid waste: a review. *Energy* 197: 117253. https://doi.org/10.1016/j.energy.2020.117253.

5 Pradana, Y.S. and Budiman, A. (2015). Bio-syngas derived from Indonesian oil palm empty fruit bunch (EFB) using middle-scale gasification. *J. Eng. Sci. Technol.* 10 (8): 1–8.

6 Hofbauer, H. and Materazzi, M. (2019). Waste gasification processes for SNG production. In: *Substitute Natural Gas from Waste. Technical Assessment and Industrial Applications of Biochemical and Thermochemical Processes*, 105–160. https://doi.org/10.1016/B978-0-12-815554-7.00007-6.

7 Fan, L., Zhang, Y., Liu, S. et al. (2017). Bio-oil from fast pyrolysis of lignin: effects of process and upgrading parameters. *Bioresour. Technol.* 241: 1118–1126. https://doi.org/10.1016/j.biortech.2017.05.129.

8 Mattos, C., Veloso, M.C.C., Romeiro, G.A., and Folly, E. (2019). Biocidal applications trends of bio-oils from pyrolysis: characterization of several conditions and biomass, a review. *J. Anal. Appl. Pyrolysis* 139: 1–12. https://doi.org/10.1016/j.jaap.2018.12.029.

9 Pathak, G., Das, D., Rajkumari, K., and Rokhum, L. (2018). Exploiting waste: towards a sustainable production of biodiesel using: *Musa acuminata* peel ash as a heterogeneous catalyst. *Green Chem.* 20 (10): 2365–2373. https://doi.org/10.1039/c8gc00071a.

10 Karmakar, B., Lalthazuala, S., and Halder, G. (2021). Injecting superheated C1 and C3 alcohol supports towards non-catalytic semi-continuous conversion of *Hevea brasiliensis* oil into biodiesel. *Fuel* 122777: https://doi.org/10.1016/j.fuel.2021.122777.

11 Nasreen, S., Liu, H., Khan, R. et al. (2015). Transesterification of soybean oil catalyzed by Sr-doped cinder. *Energy Convers. Manage.* 95: 272–280. https://doi.org/10.1016/j.enconman.2015.02.006.

12 AlSharifi, M. and Znad, H. (2020). Transesterification of waste canola oil by lithium/zinc composite supported on waste chicken bone as an effective catalyst. *Renewable Energy* 151: 740–749. https://doi.org/10.1016/j.renene.2019.11.071.

13 Sebayang, D., Agustian, E.G.I., and Praptijanto, A. (2010). Transesterification of biodiesel from waste cooking oil using ultrasonic technique. In: *International Conference on Environment*, 13–15.

14 Carere, C.R., Sparling, R., Cicek, N., and Levin, D.B. (2008). Third generation biofuels via direct cellulose fermentation. *Int. J. Mol. Sci.* 9 (7): 1342–1360. https://doi.org/10.3390/ijms9071342.

15 Tsita, K.G., Kiartzis, S.J., Ntavos, N.K., and Pilavachi, P.A. (2020). Next generation biofuels derived from thermal and chemical conversion of the Greek transport sector. *Therm. Sci. Eng. Prog.* 17: 100387. https://doi.org/10.1016/j.tsep.2019.100387.

16 Cruz, G., Silva, A.V.S., Da Silva, J.B.S. et al. (2020). Biofuels from oilseed fruits using different thermochemical processes: opportunities and challenges. *Biofuels, Bioprod. Biorefin.* 14 (3): 696–719. https://doi.org/10.1002/bbb.2089.

17 Ward, A.J., Lewis, D.M., and Green, F.B. (2014). Anaerobic digestion of algae biomass: a review. *Algal. Res.* 5: 204–214. https://doi.org/10.1016/j.algal.2014.02.001.

18 Jackson, S.A., Kang, X., O'Shea, R. et al. (2020). Anaerobic digestion performance and microbial community structures in biogas production from whiskey distillers organic by-products. *Bioresour. Technol. Rep.* 12: 100565. https://doi.org/10.1016/j.biteb.2020.100565.

19 Yin, R., Liu, R., Mei, Y. et al. (2013). Characterization of bio-oil and bio-char obtained from sweet sorghum bagasse fast pyrolysis with fractional condensers. *Fuel* 112: 96–104. https://doi.org/10.1016/j.fuel.2013.04.090.

20 Cubio, G.M., Tuates, A.M., Labitad, E. et al. (2021). Development of a thermochemical conversion unit for corn residues. *Nihon Enerugi Gakkaishi/J. Jpn. Inst. Energy* 99 (12): 243–250. https://doi.org/10.3775/JIE.99.243.

21 Jachmanián, I. and Mukherjee, K.D. (1996). Esterification and interesterification reactions catalyzed by acetone powder from germinating rapeseed. *J. Am. Oil Chem. Soc.* 73 (11): 1527–1532. https://doi.org/10.1007/BF02523520.

22 Mohd Hassim, N.A., Ismail, N.H., and Mat Dian, N.L.H. (2018). Enzymatic interesterification of palm fractions for the production of cocoa butter alternatives. *J. Oil Palm Res.* 30 (4): 537–547. https://doi.org/10.21894/jopr.2018.0038.

23 Wu, L., Wei, T.Y., Tong, Z.F. et al. (2016). Bentonite-enhanced biodiesel production by NaOH-catalyzed transesterification of soybean oil with methanol. *Fuel Process. Technol.* 144: 334–340. https://doi.org/10.1016/j.fuproc.2015.12.017.

24 Jain, S., Sharma, M.P., and Rajvanshi, S. (2011). Acid base catalyzed transesterification kinetics of waste cooking oil. *Fuel Process. Technol.* 92 (1): 32–38. https://doi.org/10.1016/j.fuproc.2010.08.017.

25 Jain, S. and Sharma, M.P. (2010). Kinetics of acid base catalyzed transesterification of *Jatropha curcas* oil. *Bioresour. Technol.* 101 (20): 7701–7706. https://doi.org/10.1016/j.biortech.2010.05.034.

26 Tacias-Pascacio, V.G., Torrestiana-Sánchez, B., Dal Magro, L. et al. (2019). Comparison of acid, basic and enzymatic catalysis on the production of biodiesel after RSM optimization. *Renewable Energy* 135 (C): 1–9. https://doi.org/10.1016/j.renene.2018.11.107.

27 Pullen, J. and Saeed, K. (2015). Investigation of the factors affecting the progress of base-catalyzed transesterification of rapeseed oil to biodiesel FAME. *Fuel Process. Technol.* 130: 127–135. https://doi.org/10.1016/j.fuproc.2014.09.013.

28 Nigade, S.S., Jadhav, S.D., and Chandgude, A.K. (2015). Development of transesterification system with acid and base homogeneous catalysts for *Mangifera indica* seed oil to *Mangifera indica* methyl ester (MOME biodiesel). *Int. J. Energy Power Eng.* 4 (5-1): 48–53.

29 Danish, M., kale, P., Ahmad, T. et al. (2020). Conversion of flaxseed oil into biodiesel using KOH catalyst: optimization and characterization dataset. *Data Brief* 29: 105225. https://doi.org/10.1016/j.dib.2020.105225.

30 Changmai, B., Sudarsanam, P., and Rokhum, L. (2020). Biodiesel production using a renewable mesoporous solid catalyst. *Ind. Crops Prod.* 145: 111911. https://doi.org/10.1016/j.indcrop.2019.111911.

31 Farouk, H., Jaafar, M.N.M., and Atabani, A.E. (2014). A study of biodiesel production from crude jatropha oil (CJO) with high level of free fatty acids. *J. Teknol. (Sciences Engineering)* 69 (3): https://doi.org/10.11113/jt.v69.3145.

32 Wang, Z., Wu, C., Yuan, Z. et al. (2007). Repeated use of methanol and sulfuric acid to pretreat jatropha oil for biodiesel production. *ISES Sol. World Congr. 2007* ISES 2007 2413–2417. https://doi.org/10.1007/978-3-540-75997-3_488.

33 Zhang, L., Guo, W., Liu, D. et al. (2008). Low boiling point organic amine-catalyzed transesterification for biodiesel production. *Energy Fuels* 22 (2): 1353–1357. https://doi.org/10.1021/ef700636u.

34 Cerro-Alarcón, M., Corma, A., Iborra, S., and Gómez, J.P. (2008). Biomass to fuels: a water-free process for biodiesel production with phosphazene catalysts. *Appl. Catal., A* 346 (1, 2): 52–57. https://doi.org/10.1016/j.apcata.2008.05.002.

35 Darnoko, D. and Cheryan, M. (2000). Kinetics of palm oil transesterification in a batch reactor. *J. Am. Oil Chem. Soc.* 77 (12): 1263–1267. https://doi.org/10.1007/s11746-000-0198-y.

36 Ezzati, R., Ranjbar, S., and Soltanabadi, A. (2021). Kinetics models of transesterification reaction for biodiesel production: a theoretical analysis. Renew. *Energy* 168 (C): 280–296. https://doi.org/10.1016/j.renene.2020.12.055.

37 Eze, V.C., Harvey, A.P., and Phan, A.N. (2015). Determination of the kinetics of biodiesel saponification in alcoholic hydroxide solutions. *Fuel* 140: 724–730. https://doi.org/10.1016/j.fuel.2014.10.001.

38 Haryanto, A., Gita, A.C., Saputra, T.W., and Telaumbanua, M. (2020). Kinetics of biodiesel synthesis using used frying oil through transesterification reaction. *Aceh Int. J. Sci. Technol.* 9 (1): 1–11. https://doi.org/10.13170/aijst.9.1.13297.

39 Borugadda, V.B. and Goud, V.V. (2012). Biodiesel production from renewable feedstocks: status and opportunities. *Renewable Sustainable Energy Rev.* 16 (7): 4763–4784. https://doi.org/10.1016/j.rser.2012.04.010.

40 Stamenković, O.S., Veličković, A.V., and Veljković, V.B. (2011). The production of biodiesel from vegetable oils by ethanolysis: current state and perspectives. *Fuel* 90 (11): 3141–3155. https://doi.org/10.1016/j.fuel.2011.06.049.

41 Ho, K.C., Chen, C.L., Hsiao, P.X. et al. (2014). Biodiesel production from waste cooking oil by two-step catalytic conversion. *Energy Procedia* 61: 302–305. https://doi.org/10.1016/j.egypro.2014.11.1086.

42 Melero, J.A., Bautista, L.F., Morales, G. et al. (2015). Acid-catalyzed production of biodiesel over arenesulfonic SBA-15: insights into the role of water in the reaction network. *Renewable Energy* 75: 425–432. https://doi.org/10.1016/j.renene.2014.10.027.

43 Lalthazuala, S., Changmai, B., Kress, T., and Wheatley, A.E.H. (2022). A one-pot route to tunable sugar-derived sulfonated carbon catalysts for sustainable production of biodiesel by fatty acid esteri fi cation. *Renewable Energy* 184: 908–919. https://doi.org/10.1016/j.renene.2021.12.001.

44 Patle, D.S., Sharma, S., Gadhamsetti, A.P. et al. (2018). Ultrasonication-assisted and benzimidazolium-based Brønsted acid ionic liquid-catalyzed transesterification of castor oil. *ACS Omega* 3 (11): 15455–15463. https://doi.org/10.1021/acsomega.8b02021.

45 Li, J. and Guo, Z. (2017). Catalytic biodiesel production mediated by amino acid-based protic salts. *ChemSusChem* 10 (8): 1792–1802. https://doi.org/10.1002/cssc.201700026.

46 Hatano, M., Tabata, Y., Yoshida, Y. et al. (2018). Metal-free transesterification catalyzed by tetramethylammonium methyl carbonate. *Green Chem.* 20 (6): 1193–1198. https://doi.org/10.1039/c7gc03858e.

47 Chen, X., Li, Z., Chun, Y. et al. (2020). Effect of the formation of diglycerides/monoglycerides on the kinetic curve in oil transesterification with methanol catalyzed by calcium oxide. *ACS Omega* 5 (9): 4646–4656. https://doi.org/10.1021/acsomega.9b04431.

48 Serafini, M.S.A. and Tonetto, G.M. (2019). Synthesis of fatty acid methyl esters from pomace oil catalyzed by zinc stearate: a kinetic study of the transesterification and esterification reactions. *Catalysts* 9 (12): 978. https://doi.org/10.3390/catal9120978.

49 Du, J., Wu, D., Hou, X., and Feng, C. (2010). Kinetic studies on lipase-catalyzed transesterification of phosphatidylcholine with ?-linolenic acid ethyl ester. *Internet J. Chem.* 2 (2): 77. https://doi.org/10.5539/ijc.v2n2p77.

50 Devi, N.A., Radhika, G.B., and Bhargavi, R.J. (2017). Lipase catalyzed transesterification of ethyl butyrate synthesis in n-hexane — a kinetic study. *J. Food Sci. Technol.* 54 (9): 2871–2877. https://doi.org/10.1007/s13197-017-2725-2.

51 Gunawan, E.R., Suhendra, D., and Miftahatul Hasanah, B.R. (2017). Enzymatic kinetics of fatty acids methyl ester from kapok seed oil (*Ceiba petandra*) using lipozyme TL IM. *Am. J. Appl. Sci.* 14 (12): 1146–1151. https://doi.org/10.3844/ajassp.2017.1146.1151.

52 Xiong, J., Huang, Y., Zhang, H., and Hou, L. (2014). Lipase-catalyzed transesterification synthesis of geranyl acetate in organic solvents and its kinetics. *Food Sci. Technol. Res.* 20 (2): 207–216. https://doi.org/10.3136/fstr.20.207.

53 Ülger, C. and Takaç, S. (2017). Kinetics of lipase-catalysed methyl gallate production in the presence of deep eutectic solvent. *Biocatal. Biotransform.* 35 (6): 407–416. https://doi.org/10.1080/10242422.2017.1359573.

54 Annapurna Devi, N. and Bhanu Radhika, G. (2018). Kinetic study of ethyl hexanoate synthesis using surface coated lipase from *Candida rugosa*. *Iran J. Chem. Chem. Eng.* 37 (4): 81–92.

55 Andrade, T.A., Errico, M., and Christensen, K.V. (2017). Influence of the reaction conditions on the enzyme catalyzed transesterification of castor oil: a possible step in biodiesel production. *Bioresour. Technol.* 243: 366–374. https://doi.org/10.1016/j.biortech.2017.06.118.

56 Christopher, L.P., Kumar, H., and Zambare, V.P. (2014). Enzymatic biodiesel: challenges and opportunities. *Appl. Energy* 119: 497–520. https://doi.org/10.1016/j.apenergy.2014.01.017.

57 Farobie, O. and Matsumura, Y. (2017). State of the art of biodiesel production under supercritical conditions. *Prog. Energy Combust. Sci.* 63: 173–203. https://doi.org/10.1016/j.pecs.2017.08.001.

58 Maçaira, J., Santana, A., Costa, A. et al. (2014). Process intensification using CO_2 as cosolvent under supercritical conditions applied to the design of biodiesel production. *Ind. Eng. Chem. Res.* 53 (10): 3985–3995. https://doi.org/10.1021/ie402657e.

59 Singh, D., Sharma, D., Soni, S.L. et al. (2020). A review on feedstocks, production processes, and yield for different generations of biodiesel. *Fuel* 262: 116553. https://doi.org/10.1016/j.fuel.2019.116553.

60 Gaduš, J., Giertl, T., and Kažimírová, V. (2016). Biogas production using anaerobic digestion of industrial waste. *Appl. Mech. Mater.* 832: 55–62. https://doi.org/10.4028/www.scientific.net/amm.832.55.

61 Shakib, N. and Rashid, M. (2019). Biogas production optimization from POME by using anaerobic digestion process. *J. Appl. Sci. Process Eng.* 6 (2): 369–377. https://doi.org/10.33736/jaspe.1711.2019.

62 Hijazi, O., Munro, S., Zerhusen, B., and Effenberger, M. (2016). Review of life cycle assessment for biogas production in Europe. *Renewable Sustainable Energy Rev.* 54: 1291–1300. https://doi.org/10.1016/j.rser.2015.10.013.

63 Weiland, P. (2003). Production and energetic use of biogas from energy crops and wastes in Germany. *Appl. Biochem. Biotechnol.(Part A Enzym. Eng. Biotechnol.)* 109 (1): 263–274. https://doi.org/10.1385/ABAB:109:1-3:263.

64 López, I., Borzacconi, L., and Passeggi, M. (2018). Anaerobic treatment of sugar cane vinasse: treatability and real-scale operation. *J. Chem. Technol. Biotechnol.* 93 (5): 1320–1327. https://doi.org/10.1002/jctb.5493.

65 Hussain Ansari, S., Khan, M.S., Haider, M.S. et al. (2011). study on economic feasibility of biogas plant for a small town. *SciInt (Lahore)* 23 (4): 325–326.

66 Jayaraj, S., Deepanraj, B., and Sivasubramanian, V. (2014). Study on the effect of pH on biogas production form food waste by anaerobic digestion. *9th Int. Green Energy Conf.* 25–28.

67 Van, D.P., Fujiwara, T., Tho, B.L. et al. (2020). A review of anaerobic digestion systems for biodegradable waste: configurations, operating parameters, and current trends. *Environ. Eng. Res.* 25 (1): 1–17. https://doi.org/10.4491/eer.2018.334.

68 Surendran, C.P.A. and Shanmugam, P. (2021). Correlation between empirical formulae based stoichiometric and experimental methane potential and calorific energy values for vegetable solid wastes. *Energy Rep.* 7: 19–31. https://doi.org/10.1016/j.egyr.2020.10.071.

69 Li, Y., Alaimo, C.P., Kim, M. et al. (2019). Composition and toxicity of biogas produced from different feedstocks in California. *Environ. Sci. Technol.* 53 (19): 11569–11579. https://doi.org/10.1021/acs.est.9b03003.

70 Greco, A., Mira, D., and Jiang, X. (2017). Effects of fuel composition on biogas combustion in premixed laminar flames. *Energy Procedia* 105: 1058–1062. https://doi.org/10.1016/j.egypro.2017.03.457.

71 Goswami, R., Chattopadhyay, P., Shome, A. et al. (2016). An overview of physico-chemical mechanisms of biogas production by microbial communities: a step towards sustainable waste management. *3 Biotech* 6 (1): 1–12. https://doi.org/10.1007/s13205-016-0395-9.

72 Huang, Q., Liu, Y., and Dhar, B.R. (2020). A critical review of microbial electrolysis cells coupled with anaerobic digester for enhanced biomethane recovery from high-strength feedstocks. *Crit. Rev. Env. Sci. Technol.* 52 (1): 50–89. https://doi.org/10.1080/10643389.2020.1813065.

73 Farzad, S., Mandegari, M.A., and Görgens, J.F. (2016). A critical review on biomass gasification, co-gasification, and their environmental assessments. *Biofuel Res. J.* 3 (4): 483–495. https://doi.org/10.18331/BRJ2016.3.4.3.

74 Dawood, F., Anda, M., and Shafiullah, G.M. (2020). Hydrogen production for energy: an overview. *Int. J. Hydrogen Energy* 45 (7): 3847–3869. https://doi.org/10.1016/j.ijhydene.2019.12.059.

75 Battista, F., Mancini, G., Ruggeri, B., and Fino, D. (2016). Selection of the best pretreatment for hydrogen and bioethanol production from olive oil waste products. *Renewable Energy* 88: 401–407. https://doi.org/10.1016/j.renene.2015.11.055.

76 Cheng, Y., Qiao, M., and Zong, B. (2017). Fischer-Tropsch synthesis. *Encycl. Sustain. Technol.* 403–410. https://doi.org/10.1016/B978-0-12-409548-9.10107-1.

77 Li, T., Wang, H., Yang, Y. et al. (2014). Study on an iron-nickel bimetallic Fischer-Tropsch synthesis catalyst. *Fuel Process. Technol.* 118: 117–124. https://doi.org/10.1016/j.fuproc.2013.08.015.

78 Pan, Z., Parvari, M., and Bukur, D.B. (2014). Fischer-Tropsch synthesis on Co/ZnO - two step activation procedure for improved performance. *Appl. Catal., A* 480: 79–85. https://doi.org/10.1016/j.apcata.2014.04.040.

79 Shah, Y.T. and Perrotta, A.J. (1976). Catalysts for Fischer-Tropsch and isosynthesis. *Ind. Eng. Chem. Prod. Res. Dev.* 15 (2): 123–131. https://doi.org/10.1021/i360058a005.

80 Zhao, X., Wei, L., Cheng, S., and Julson, J. (2017). Review of heterogeneous catalysts for catalytically upgrading vegetable oils into hydrocarbon biofuels. *Catalysts* 7 (3): 83. https://doi.org/10.3390/catal7030083.

81 Hassan, N.S., Jalil, A.A., Hitam, C.N.C. et al. (2020). Biofuels and renewable chemicals production by catalytic pyrolysis of cellulose: a review. *Environ. Chem. Lett.* 18 (5): 1625–1648. https://doi.org/10.1007/s10311-020-01040-7.

82 Liu, S., He, J., Lu, D., and Sun, J. (2020). Optimal integration of methanol-to-gasoline process with organic rankine cycle. *Chem. Eng. Res. Des.* 154: 182–191. https://doi.org/10.1016/j.cherd.2019.11.036.

83 Bidir, M.G., Millerjothi, N.K., Adaramola, M.S., and Hagos, F.Y. (2021). The role of nanoparticles on biofuel production and as an additive in ternary blend fuelled diesel engine: a review. *Energy Rep.* 7: 3614–3627. https://doi.org/10.1016/j.egyr.2021.05.084.

84 Sekoai, P.T., Ouma, C.N.M., du Preez, S.P. et al. (2019). Application of nanoparticles in biofuels: an overview. *Fuel* 237: 380–397. https://doi.org/10.1016/j.fuel.2018.10.030.

85 Zhang, J., Zhu, Y., Chen, C. et al. (2012). Carbon nanotubes coated with platinum nanoparticles as anode of biofuel cell. *Particuology* 10 (4): 450–455. https://doi.org/10.1016/j.partic.2011.11.014.

86 Pakapongpan, S., Tuantranont, A., and Poo-Arporn, R.P. (2017). Magnetic nanoparticle-reduced graphene oxide nanocomposite as a novel bioelectrode for mediatorless-membraneless glucose enzymatic biofuel cells. *Sci. Rep.* 7 (1): 1–12. https://doi.org/10.1038/s41598-017-12417-0.

87 Muhammad, A. and Badshah, M. (2019). Magnetic nanoparticles: eco-friendly application in biofuel production. *Nanotechnol. Life Sci.* 109-129: https://doi.org/10.1007/978-3-030-16439-3_7.

88 Teymourian, H., Salimi, A., and Khezrian, S. (2013). Fe_3O_4 magnetic nanoparticles/reduced graphene oxide nanosheets as a novel electrochemical and bioeletrochemical sensing platform. *Biosens. Bioelectron.* 49: 1. https://doi.org/10.1016/j.bios.2013.04.034.

89 Ji, Y., Gai, P., Feng, J. et al. (2017). A Fe_3O_4-carbon nanofiber/gold nanoparticle hybrid for enzymatic biofuel cells with larger power output. *J. Mater. Chem. A* 5 (22): 11026–11031. https://doi.org/10.1039/c7ta01931a.

90 Guo, F., Fang, Z., Xu, C.C., and Smith, R.L. (2012). Solid acid mediated hydrolysis of biomass for producing biofuels. *Prog. Energy Combust. Sci.* 38 (5): 672–690. https://doi.org/10.1016/j.pecs.2012.04.001.

91 Kowthaman, C.N. and Varadappan, A.M.S. (2019). Synthesis, characterization, and optimization of Schizochytrium biodiesel production using Na+-doped nanohydroxyapatite. *Int. J. Energy Res.* 43 (8): 3182–3200. https://doi.org/10.1002/er.4387.

92 Rahimi, T., Kahrizi, D., Feyzi, M. et al. (2021). Catalytic performance of MgO/Fe_2O_3-SiO_2 core-shell magnetic nanocatalyst for biodiesel production of *Camelina sativa* seed oil: optimization by RSM-CCD method. *Ind. Crops Prod.* 159: 113065. https://doi.org/10.1016/j.indcrop.2020.113065.

93 Mansir, N., Teo, S.H., Mijan, N.A., and Taufiq-Yap, Y.H. (2021). Efficient reaction for biodiesel manufacturing using bi-functional oxide catalyst. *Catal. Commun.* 149: 106201. https://doi.org/10.1016/j.catcom.2020.106201.

94 Klein-Marcuschamer, D., Oleskowicz-Popiel, P., Simmons, B.A., and Blanch, H.W. (2012). The challenge of enzyme cost in the production of lignocellulosic biofuels. *Biotechnol. Bioeng.* https://doi.org/10.1002/bit.24370.

95 de Jong, S., Hoefnagels, R., Wetterlund, E. et al. (2017). Cost optimization of biofuel production – the impact of scale, integration, transport and supply chain configurations. *Appl. Energy* 109 (4): 1083–1087. https://doi.org/10.1016/j.apenergy.2017.03.109.

96 Zhang, J. and Cross, M.L. (2020). Carbon policies, fossil fuel price, and the impact on employment. *Clean Technol. Environ. Policy* 22 (5): 1085–1095. https://doi.org/10.1007/s10098-020-01850-x.

97 Abomohra, A.E.F., Elsayed, M., Esakkimuthu, S. et al. (2020). Potential of fat, oil and grease (FOG) for biodiesel production: a critical review on the recent progress and future perspectives. *Prog. Energy Combust. Sci.* 81: 100868. https://doi.org/10.1016/j.pecs.2020.100868.

98 Raftery, J.P. and Karim, M.N. (2017). Economic viability of consolidated bioprocessing utilizing multiple biomass substrates for commercial-scale cellulosic bioethanol production. *Biomass Bioenergy* 103: 35–46. https://doi.org/10.1016/j.biombioe.2017.05.012.

99 Abdul Rahman, A.Z. and Saidina Amin, N.A. (2019). Environmental sustainability assessment of palm lignocellulosic biomass pretreatment methods. *Chem. Eng. Trans.* 72: 13–18. https://doi.org/10.3303/CET1972003.

100 Solarin, S.A. (2020). An environmental impact assessment of fossil fuel subsidies in emerging and developing economies. *Environ. Impact Assess. Rev.* 85: 106443. https://doi.org/10.1016/j.eiar.2020.106443.

101 Navas-Anguita, Z., García-Gusano, D., Dufour, J., and Iribarren, D. (2020). Prospective techno-economic and environmental assessment of a national hydrogen production mix for road transport. *Appl. Energy* 259: 114121. https://doi.org/10.1016/j.apenergy.2019.114121.

102 Jeswani, H.K., Chilvers, A., and Azapagic, A. (2020). Environmental sustainability of biofuels: a review: environmental sustainability of biofuels. *Proc. R. Soc. A Math. Phys. Eng. Sci.* 476 (2243): 20200351. https://doi.org/10.1098/rspa.2020.0351.

103 Faleh, N., Khila, Z., Wahada, Z. et al. (2018). Exergo-environmental life cycle assessment of biodiesel production from mutton tallow transesterification. *Renewable Energy* 127: 74–83. https://doi.org/10.1016/j.renene.2018.04.046.

104 Branco-Vieira, M., Costa, D.M.B., Mata, T.M. et al. (2020). Environmental assessment of industrial production of microalgal biodiesel in central-South Chile. *J. Cleaner Prod.* 266: 121756. https://doi.org/10.1016/j.jclepro.2020.121756.

105 Wahyono, Y., Hadiyanto, H., Budihardjo, M.A., and Adiansyah, J.S. (2020). Assessing the environmental performance of palm oil biodiesel production in Indonesia: a life cycle assessment approach. *Energies* 13 (12): 3248. https://doi.org/10.3390/en13123248.

106 Chen, H., Xie, B., Ma, J., and Chen, Y. (2018). NOx emission of biodiesel compared to diesel: higher or lower? *Appl. Therm. Eng.* 137: 584–593. https://doi.org/10.1016/j.applthermaleng.2018.04.022.

107 Chirico, R., Clairotte, M., Adam, T.W. et al. (2014). Emissions of organic aerosol mass, ack carbon, particle number, and regulated and unregulated gases from scooters and light and heavy duty vehicles with different fuels. *Atmos. Chem. Phys. Discuss.* 14 (11): 16591–16639. https://doi.org/10.5194/acpd-14-16591-2014.

6

Microalgae – Biofuel Production Trends

6.1 Introduction

A significant amount of CO_2 is released in the atmosphere causing greenhouse effect. It has been observed that the transport sector is the major source of harmful emissions. A good substitute for fossil fuel was the vital need for growing technology. Biodiesel is considered to be a potential solution on these environmental problems. Research is continuously going on in the production of biofuels in the form of biochar, biodiesel, biogas, biohydrogen, syngas, etc., in solid, liquid as well as gaseous forms as per the need. These are derived from plant-based feedstock. First-generation biofuel requires edible feedstock, which have a direct impact on food prices. Likewise, second-generation biofuel was developed from non-edible feedstock, but again it competes with the requirement of land, water and nutrients for edible crop production. To address such issues and meet the global demand of energy, researchers turned their focus towards microalgae and macroalgae, which are known to be the third generation of biofuels [1, 2].

Growth of algae is rapid. If unchecked, its overgrowth can cause harmful effects on marine life. Some of these algae blooms are toxic in nature. Though on smaller scale, it is used as pest in fish aquarium. but in oceans its growth has a negative impact on human life, animals, environment, marine life, water purification technology, tourism and economy. The nutrients discharged from household sewage or industrial waste, slow moving water, change in temperature or pH are some of the factors responsible for algal blooms [3, 4]. These algal blooms reduce the oxygen level in water bodies and also restrict sunlight to reach to marine life. Harmful toxins emitted by algal blooms have adverse effects on human and animal life when exposed to such water. It also increases the cost of water purification technology in industries, which rely greatly on the water body reservoirs.

Algae have a great potential of storing high amounts of nutrients such as carbohydrates and proteins, which can easily be converted into biofuel. Algae, through photosynthesis respiration, are able to convert CO_2 to O_2 and can generate large amounts of energy content in the form of sugar, protein and lipid. The harmful algae bloom now can offer the best solution for energy crisis due to the increasing dependency on energy. The successful replacement of earlier generation biodiesel with that from microalgae needs a detailed comparison of the biodiesel productivity of microalgae with that of others. Figure 6.1

Sustainability in Biofuel Production Technology, First Edition. Pratibha S. Agrawal, Pramod N. Belkhode, and Samuel Lalthazuala Rokhum.

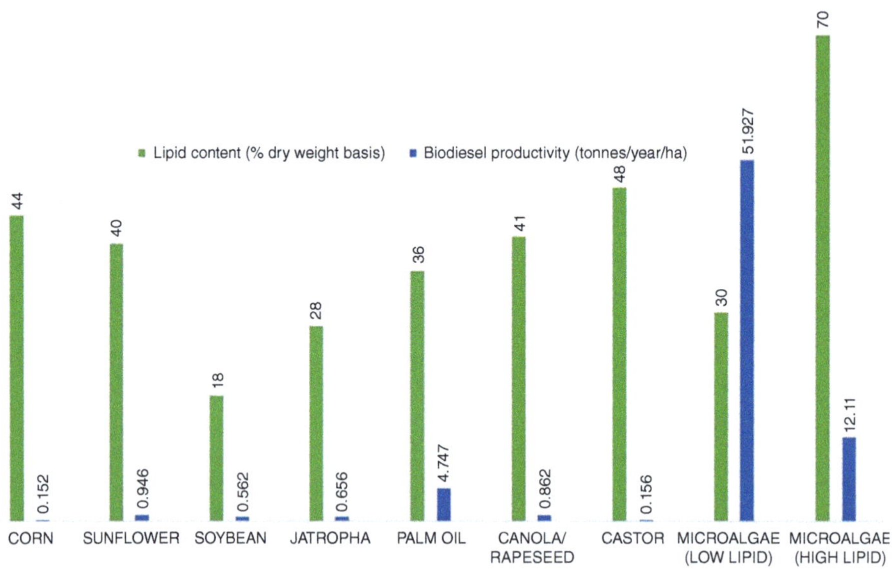

Figure 6.1 Different generation biodiesel production processes compared with microalgae. *Source:* Data from [5].

shows different generation biodiesel production compared with microalgae, which clearly shows the high productivity of biodiesel from microalgae.

Oil content found in microalgae resembles to vegetable oils in physicochemical properties [6, 7]. It is a goof alternative for lipid production that varies in different microalgae samples with the type and conditions of the surrounding in which they grow. Moreover, microalgae significantly contribute to decrease CO_2 pollution, which is an environmental benefit.

The use of fossil fuel in transport has made a major contribution to greenhouse gases causing global warming. To check this problem along with the shortage of the existing fuels, carbon neutral biofuels have been discovered. Compared to all forms of energy, liquid biofuels have the capacity to store solar energy and make its use directly. Photosynthetic organism plays a major role in capturing and storing energy from sunlight in the form of chemical energy. Plants absorb carbon from atmosphere using solar light and CO_2. Photosynthesised carbon is now converted to various compounds like carbohydrates and thus helps in balancing the atmospheric carbon. Highly versatile photosynthetic microorganism is always of growing interest as it can grow on non-arable land using wastewater and it tends to capture solar energy and convert atmospheric and waste CO_2 into various highly energetic chemical products.

The third-generation biofuel production from photosynthetic organisms (cyanobacteria and algae) is of growing interest due to CO_2 fixation with sunlight, but the slow growth rate of cell is creating hurdle in the productivity efficiency. Microalgae are unicellular photosynthetic microorganisms, efficiently trapping and converting solar energy to chemical energy. These are rich in oil content, especially if grown under nitrogen starved conditions. It can easily convert CO_2 into carbon-rich lipids, which can be converted into biodiesel.

Cultivation of algae does not entail useful land, restrict separate addition of nutrients and does not require fresh water, and thus proven to be safer feedstock from the point of view of world food production. It has also been proven to be a potential candidate in the production of biodiesel, bioethanol and biohydrogen by many researchers. However, the process requires proper cultivation and harvesting of microalgae which is a costlier step compared to traditional methods of biodiesel production making it less preferable [8]. However, the integration of cultivation method with wastewater treatment plant can contribute to reduce the cost as wastewater is rich in nutrients and other microelements. Carbon present in wastewater can be feed to microalgae, which will convert it into biomass. Similarly, other nutrients such as nitrogen and phosphorous can also be used by microalgae for their growth. In addition, such wastewater if fed to microalgae needs not to be treated separately from environmental policy concerns as it will contain less amount of carbon now. Such integrated technology will make the biodiesel production technique commercially sustainable. This chapter deals with the developing status of biodiesel production using microalgae. Its high diesel productivity/ha/year, fast growth speed over a limited period of time, supporting technology to wastewater treatment plants and the controlled methods of cultivation make it a more promising source of biodiesel production in the future.

6.2 Technology for Microalgae Cultivation

Microalgae are found to grow faster in dumped water bodies. They consume sunlight carbon dioxide and water for their bloom. Usually, it can be seen as green, blue green or sometimes red and brown. The lipid content in the microalgae may vary from 5% to 50% or even 75% in some cases. The cultivation method is comparatively easy as it can sustain variations in the water contaminations, temperature, pH and intensity of the sunlight. The methods of cultivations are divided into four types depending on the source of energy and carbon used for microalgae bloom. These are illustrated in Figure 6.2 and discussed below.

6.2.1 Autotropic/Phototropic Cultivation

This is the simplest method of cultivation in which algae use sunlight and CO_2 as carbon source, which produces chemical energy by photosynthesis. It is the cheapest method of cultivation. CO_2, which is a greenhouse gas, is successfully consumed in algae growth for biofuel production. This process is energetically as well as environmentally efficient one, and hence is preferably used. However, at the same time, variable intensity of light, exposure time to sunlight and available limited CO_2 supply affect the lipid content, which is reflected in terms of biomass contents. A continuous supply of CO_2 could increase the lipid content and hence the biomass productivity in turn.

It can be cultivated in two ways: open and closed, based on the role of atmosphere [9]. Atmospheric factors greatly influence the lipid content and algae growth. Algae if cultivated in an open pond is termed as open cultivation method. It can be grown in typically artificially built open ponds in which water can be kept in motion using wheels. However,

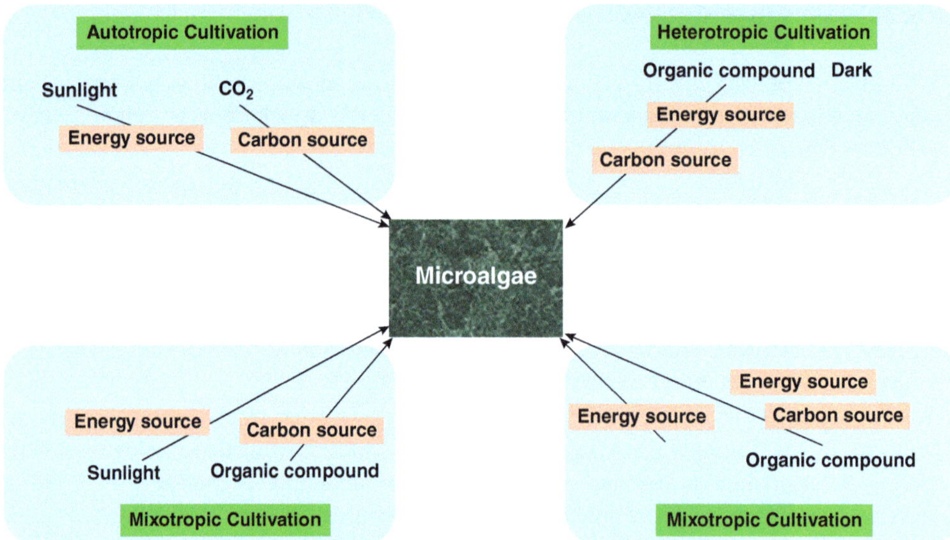

Figure 6.2 Cultivation methods for microalgae.

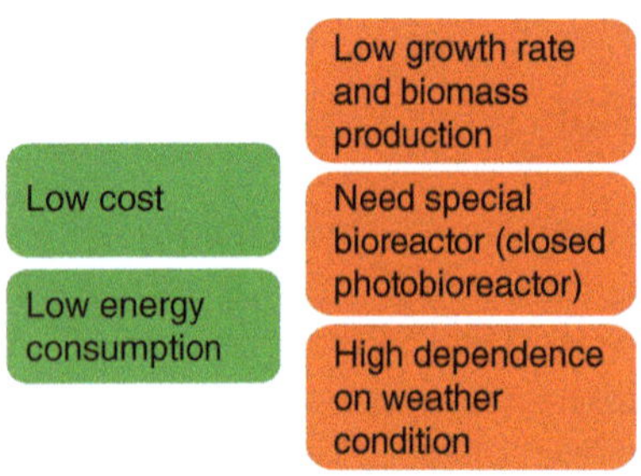

Figure 6.3 The advantages and disadvantages of the autotropic/phototropic cultivation method.

if the algae are separated from the atmosphere and allowed to grow in a closed container using various types of bioreactors, it is termed as a closed system of cultivation. In this process, CO_2 can be induced or pumped from outside. The benefits and disadvantages of the cultivation method are presented in Figure 6.3.

Though open ponds are inexpensive, intensive care should be taken to avoid atmospheric contamination and evaporation of water. It increases the cost; otherwise, the biomass quality will be degraded. It is always susceptible to the changing weather condition and is greatly exposed to contaminations. In contrast, closed methods require initial knowledge and funding for set-up, but it will produce reliable quality of biomass. All the atmospheric affecting parameters can be duly controlled in this way, producing biomass even of food quality. It can be optimised as per the desired quality of the biomass. Such type of process is usually followed when high quality of algae, for example, for food or pharmaceutical application is required.

Thus, the following are the point of superiority of closed methods over the open ones [10]:

- Risk of contamination is more in open systems and can be very well addressed in closed ones.
- Exposure to light can be equal and in the required amount all over the pond.

- Productivity in terms of both quality and quantity can be improved many folds in closed systems.
- Process is well controlled in closed systems.
- Reliable and stable products, irrespective of atmospheric conditions, will be obtained in closed ones.
- Possibility of blooming in closed systems is for all 24 hours due to artificial lighting and other required feed, which can be managed well.

6.2.2 Heterotropic Cultivation

Some algae cannot only grow under phototropic conditions but also able to grow using an external source of organic carbon in dark. Both energy and carbon source are supplied by these external carbon sources. It can overcome the disadvantages of uneven distribution of sunlight for the growth and thus can produce biomass with improved quality. The basic processes involved are presented in Figure 6.4.

Wen et al. reported 40% increase in the lipid content when the method is shifted from phototrophic to heterotrophic system [11]. The only difficulty in this method is the addition of source of organic carbon makes the method costly. The advantages and disadvantages of the heterotropic cultivation method are presented in Figure 6.5.

Biomass with lower protein contents can be obtained, which will have lower market value. If such a method is integrated with industries, the industrial waste can be the largest source of carbon, thereby reducing the purchase cost of carbon source and minimising the waste produced from industries. This way, microalgae can be cultured on new innovative low-cost carbon sources and hence heterotopic methods can become effective in cost consideration.

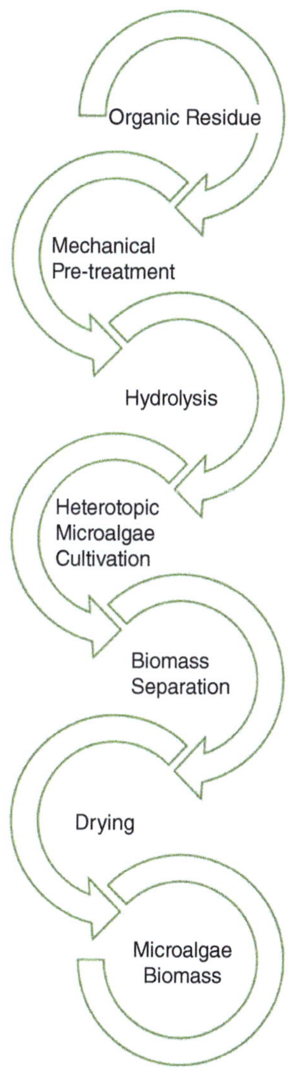

Figure 6.4 Heterotropic microalgae cultivation.

6.2.3 Mixotropic Cultivation

In the mixotrophic cultivation method, light-induced photosynthesis is used along with the organic and inorganic CO_2 compounds as a source of carbon that is required for its growth. By adopting this method, microalgae can be effectively cultured on both phototropic and heterotropic methods [12, 13]. As organic compounds are also used, CO_2, which is released during respiration, can be absorbed, and thus cultivation becomes environmentally reliable. The advantages and disadvantages of mixotropic cultivation method are presented in Figure 6.6.

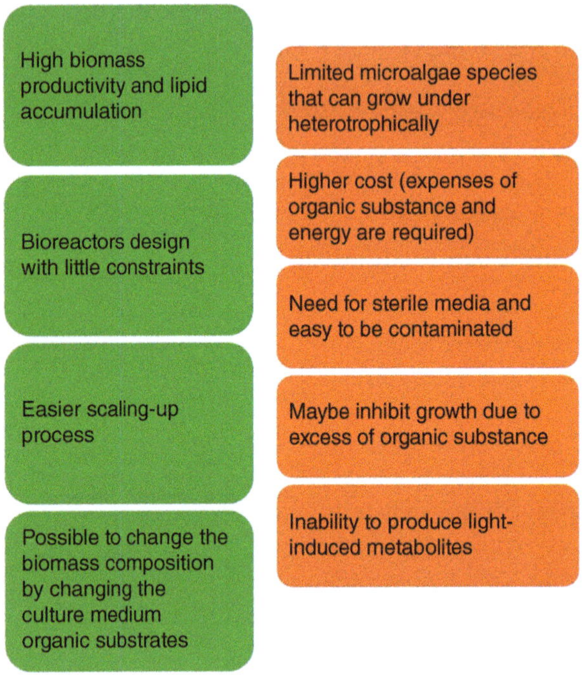

Figure 6.5 The advantages and disadvantages of heterotropic microalgae cultivation.

6.2.4 Photoheterotropic Cultivation

In this method, algae are cultivated using both light and organic compounds as a source of energy. These methods are rarely used due to contamination risk and increased cost. It favours the speedy growth of algae with valuable metabolites obtained as side products.

The advantages and disadvantages of photoheterotropic cultivations are shown in Figure 6.7.

6.2.5 Large-scale and Lab-scale Microalgal Cultivation for Biomass Production

In large-scale microalgal cultivation methods, open and closed photo-bioreactors (PBRs) are used. The open cultivation method, being cheaper one, is used for large-scale production. In the open method, the sunlight as well as atmospheric CO_2 are used as sources of energy and carbon, and nutrients are added externally for producing biomass on a quantitative scale, and closed photo-bioreactors are used for producing value-added biomass on a qualitative scale. The open cultivation method is provided with a paddle wheel to mix well the algae and nutrients [14].

The PBR method requires several supporting techniques for increasing the quality of biomass. Usually, researchers use this technique in laboratory in the controlled environment to void any atmospheric contamination. Several researchers are working for improving the operational parameters such as pH, temperature and catalyst and also reducing the cost of process to obtain the product on a larger scale without compromising its quality.

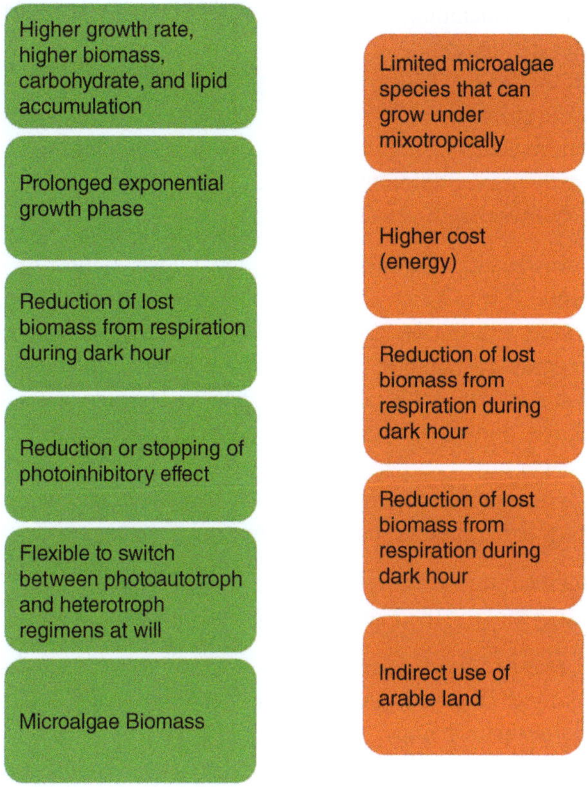

Figure 6.6 Advantages and disadvantages of mixotropic cultivation method.

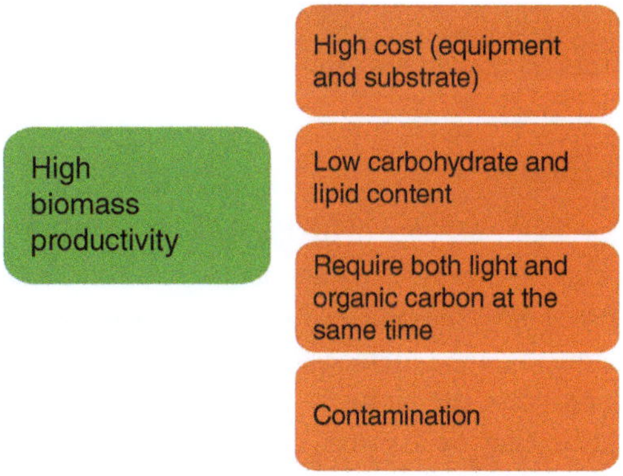

Figure 6.7 Advantages and disadvantages of photoheterotropic cultivation method.

Thus, microalgae cultivation for biofuel production has shown numerous advantages as listed below [15, 16]:

i) they do not compete with human food demands,
ii) they have high lipid contents with carbohydrates, fats and proteins,

iii) they can be grown in any water bodies including wastewater, saving the agricultural land,
iv) they do not need external water for blooming,
v) they can be grown throughout the year,
vi) they contribute to decreasing the atmospheric CO_2,
vii) more than half of the O_2 in nature can be generated in photosynthesis respiration,
viii) its cultivation does not require the use/addition of herbicides or pesticides and
ix) they are thus contributing to an energy-rich cleaner environment.

Industrial evolution increased wastewater discharge day by day, which many times contains microalgal growth nutrients. If this water is utilised for microalgal cultivation, then the addition of external nutrients will not be required and the waste wager will be utilised in a useful way. Such an integrated technique will reduce the cost of cultivation as well as the efforts made for wastewater treatments. Such a commercial sustainable method will promote high energy production, reduce CO_2 emission checking the greenhouse emission and make the method cheaper one.

After cultivation, the biomass is harvested, which is again an expensive technique. A greater amount of work and energy is required for harvesting for two main reasons:

i) Low density of algae in cultural medium.
ii) Algae cells carry a negative charge making their suspension in the medium.

Various harvesting methods such as centrifugation, sedimentation, filtration and flocculations can be used. Out of these methods, the flocculation method is widely used due to simpler, cheaper and high efficiency of the technique. There is still a chance of contamination that can occur in biomass, reducing its market value and biodiesel productivity. Due care should be taken during all those processes.

6.3 Biofuels from Microalgae

Various types of biofuels can be produced from microalgae as shown in Figure 6.8.

6.3.1 Pyrolysis of Microalgae to Biochar/Bio-oil

Pyrolysis of algae is a thermochemical process in the absence of oxygen in the temperature range of 300–700 °C to produce bio-oil, biochar and gaseous component. Bio-oil and biochar can be used as important catalysts in many reactions and also as fuel for heat generation. The quality of the product depends on the type of feedstock and the operating temperature. Depending on the reaction conditions, the pyrolysis process can be categorised in various types, such as slow, fast, flash, catalytic assisted, microwave assisted and hydrolytic pyrolysis. Amongst the other thermochemical processes, pyrolysis is preferred due to simplicity and the speed of the reaction.

6.3.1.1 Slow Pyrolysis

The difference between slow, fast and flash pyrolysis is the heating rate, time spent by vapours and temperature. In slow pyrolysis, temperature is allowed to increase gradually

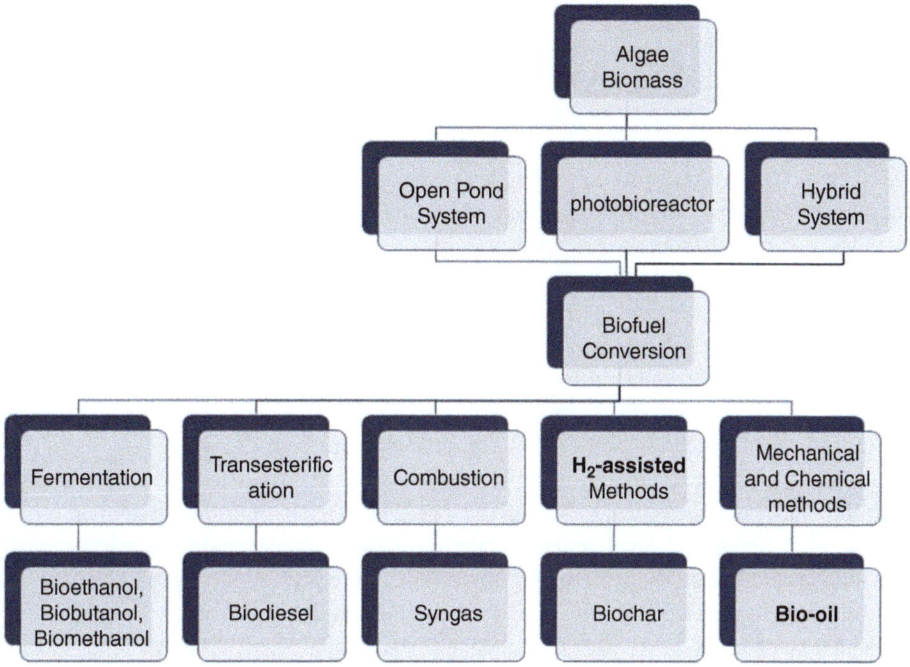

Figure 6.8 Biofuels from microalgae.

(1 °C/s) maintaining the slow heating rate. This allows vapours to spent more time in the reactor (10–60 minutes). The microalgae, which is placed in the reactor, is heated using the electric furnace at a slow rate. As this reaction is carried out in the absence of oxygen, the sweeping gas nitrogen is used to avoid the presence of oxygen in the reactor. Slow pyrolysis favours biochar formation, where the formation of other by-products is reduced. In this process, the biomass is allowed to decompose at 400–500 °C in different steps. Initially, as temperature increases and approaches to 100 °C, breaking of bond is observed with the removal of water molecules. On further heating, lipids, proteins and carbohydrates are decomposed as a result of pyrolysis. Finally, carbon-rich residues are formed at higher temperatures. Between the temperature range 450–500 °C, 50–70% w/w organic liquid is obtained [17]. Temperature and catalyst can change the reaction conditions. Catalyst addition can change the heating value of the liquid. By the addition of *Nannochloropsis* sp. catalyst, the formation of the yield rate increased by 48%, but the favourable charcoal product still limits its addition [18]. Rizzo et al. [19] found that at reduced pH, *Chlorella* sp. increases the yield with a high heating value and a lower density. It is observed that as the temperature elevated above 500 °C, the yield is decreased. A wastewater grown algae also produces the high yield but with high ash content and less stability. Thus, with respect to cost and energy, biochar production from microalgae is most economical and significant process [20]. It was also observed that proper pre-treatment of the microalgae will reduce the decomposition temperature. Thus, the limitations associated with the cracking process, slow reaction and high temperature need make the process less popular.

6.3.1.2 Fast/Flash Pyrolysis

A rapid heating process is used when the S and N content is low and high bio-oil production is required. It produces bio-oil (60–75%), biochar (15–25%) and other gaseous by-products (10–20%). Biomass is fed into the electrically powered reactor, in which feeders, condenser and carrier gas play a crucial role in the process. It operates at a high heating rate ($1000\,°C\,s^{-1}$) [21]. In this process, a higher pyrolysis temperature is reached before the decomposition starts. The process is easy to handle. The properties of biochar or bio-oil depends on the feedstock used. Kim et al. observed that using *Chlorella protothecoides*, *M. aeruginosa*, and *Scenedesmus* sp., the yield is increased noticeably [22]. Reactions with the highest temperature between 500 and 900 °C are observed. Fluidised beds are generally used in the process. Cultivation using heterotrophic techniques is found to be 3.4 times higher than the autotropic cultivation method. In addition, if *C. protothecoides* and *Microcystis aeruginosa* are used, the oil yield increases by 29–33% [23]. Insufficient heat transfer in the reactor can be sometimes responsible to decrease the yield.

Many researchers studied that the yield of product can be improved using a suitable catalyst. Aysu and Naqvi et al. [24, 25] observed the effect of carrier gas and the catalyst on the pyrolysis of green algae. Nickel- and molybdenum-based catalysts make the reactant easily accessible to catalytic surface, thereby brings out decarboxylation of oxygenated chemicals and obtains the increased amounts of hydrocarbons and aliphatic alkenes.

Elevated pressure of the reactor may reduce the volatile release, thereby higher biochar may be obtained. The flue gas from the reactors is utilised for drying the raw materials. The type of cultivations also affects the bio-oil yield. The production yield in the fast technique is typically higher than that in slow pyrolysis due to the increased heating rate.

6.3.1.3 Flash Pyrolysis

It is a special type of fast pyrolysis. The conditions of higher reaction temperature, rapid heating and less vapor residence time are collectively considered in fast pyrolysis. Another difference is very fine particles used in the flash pyrolysis as compared to slow and fast pyrolysis. All this results in a higher biomass to bio-oil conversion rate [26, 27]. Various physical properties, such as pH, stability, viscosity, solubility, etc., also affect the effectiveness of the yield. Thus, the controlled pyrolysis using catalysts gradually becomes more favourable.

6.3.1.4 Catalytic Pyrolysis

The oxygen content decreases the process efficiency of the slow and fast pyrolysis process, which is needed to be at least removed or reduced. The catalysts are added for reducing the oxygen and nitrogen content in the bio-oil and generating upgraded bio-oil. The electrical-powered tubular fixed bed reactor temperature can be maintained between 300 and 600 °C. Various inert gases are used for the condensation process. This improves the quality of the main product that is more stable and less acidic. It has been observed that acidic catalysts promote high biochar, and base catalysts favour higher bio-oil yields [28, 29].

Amongst all catalysts, higher aromatic-supported hydrocarbons, nickel-based catalysts [30, 31], zeolite-based catalyst [32, 33], ceria-based catalysts [34], zirconia-supported catalysts [35] and mixed modified catalysts [36–38] with metal ions are more preferred. The study reveals that the most critical factor is the pyrolysis temperature to support the cleavage and further conversion.

6.3.1.5 Microwave-Assisted Pyrolysis (MAP)

Microwaves are the electromagnetic radiations in the radio frequency range from 0.3 to 300 GHz. Microwave-assisted pyrolysis (MAP) uses microwave irradiation (400–800 °C) for pyrolysing microalgae biomass to produce higher yields of bio-oil and biochar due to rapid and even heating. Bulky mass conversion, uniform heating, easily controllable and non-agitation technique were soon accepted globally. The only limitation with the MAP technique is poor absorption, for which absorbers such as metallic oxides, activated carbons and chars [39] are used in the reactor to significantly enhance the temperature and resulting in an increase in the yield of bio-fuel up to 87% [40–42]. Co-pyrolysis, implantation of the composite catalyst and activated carbons favour product formation. Similarly, it enables a higher yield rate. Zhou et al. [43] observed that the major yield is bio-oil containing acids, hydrocarbons and organic nitrogen compounds. The production of polycyclic aromatic hydrocarbons is hindered at higher temperatures (above 450 °C).

Parvez et al. reported that though microwave pyrolysis has proven thermodynamically superior to electrically heated processes, it cannot be implemented commercially [44]. The limitation of the process is extra heating is required if the product contains a trace amount of water. In addition, if the biomass is in large amount, then even heating using microwave radiation becomes little bit difficult.

6.3.1.6 Hydropyrolysis

It is a special type of temperature-dependent pyrolysis in the high-pressure hydrogen atmosphere where nitrogen is used as a carrier gas. This results in the increased production of hydrocarbons with improved structural preservation. Such a process is useful to produce bio-oil with reduced oxygen content.

The optimised conditions reported by Malik et al. are 310 °C, 3 MPa, and 60 minute yielding CH_4, unreacted H_2, CO and CO_2 [45, 46]. Wang et al. [47] reported the catalytic hydropyrolysis and co-hydropyrolysis of algae. They reported that the co-hydropyrolysis of algae led to increased energy recovery (80.19–91.26%) compared to that with the hydropyrolysis of algae (53.25–79.58%). Furthermore, the addition of catalyst improves the product in terms of higher contents of carbon and hydrogen and simultaneously reducing the oxygen, sulphur and nitrogen contents [48].

The lifespan, catalytic usability and efficiency of the catalyst will be affected as the feed pressure is low. Fast hydro pyrolysis is also recommended for a low bio-oil yield process. It increases the yield rate. To select the appropriate process for bio-oil production, the decomposition of microalgae should be understood in detail along with the fuel production process. Much more efforts are required to optimise the production of higher bio-oil in the near future.

6.3.2 Biodiesel from Microalgae

Microalgae can accumulate very high lipid (5–60%) on a dry weight content. Except the high viscosity, those lipids resemble the vegetable oil in physic chemical properties. Thus, it can be a potential candidate for biodiesel production. Microalgal triglycerides (MTG) can easily undergo the transesterification reaction in which viscosity will also be reduced to produce biodegradable, renewable alternative energy sources. The derived biodiesel has a comparable heating value of 39–41 MJ kg^{-1} with the petroleum used in transportation. It has also been observed that microalgae produce high amounts of lipids under the stressful condition of nitrogen deficiency, high light intensity, salinity, etc. The quality of lipids depends on the cultivation and growing conditions of the algae, which is reflected in the quality of biodiesel produced. Nascimento et al. [49], after screening 12 microalgae strains, reported that *Kirchneriellalunaris*, *Ankistrodesmus fusiformis*, *Chlamydocapsabacillus Ankistrodesmusfalcatus* showed the highest levels of polyunsaturated FAME, which produce biodiesels with the lowest oxidation stability, the lowest (42.47–50.52) cetane number and the highest (101.33–136.97) iodine values, whereas Chlamydomonas sp. and *Scenedesmus obliquus* showed the higher levels of saturated FAME-producing biodiesel with higher oxidation stability, higher cetane numbers (63.63–64.94) and lower iodine values (27.34–35.28). Lipids with the higher levels of saturated FAME (*Chlorella vulgaris*) and with the higher levels of monounsaturated FAME (*Amphora* sp.) result in the biodiesel of enhanced quality. Such biodiesel with a higher value of cetane numbers helps in timely ignition and complete combustion of fuel. *Scenedesmus abundans* with the higher levels of monounsaturated FAME produces biodiesel that meets the standards of EN 14214, SANS1935 and DIN 51606 of European, South African and Germany's standard, respectively. According to the Chinese National Standards, the oxidative stability of such biodiesel is lower, which can be increased by catalytic hydrogenation (Pd/C).

6.3.3 Bioethanol Production from Microalgae

This is the most promising third-generation bioethanol that is divided into the following three technologies:

i) traditional fermentation of pre-treated carbohydrate-rich microalgae biomass (enzymatic hydrolysis and yeast fermentation),
ii) dark fermentation of reserved carbohydrates (photosynthesis is redirected to produce hydrogen, acids and alcohols) and
iii) using genetic engineering direct, photo-fermentation from carbon dioxide and water in a single step to bioethanol using light energy.

6.3.3.1 Fermentative Bioethanol Production

The carbohydrate content of microalgae is mainly focused on feedstock. After trapping the light energy, microalgae convert it into chemical energy by photosynthetic reactions. These reactions can be light or dark one. In light reactions, water breaks into protons and electrons using solar energy. These electrons and protons are then used to generate energy carriers (NADPH and ATP) that are required to support the metabolic reactions. In dark reactions, carbon dioxide is reduced to carbohydrates using NADPH and ATP [50].

The microalgae cell is properly separated first by physical or chemical methods to access the polyglucans. Polyglucans contain the major carbohydrate content. They are either hoarded in the plastids as reserve components (starch and glycogen), providing the energy for metabolic processes and thus helpful in temporary survival in dark conditions, or become the key structural component of the cell wall (cellulose and sulphated polysaccharides). The types and characteristics of reserve carbohydrates are dependent on the species. The key structural component of the cell walls also proved itself suitable feedstock for bioethanol production, and again different microorganisms have different cell wall structures [51]. These carbon storage compounds, such as starch, undergo hydrolytic degradation to form glucose that can be fermented to obtain bioethanol.

6.3.3.2 Bioethanol Production by Dark Fermentation

Dark fermentation occurs in the absence of light where starch reserves are hydrolysed to sugar. The main advantage of the dark fermentation is to generate ATP that is necessarily required for metabolic processes.

Starch can be converted into pyruvate as a major intermediate compound and then into a variety of end involving various fermentative pathways. The final products vary with different algal species and with changes in environmental conditions.

6.3.3.3 Photofermentative Bioethanol Production

Cyanobacteria are able to produce ethanol directly in photosynthetic process. It is called the "photofermentative" or "photanol" route of bioethanol production. In this genetic engineering-established metabolic pathway, the carbon dioxide is fixed in the Calvin-Benson cycle using the reducing power of photosynthesis to generate pyruvate that is converted into acetaldehyde and ethanol. Though the finding on the initial growing technology step proved to be fascinating, the slow ethanol production rate and lower volume engaged researchers further in improving the process more sustainably.

6.3.4 Biohydrogen and Bio-Syngas Production from Microalgae

This is favourable production for which technologies used are pyrolysis, gasification, reforming and combustion. The choice of the technique is not simpler, but it is a big challenge for chemical engineers to go for process development.

Pyrolysis

The simplest, cheaper technique is pyrolysis. It is also the basic step for other gasification and combustion. Valuable gases, such as H_2 and CO, are produced in this process and are effectively used in fuel cells. It is a thermal process of conversion of biomass into solid (coal and ash), liquid (oxygenated hydrocarbons and water) or gas (H_2, CO, CO_2, CH_4 and other low molecular weight hydrocarbons) in the absence of air. Temperature, heating rate, residence time and types of catalysts are the efficiency deciding factors. Hydrogen production is efficient at high temperature, high heating rate and long volatile phase residence time.

This is an endothermic process, so decreasing the temperature decreases the gas yield, and the yield producing liquid products increases. Slow pyrolysis is usually not considered for hydrogen production. The fast pyrolysis can be efficient in producing all gas, liquid and solid phases. In addition, it is carried out with negligible toxic emissions.

Gasification

Gasification is a thermochemical technology involving the structural changes of the biomass at higher temperatures (500–900 °C) in the presence of air/oxygen/steam/CO_2 as a gasifying agent. Feedstock with less than 35% moisture content is destructed to produce hydrogen-rich gas. It is a partial combustion, which uses lesser amount of air than that required by stoichiometric. Hydrogen or syngas or CO_2 can be produced from the thermochemical processes of biomass. Like pyrolysis, it is not necessary to use carrier gas in this process. It is a renewable technique to decrease CO_2 emissions. Increasing the temperature with steam is a steam gasification, which is an endothermic reaction causing additional cost and an increase in the gas yield.

Reforming

It is the most widely used method to produce hydrogen and syngas. Catalytic reforming of hydrocarbons is a process that involves the breaking of H—C and C—C bonds. Hydrogen is currently formed from petroleum reforming and electrolysis processes. Steam reforming is performed at lower temperatures (150–700 °C) in the presence of catalysts to produce H_2, CO and CO_2 gases. It is important to control or remove the coke formed, which accumulates on the reactor and the catalyst surface, thereby decreasing its activity.

Combustion

It is the direct burning of biomass in air to convert it into chemical energy. It is not suitable for hydrogen production due to low energy efficiency and high emission pollutants. However, it converts hydrogen-containing waste material into a hydrogen-rich stream.

The pyrolysis and gasification methods are economically sustainable, and the cost analysis showed that the pyrolysis method is more effective and cheaper one. Syngas is a vital building block and acts as a raw material of methanol for the industries. It is produced from biomass pyrolysis and gasification of natural gas, coal or heavy oil residues. To make it more sustainable, integration with existing agricultural and forest industries, energy industry or chemical industry in a biorefinery will make it economical. Because of the production need for hydrogen, an integrated process in which hydrogen can be supplied from waste and residual (e.g. glycerol from biodiesel) is considered the only economically viable approach to operate a biorefinery.

6.4 Role of Nanoadditives in Algae-based Biofuel Production

Traditional microalgae-based-biofuel production is not accepted globally on an industrial scale due to high cost and energy required. Nanotechnology can be helpful in addressing these issues due to its stable catalytic performances, recyclability and environmentally friendly characteristics. Moreover, it can be implemented at any stage of microalgae cultivation for biodiesel use as a fuel.

It can be used at the cultivation stage to improve the lipid mass and properties, or at the biodiesel production stage to generate bioethanol and biodiesel effectively or in the fuel application, which will lead to complete combustion process. The advantages of nanoadditives in different stages are shown in Figure 6.9.

Nanoparticle addition to microalgae biofuel production intensified and improved the overall yield in every stage.

In photobioreactor (PBR), as the algae began to grow, the equivalent distribution of light intensity all over the algae becomes impossible. Shading of grown-up algae that also covers bioreactor makes it impossible to reach the light to the algae. The position of light cannot be set just near the reactor as the heat generated may increase the temperature that is again undesirable. In such cases, nanoparticle components of light-emitting diodes (LEDs) is the

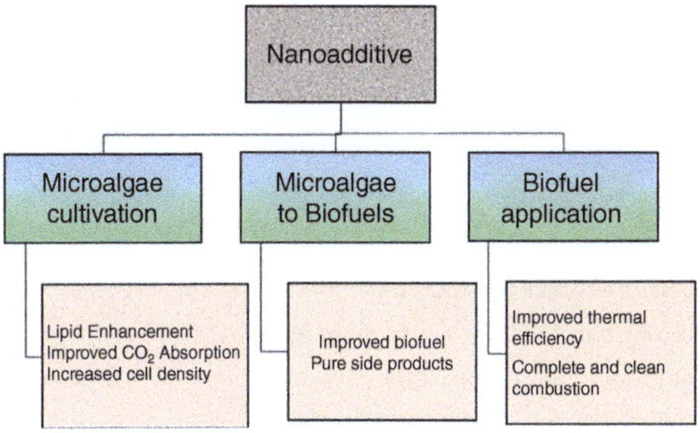

Figure 6.9 Nanoadditive applications in microalgae cultivation to biofuel application.

best option, which due to smaller size can be set into PBR. In addition, LED generated a minimum amount of heat. The use of gallium aluminium arsenide in LED is observed to improve the algal cultivations [52]. Yeh et al. [53] observed that red LED and Katsuda et al. [54] observed flashing blue LED can improve the indoor cultivation to a considerable extent. The use of cost-effective optical fibres are the better option in this regard as studied by Henshaw and Lee et al. [55, 56]. The use of silver nanoparticles coupled with plasmons has been shown to increase the photosynthetic activity in cultivation [57], and the chances of photoinhibition can be controlled by checking the size and concentration of these nanoparticles [58]. In addition to the efforts taken for illumination, nano- and microbubbles are also used in the reactor for ensuring proper mixing. Such bubbles remain in the system for a longer time, uniformly helping in CO_2 delivery and O_2 stripping in the system. This increases the photosynthesis rate and also results in an increase in the biomass yield [59, 60].

Besides cultivation, the major step is biomass conversion in which nanoparticle showed improved assistance. Various nanomaterials [61], nanophotocatalysts [62], metal oxides [63] and mesoporous materials [64] have shown their use in cost reduction and reducing energy consumption. Increased surface area of nanomaterials helps in loading of enzymes [65, 66] and diffusion of substrate to enzymes [67], which results in increased biofuel reduction. Easier catalyst recovery and reusability [68] increase the efficiency of catalyst and make the process cost effective. CaO nanoparticles also greatly contribute to biomass production and provide >99% conversion [69]. Zeolites are mesoporous with cage structure, so they can act as an adsorbent for biofuel separation [70]. It also helps in removing trace quantity of water, which is undesirable in the transesterification process and thus helps in improving the efficiency [71]. Nanofarming using sponge-like material for algae production is also used extensively nowadays [72]. These particles act as absorbents to selectively remove the lipids from the algal cell membranes, reducing the further cost of operation and producing a fair chance of in situ transesterification using calcium/strontium oxide nanoparticles [73].

When biodiesel is used as fuel, the doping of nanoparticles in different ratios was blended by physical or chemical methods and showed improved and enhanced properties as shown in Table 6.1.

Table 6.1 Nanoparticle blended biofuels applications.

Nanoparticle as additives	Application output	References
Pt–Pd	Improved glucose property	[74]
Zr/CeO_2	Facilitating the conversion of cellulose into high-quality biofuels	[75]
Mg or Al or Ca	Lipid-extraction efficiency increased	[76]
silica-based Si, Fe	Enhanced carbon and excess nutrient in feedstock	[77]
ZnO	Better nutrient removal from sewage wastewater	[78]
Na + ZSM-5	Higher temperature is more favourable to aromatisation	[79]
Ni-doped ZnO	Less emission	[80]
CeO_2	Reduction in exhaust emissions	[81]
$CaO/KOH–Fe_3O_4$ and Fe_3O_4-KF/KOH	Higher mass yield in the transesterification reaction	[82]
Fe_3O_4	Enhanced enzyme loading and productivity	[83]
CaO	Effective catalyst for transesterification	[84]
ZrO_2	Decreased hydrocarbon and CO	[85]
TiO_2	Reduction in exhaust emissions	[86]
Al_2O_3	Improved brake thermal efficiency	[87]

Thus, nanoadditive uses at different stages from microalgae culture to biodiesel application proved to be a commercial and positive approach for fulfilling future demands.

6.5 Cost Analysis of Microalgae-based Biofuel Production

For commercialising, the microalgae-based fuel techniques, various factors should be considered such as feedstock availability, quality and quantity of feedstock, biomass production cost, plant location, capacity and design, etc. If proper care of irradiation, mixing and other factors is taken, then the microalgae will offer a promising feedstock for energy generation.

The cost analysis of biofuels until 2035 can be performed on the basis of raw feedstock market prices (Figure 6.10).

Fuel prices beyond 2035 were estimated on the basis of \$/gge or \$/kWh, which shows the consumer's annual fuel costs.

Few factors worth mentioning while interpreting the graph are as follows:

- The fuel costs are untaxed. Not the current as well as future tax prediction was made.
- The fuel cost estimates in 2030 will be changed if energy prices increase.

For making the process viable, biomass conversion plants would have to be started along with specialised equipment for the conversion facilities. Economic review has shown that the conversion plants should be near the feedstock pond.

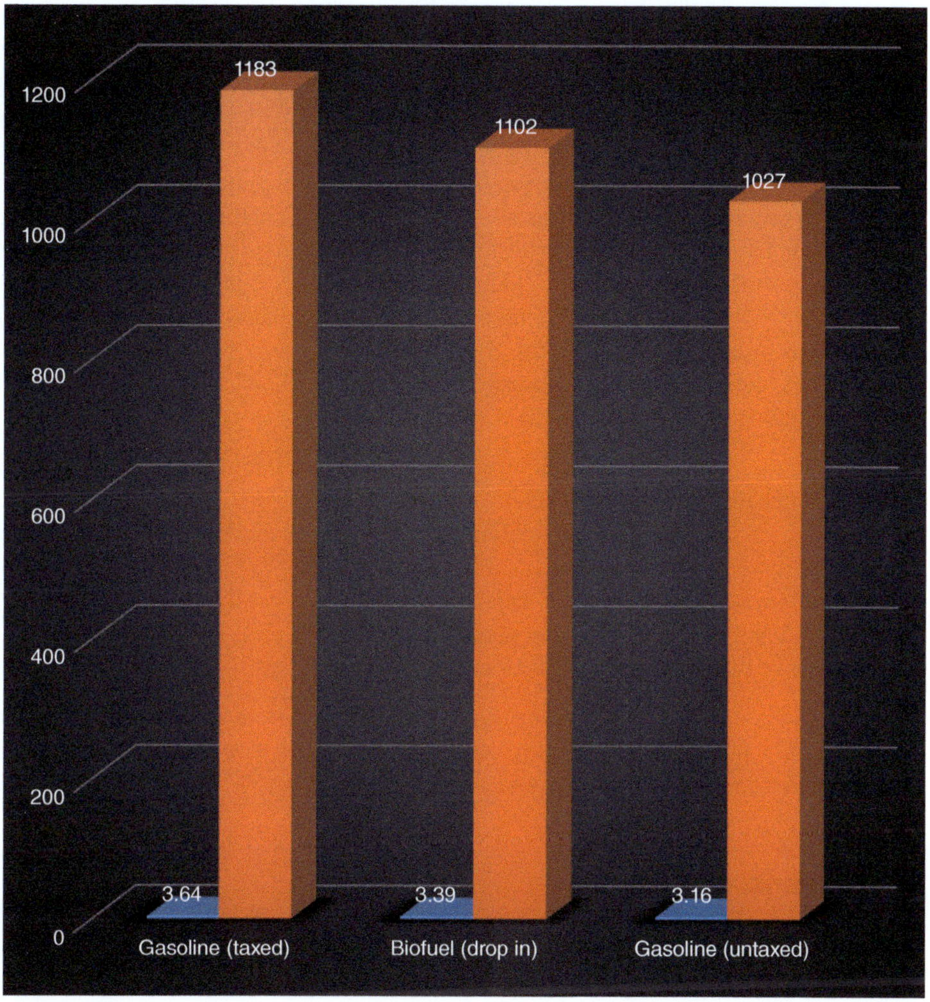

Figure 6.10 Fuel cost ($/gge or kWh) and annual consumer fuel cost ($/yr) represented by blue and brown bars.

6.6 Challenges and Opportunities in Microalgae-based Biofuel Production

Microalgae are proved to be the best alternative for fossil fuel to meet the ever-growing industrial demands. Microalgae production from the laboratory scale to the commercial scale is still facing many challenges, such as improvement in algal cultivation, increase in biomass productivity, improvement in biodiesel productivity, minimization of the energy requirement and high cost techniques.

Usually, the closed PBR cultivation technique is more expensive than open pond cultivation, but at the same time, the open pond method has limitations of contamination, variable light intensity and temperature. Due to all such limitations, commercialisation is still lacking due to an insufficient supply of water, CO_2 and other nutrient at large levels. The search for value-added coproduct should be realised along with the biofuel production technology. The use of unused, wastewater resources should be properly channelled along with this technique.

Many areas are still lacking for making the process commercial. It is still standing on the laboratory level. A vast amount of research is still required for good characterisation and modification of nanoparticles in terms of shape, size, catalytic activity as per the demand of the reaction for optimum productivity. Engine performance and gas emission characteristics must be well understood before judging the role of the appropriate catalyst. Many interdisciplinary research studies are required. Scientist should work hand in hand for making the technique sustainable at the global level.

Researchers are engaged in continuously addressing the following issues related to the commercialisation of biodiesel from microalgae:

1) to improve the biomass feedstock along with its lipid content through genetic engineering,
2) technical designing of low-cost PBRs,
3) integration of the wastewater treatment technology with algal cultivation to reduce the cost,
4) hosting innovative microalgal harvesting methods,
5) use of modified catalysts for biofuel production at the desired stage and
6) inventions of integrated microalgal biorefinery methodology.

Besides all these factors, public security, impact on life, environmental concern and the toxicity of side products obtained are also desirable to be scrutinised widely before commercialisation.

6.7 Summary

Microalgae for biofuel production are desired all over the world. Though it is energy-efficient and environmentally friendly, professionals are still looking ahead for decreasing the cost and making the technique to be accepted globally. The use of nanoadditives in various forms at different phases in algae to biofuel production can be an innovative step. Sustainability analysis still requires future research plans in proper direction. It is clear that the technique is stuck at high energy demand and increased cost of production. This is a third-generation biodiesel, thereby alarming the research area for improvement in it. Wastewater treatment can be successfully combined with CO_2 mitigation, use of value-added by-products and production of biofuel along with the addition of cost-effective nanomaterial. Thus, a new and innovative biorefinery-based low-cost technology should be developed to make the process feasible and globally accepted.

References

1 Ananthi, V., Raja, R., Carvalho, I.S. et al. (2021). A realistic scenario on microalgae based biodiesel production: third generation biofuel. *Fuel* 284: 118965. https://doi.org/10.1016/j.fuel.2020.118965.

2 Chhandama, M.V.L., Satyan, K.B., Changmai, B. et al. (2021). Microalgae as a feedstock for the production of biodiesel: a review. *Bioresour. Technol. Rep.* 15: 100771. https://doi.org/10.1016/j.biteb.2021.100771.

3 Han, S.F., Jin, W.B., Tu, R.J., and Wu, W.M. (2015). Biofuel production from microalgae as feedstock: current status and potential. *Crit. Rev. Biotechnol.* 35 (2): 255–268. https://doi.org/10.3109/07388551.2013.835301.

4 Gong, Y. and Jiang, M. (2011). Biodiesel production with microalgae as feedstock: from strains to biodiesel. *Biotechnol. Lett* 33 (7): 1269–1284. https://doi.org/10.1007/s10529-011-0574-z.

5 Veeramuthu, A. and Ngamcharussrivichai, C. (2021). Potential of microalgal biodiesel: challenges and applications. *Renew. Ener. Technol. Appl.* 167-180: https://doi.org/10.5772/intechopen.91651.

6 Chen, J., Li, J., Dong, W. et al. (2018). The potential of microalgae in biodiesel production. *Renewable Sustainable Energy Rev.* 90: 336–346. https://doi.org/10.1016/j.rser.2018.03.073.

7 Hossain, N., Hasan, M.H., Mahlia, T.M.I. et al. (2020). Feasibility of microalgae as feedstock for alternative fuel in Malaysia: a review. *Energy. Strateg. Rev.* 32: 100536. https://doi.org/10.1016/j.esr.2020.100536.

8 de Freitas, C.D., Tundisi, L.L., Cerqueira, K.S. et al. (2019). Microalgae: cultivation aspects and bioactive compounds. *Braz. Arch. Biol. Technol.* 62: https://doi.org/10.1590/1678-4324-2019180343.

9 Vandenbrink, J.P., Herranz, R., Medina, F.J. et al. (2016). A novel blue-light phototropic response is revealed in roots of Arabidopsis thaliana in microgravity. *Planta.* 244 (6): 1201–1215. https://doi.org/10.1007/s00425-016-2581-8.

10 Troell, M., Halling, C., Nilsson, A. et al. (1997). Integrated marine cultivation of Gracilaria chilensis (Gracilariales, Rhodophyta) and salmon cages for reduced environmental impact and increased economic output. *Aquacult.* 156 (1–2): 45–61. https://doi.org/10.1016/S0044-8486(97)00080-X.

11 Wen, X., Tao, H., Peng, X. et al. (2019). Sequential phototrophic-mixotrophic cultivation of oleaginous microalga Graesiella sp. WBG-1 in a 1000 m 2 open raceway pond. *Biotechnol. Biofuels* 12 (1): 1–4. https://doi.org/10.1186/s13068-019-1367-1.

12 Kulkarni, S.D., Auti, T., and Saraf, S. (2016). Bioremediation study of dairy effluent by using spirulina platensis. *Res. J. Life Sci. Bioinform. Pharm. Chem. Sci.* 1 (6): 325.

13 Kandimalla, P., Desi, S., and Vurimindi, H. (2016). Mixotrophic cultivation of microalgae using industrial flue gases for biodiesel production. *Environ. Sci. Pollut. Res.* 23 (10): 9345–9354. https://doi.org/10.1007/s11356-015-5264-2.

14 Choudri, B.S., Charabi, Y., Baawain, M., and Ahmed, M. (2017). Bioenergy from biofuel residues and wastes. *Water Environ. Res.* 89 (10): 1441–1460. https://doi.org/10.2175/106143017x15023776270511.

15 Chowdury, K.H., Nahar, N., and Deb, U.K. (2020). The growth factors involved in microalgae cultivation for biofuel production: a review. *Comput. Water. Ener. Environ. Eng.* 9 (4): 185–215. https://doi.org/10.4236/cweee.2020.94012.

16 Zewdie, D.T. and Ali, A.Y. (2020). Cultivation of microalgae for biofuel production: coupling with sugarcane-processing factories. *Energy. Sustain. Soc.* 10 (1): 1–16. https://doi.org/10.1186/s13705-020-00262-5.

17 Oasmaa, A., Källi, A., Lindfors, C. et al. (2012). Guidelines for transportation, handling, and use of fast pyrolysis bio-oil. 1. Flammability and toxicity. *Energy Fuels* 26 (6): 3864–3873. https://doi.org/10.1021/ef300418d.

18 Pan, P., Hu, C., Yang, W. et al. (2010). The direct pyrolysis and catalytic pyrolysis of Nannochloropsis sp. residue for renewable bio-oils. *Bioresour. Technol.* 101 (12): 4593–4599. https://doi.org/10.1016/j.biortech.2010.01.070.

19 Rizzo, A.M., Prussi, M., Bettucci, L. et al. (2013). Characterization of microalga chlorella as a fuel and its thermogravimetric behavior. *Appl. Energy* 102: 24–31. https://doi.org/10.1016/j.apenergy.2012.08.039.

20 Akhtar, A., Jiříček, I., Ivanova, T. et al. (2019). Carbon conversion and stabilisation of date palm and high rate algal pond (microalgae) biomass through slow pyrolysis. *Int. J. Energy Res.* 43 (9): 4403–4416. https://doi.org/10.1002/er.4565.

21 Miao, X., Wu, Q., and Yang, C. (2004). Fast pyrolysis of microalgae to produce renewable fuels. *J. Anal. Appl. Pyrolysis* 71 (2): 855–863. https://doi.org/10.1016/j.jaap.2003.11.004.

22 Kim, S.W., Koo, B.S., and Lee, D.H. (2014). A comparative study of bio-oils from pyrolysis of microalgae and oil seed waste in a fluidized bed. *Bioresour. Technol.* 162: 96–102. https://doi.org/10.1016/j.biortech.2014.03.136.

23 Chiaramonti, D., Prussi, M., Buffi, M. et al. (2017). Review and experimental study on pyrolysis and hydrothermal liquefaction of microalgae for biofuel production. *Appl. Energy* 185: 963–972. https://doi.org/10.1016/j.apenergy.2015.12.001.

24 Aysu, T., Ola, O., Maroto-Valer, M.M., and Sanna, A. (2017). Effects of titania based catalysts on in-situ pyrolysis of pavlova microalgae. *Fuel Process. Technol.* 166: 291–298. https://doi.org/10.1016/j.fuproc.2017.05.001.

25 Naqvi, S.R., Naqvi, M., Inayat, A., and Blanco-Sanchez, P. (2021). Impact of layered and delaminated zeolites on catalytic fast pyrolysis of microalgae using fixed-bed reactor and Py-GC/MS. *J. Anal. Appl. Pyrolysis* 155: 105025. https://doi.org/10.1016/j.jaap.2021.105025.

26 Matayeva, A., Basile, F., Cavani, F. et al. (2019). Development of upgraded bio-oil via liquefaction and pyrolysis. *Stud. Surf. Sci. Catal.* 178: 231–256. https://doi.org/10.1016/B978-0-444-64127-4.00012-4.

27 Wang, A., Austin, D., and Song, H. (2019). Investigations of thermochemical upgrading of biomass and its model compounds: opportunities for methane utilization. *Fuel* 246: 443–453. https://doi.org/10.1016/j.fuel.2019.03.015.

28 Conti, R., Pezzolesi, L., Pistocchi, R. et al. (2016). Photobioreactor cultivation and catalytic pyrolysis of the microalga Desmodesmus communis (Chlorophyceae) for hydrocarbons production by HZSM-5 zeolite cracking. *Bioresour. Technol.* 222: 148–155. https://doi.org/10.1016/j.biortech.2016.10.002.

29 Tripathi, M., Mubarak, N.M., Sahu, J.N., and Ganesan, P. (2017). Overview on synthesis of magnetic bio char from discarded agricultural biomass. *Handb. Compos. from Renew. Mater.* 1: 435. https://doi.org/10.1002/9781119441632.ch16.

30 Choo, M.Y., Oi, L.E., Daou, T.J. et al. (2020). Deposition of NiO nanoparticles on nanosized zeolite nay for production of biofuel via hydrogen-free deoxygenation. *Mater. (Basel)* 13 (14): https://doi.org/10.3390/ma13143104.

31 Aysu, T., Abd Rahman, N.A., and Sanna, A. (2016). Catalytic pyrolysis of Tetraselmis and Isochrysis microalgae by nickel ceria based catalysts for hydrocarbon production. *Energy* 103: 205–214. https://doi.org/10.1016/j.energy.2016.02.055.

32 Zainan, N.H., Srivatsa, S.C., and Bhattacharya, S. (2015). Catalytic pyrolysis of microalgae Tetraselmis suecica and characterization study using in situ synchrotron-based infrared microscopy. *Fuel* 161: 345–354. https://doi.org/10.1016/j.fuel.2015.08.030.

33 Rahman, N.A.A., Fermoso, J., and Sanna, A. (2018). Effect of Li-LSX-zeolite on the in-situ catalytic deoxygenation and denitrogenation of Isochrysis sp. microalgae pyrolysis vapours. *Fuel Process. Technol.* 173: 253–261. https://doi.org/10.1016/j.fuproc.2018.01.020.

34 Aysu, T., Fermoso, J., and Sanna, A. (2018). Ceria on alumina support for catalytic pyrolysis of pavlova sp. microalgae to high-quality bio-oils. *J. Energy. Chem.* 27 (3): 874–882. https://doi.org/10.1016/j.jechem.2017.06.014.

35 Fonseca, J.M., Teleken, J.G., de Cinque, A.V. et al. (2018). Microalgae biodiesel production in China : a preliminary economic analysis. *Renewable Sustainable Energy Rev.* 104: 296–306.

36 Andrade, L.A., Barbosa, J.M., Barrozo, M.A.S., and Vieira, L.G.M. (2020). A comparative study of the behavior of Chlamydomonas reinhardtii and Spirulina platensis in solar catalytic pyrolysis. *Int. J. Energy Res.* 44 (7): 5397–5411. https://doi.org/10.1002/er.5289.

37 Shirazi, Y., Viamajala, S., and Varanasi, S. (2020). In situ and ex situ catalytic pyrolysis of microalgae and integration with pyrolytic fractionation. *Front. Chem.* 786: https://doi.org/10.3389/fchem.2020.00786.

38 da Silveira, A., Rossi, R., Barbosa, J.M. et al. (2021). Solar assisted catalytic thermochemical processes: pyrolysis and hydropyrolysis of Chlamydomonas reinhardtii microalgae. Renew. *Energy* 170: 669–682. https://doi.org/10.1016/j.renene.2021.02.034.

39 Du, Z., Li, Y., Wang, X. et al. (2011). Microwave-assisted pyrolysis of microalgae for biofuel production. *Bioresour. Technol.* 102 (7): 4890–4896. https://doi.org/10.1016/j.biortech.2011.01.055.

40 Li, L., Ma, X., Xu, Q., and Hu, Z. (2013). Influence of microwave power, metal oxides and metal salts on the pyrolysis of algae. *Bioresour. Technol.* 142: 469–474. https://doi.org/10.1016/j.biortech.2013.05.080.

41 Sekar, M., Mathimani, T., Alagumalai, A. et al. (2021). A review on the pyrolysis of algal biomass for biochar and bio-oil – bottlenecks and scope. *Fuel* 283: 119190. https://doi.org/10.1016/j.fuel.2020.119190.

42 Zhang, R., Li, L., Tong, D., and Hu, C. (2016). Microwave-enhanced pyrolysis of natural algae from water blooms. *Bioresour. Technol.* 212: 311–317. https://doi.org/10.1016/j.biortech.2016.04.053.

43 Zhou, N. and Dunford, N.T. (2017). Thermal degradation and microwave-assisted pyrolysis of green algae and cyanobacteria isolated from the great salt plains. *Trans. ASABE* 60 (2): 561–569. https://doi.org/10.13031/trans.12028.

44 Parvez, A.M., Afzal, M.T., Jiang, P., and Wu, T. (2020). Microwave-assisted biomass pyrolysis polygeneration process using a scaled-up reactor: product characterization, thermodynamic assessment and bio-hydrogen production. *Biomass Bioenergy* 139: 105651. https://doi.org/10.1016/j.biombioe.2020.105651.

45 Choudhary, P., Malik, A., and Pant, K.K. (2020). Exploration of a novel biorefinery based on sequential hydropyrolysis and anaerobic digestion of algal biofilm: a comprehensive characterization of products for energy and chemical production. *Sustain. Ener. Fuels.* 4 (3): 1481–1495. https://doi.org/10.1039/c9se00939f.

46 Gamliel, D.P., Bollas, G.M., and Valla, J.A. (2018). Two-stage catalytic fast hydropyrolysis of biomass for the production of drop-in biofuel. *Fuel* 216: 160–170. https://doi.org/10.1016/j.fuel.2017.12.017.

47 Yan, W.H., Wang, K., Duan, P.G. et al. (2017). Catalytic hydropyrolysis and co-hydropyrolysis of algae and used engine oil for the production of hydrocarbon-rich fuel. *Energy* 133: 1153–1162. https://doi.org/10.1016/j.energy.2017.03.034.

48 Chang, Z., Duan, P., and Xu, Y. (2015). Catalytic hydropyrolysis of microalgae: influence of operating variables on the formation and composition of bio-oil. *Bioresour. Technol.* 184: 349–354. https://doi.org/10.1016/j.biortech.2014.08.014.

49 Nascimento, I.A., Marques, S.S.I., Cabanelas, I.T.D. et al. (2013). Screening microalgae strains for biodiesel production: lipid productivity and estimation of fuel quality based on fatty acids profiles as selective criteria. *Bioenergy Res.* 6 (1): 1–3. https://doi.org/10.1007/s12155-012-9222-2.

50 Hwang, J.H., Kabra, A.N., Ji, M.K. et al. (2016). Enhancement of continuous fermentative bioethanol production using combined treatment of mixed microalgal biomass. *Algal. Res.* 17: 14–20. https://doi.org/10.1016/j.algal.2016.03.029.

51 Lamb, J.J., Sarker, S., Hjelme, D.R., and Lien, K.M. (2018). Fermentative bioethanol production using enzymatically hydrolysed *saccharina latissima*. *Adv. Microbiol.* 8 (05): 378. https://doi.org/10.4236/aim.2018.85025.

52 Lee, C.-G. and Palsson, B. (1994). High-density algal photobioreactors using light-emitting diodes. *Biotechnol. Bioeng.* 44 (10): 1161–1167. https://doi.org/10.1002/bit.260441002.

53 Yeh, N. and Chung, J.P. (2009). High-brightness LEDs-energy efficient lighting sources and their potential in indoor plant cultivation. *Renewable Sustainable Energy Rev.* 13 (8): 2175–2180. https://doi.org/10.1016/j.rser.2009.01.027.

54 Katsuda, T., Shimahara, K., Shiraishi, H. et al. (2006). Effect of flashing light from blue light emitting diodes on cell growth and astaxanthin production of Haematococcus pluvialis. *J. Biosci. Bioeng.* 102 (5): 442–446. https://doi.org/10.1263/jbb.102.442.

55 Henshaw, P.F. and Zhu, W. (2001). Biological conversion of hydrogen sulphide to elemental Sulphur in a fixed-film continuous flow photo-reactor. *Water Res.* 35 (15): 3605–3610. https://doi.org/10.1016/S0043-1354(01)00082-3.

56 Lee, K.H. and Kim, B.W. (1998). Enhanced microbial removal of H2S using Chlorobium in an optical-fiber bioreactor. *Biotechnol. Lett* 20 (5): 525–529. https://doi.org/10.1023/A:1005408817125.

57 Torkamani, S., Wani, S.N., Tang, Y.J., and Sureshkumar, R. (2010). Plasmon-enhanced microalgal growth in miniphotobioreactors. *Appl. Phys. Lett.* 7 (4): 043703. https://doi.org/10.1063/1.3467263.

58 Steele, J.M., Grady, N.K., Nordlander, P., and Halas, N.J. (2007). Plasmon hybridization in complex nanostructures. *Springer Ser. Opt. Sci.* 183–196. https://doi.org/10.1007/978-1-4020-4333-8_13.

59 Zimmerman, W.B., Hewakandamby, B.N., Tesař, V. et al. (2009). On the design and simulation of an airlift loop bioreactor with microbubble generation by fluidic oscillation. *Food Bioprod. Process.* 87 (3): 215–227. https://doi.org/10.1016/j.fbp.2009.03.006.

60 Zimmerman, W.B., Tesař, V., and Bandulasena, H.C.H. (2011). Towards energy efficient nanobubble generation with fluidic oscillation. *Curr. Opin. Colloid Interface Sci.* 16 (4): 350–356. https://doi.org/10.1016/j.cocis.2011.01.010.

61 Laudenslager, M.J., Scheffler, R.H., and Sigmund, W.M. (2010). Electrospun materials for energy harvesting, conversion, and storage: a review. *Pure Appl. Chem.* 82 (11): 2137–2156. https://doi.org/10.1351/PAC-CON-09-11-49.

62 Zhong, Z., Ang, T.P., Luo, J. et al. (2005). Synthesis of one-dimensional and porous TiO2 nanostructures by controlled hydrolysis of titanium alkoxide via coupling with an esterification reaction. *Chem. Mater.* 17 (26): 6814–6818. https://doi.org/10.1021/cm051695b.

63 Crossley, S., Faria, J., Shen, M., and Resasco, D.E. (2010). Solid nanoparticles that catalyze biofuel upgrade reactions at the water/oil interface. *Science* (80-) 327 (5961): 68–72. https://doi.org/10.1126/science.1180769.

64 Lucena, I.L., Silva, G.F., and Fernandes, F.A.N. (2008). Biodiesel production by esterification of oleic acid with methanol using a water adsorption apparatus. *Ind. Eng. Chem. Res.* 47 (18): 6885–6889. https://doi.org/10.1021/ie800547h.

65 Cruz, J.C., Pfromm, P.H., Tomich, J.M., and Rezac, M.E. (2010). Conformational changes and catalytic competency of hydrolases adsorbing on fumed silica nanoparticles: II Secondary structure. *Colloids Surf. B Biointerf.* 81 (1): 1–10. https://doi.org/10.1016/j.colsurfb.2010.06.005.

66 Wang, Y. and Hsieh, Y.L. (2008). Immobilization of lipase enzyme in polyvinyl alcohol (PVA) nanofibrous membranes. *J. Membr. Sci.* 309 (1–2): 73–81. https://doi.org/10.1016/j.memsci.2007.10.008.

67 Kim, H.J., Kang, B.S., Kim, M.J. et al. (2004). Transesterification of vegetable oil to biodiesel using heterogeneous base catalyst. *Catal. Today* 93-95: 315–320. https://doi.org/10.1016/j.cattod.2004.06.007.

68 Ganesan, A., Moore, B.D., Kelly, S.M. et al. (2009). Optical spectroscopic methods for probing the conformational stability of immobilised enzymes. *ChemPhysChem* 10 (9–10): 1492–1499. https://doi.org/10.1002/cphc.200800759.

69 Reddy, C., Reddy, V., Oshel, R., and Verkade, J.G. (2006). Room-temperature conversion of soybean oil and poultry fat to biodiesel catalyzed by nanocrystalline calcium oxides. *Energy Fuels* 20 (3): 1310–1314. https://doi.org/10.1021/ef050435d.

70 Cardona Alzate, C.A. and Sánchez Toro, O.J. (2006). Energy consumption analysis of integrated flowsheets for production of fuel ethanol from lignocellulosic biomass. *Energy* 31 (13): 2447–2459. https://doi.org/10.1016/j.energy.2005.10.020.

71 Kusdiana, D. and Saka, S. (2004). Effects of water on biodiesel fuel production by supercritical methanol treatment. *Bioresour. Technol.* 91 (3): 289–295. https://doi.org/10.1016/S0960-8524(03)00201-3.

72 Patil, J.A. and Honaguntikar, P. (2014). An overview on developments in biodiesel production from algae. *Int. J. Sci. Res.* 3 (12): 102–106.

73 Liu, X., Piao, X., Wang, Y., and Zhu, S. (2008). Calcium ethoxide as a solid base catalyst for the transesterification of soybean oil to biodiesel. *Energy Fuels* 22 (2): 1313–1317. https://doi.org/10.1021/ef700518h.

74 Hassan, K.M., Khalifa, Z., Elhaddad, G.M., and Abdel, A.M. (2020). The role of electrolytically deposited palladium and platinum metal nanoparticles dispersed onto poly(1,8-diaminonaphthalene) for enhanced glucose electrooxidation in biofuel cells. *Electrochim. Acta* 355: 136781. https://doi.org/10.1016/j.electacta.2020.136781.

75 Zhou, H., Akhtar, M.A., Wan, Y. et al. (2020). Catalytic ketonization of levoglucosan over nano-CeO2 for production of hydrocarbon precursors. *J. Anal. Appl. Pyrolysis* 152: 104973. https://doi.org/10.1016/j.jaap.2020.104973.

76 Lee, Y.C., Huh, Y.S., Farooq, W. et al. (2013). Lipid extractions from docosahexaenoic acid (DHA)-rich and oleaginous chlorella sp. biomasses by organic-nanoclays. *Bioresour. Technol.* 37: 74–81. https://doi.org/10.1016/j.biortech.2013.03.090.

77 Marella, T.K., Parine, N.R., and Tiwari, A. (2018). Potential of diatom consortium developed by nutrient enrichment for biodiesel production and simultaneous nutrient removal from waste water. *Saudi. J. Biol. Sci.* 25 (4): 704–709. https://doi.org/10.1016/j.sjbs.2017.05.011.

78 Vasistha, S., Khanra, A., and Rai, M.P. (2021). Influence of microalgae-ZnO nanoparticle association on sewage wastewater towards efficient nutrient removal and improved biodiesel application: an integrated approach. *J. Water. Process. Eng.* 39: 101711. https://doi.org/10.1016/j.jwpe.2020.101711.

79 Hosseinpour, M., Golzary, A., Saber, M., and Yoshikawa, K. (2017). Denitrogenation of biocrude oil from algal biomass in high temperature water and formic acid mixture over H+ZSM-5 nanocatalyst. *Fuel* 206: 628–637. https://doi.org/10.1016/j.fuel.2017.06.055.

80 Karthikeyan, S. and Dharma, P.T. (2018). Emission analysis of Botryococcus braunii algal biofuel using Ni-doped ZnO nano additives for IC engines. *Ener. Sour. Part A. Rec. Util. Environ. Eff.* 40 (9): 1060–1067. https://doi.org/10.1080/15567036.2018.1468517.

81 Karthikeyan, S. and Prathima, A. (2017). Microalgae biofuel with CeO2 nano additives as an eco-friendly fuel for CI engine. *Ener. Sour. Part A. Rec. Util. Environ. Eff.* 39 (13): 1332–1338. https://doi.org/10.1080/15567036.2017.1328002.

82 Farrokheh, A., Tahvildari, K., and Nozari, M. (2020). Comparison of biodiesel production using the oil of chlorella vulgaris micro-algae by electrolysis and reflux methods using CaO/KOH-Fe3O4 and KF/KOH-Fe3O4 as magnetic Nano catalysts. *Waste Biomass Valorization* 12 (6): 3315–3329. https://doi.org/10.1007/s12649-020-01229-5.

83 Nematian, T., Salehi, Z., and Shakeri, A. (2020). Conversion of bio-oil extracted from Chlorella vulgaris micro algae to biodiesel via modified superparamagnetic nano-biocatalyst. *Renewable Energy* 146: 1796–1804. https://doi.org/10.1016/j.renene.2019.08.048.

84 Siva, S. and Marimuthu, C. (2015). Production of biodiesel by transesterification of algae oil with an assistance of nano-CaO catalyst derived from egg shell. *Int. J. Chem.Tech. Res.* 7 (4): 2112–2116.

85 Karthikeyan, S. and Prathima, A. (2017). Analysis of emissions from use of an algae biofuel with nano-ZrO2. *Energy Sources, Part A Recover Util Environ Eff* 39 (5): 473–479. https://doi.org/10.1080/15567036.2016.1230799.

86 Pandian, A.K., Ramakrishnan, R.B.B., and Devarajan, Y. (2017). Emission analysis on the effect of nanoparticles on neat biodiesel in unmodified diesel engine. *Environ. Sci. Pollut. Res.* 24 (29): 23273–23278. https://doi.org/10.1007/s11356-017-9973-6.

87 Basha, J.S. and Anand, R.B. (2013). The influence of nano additive blended biodiesel fuels on the working characteristics of a diesel engine. *J. Braz. Soc. Mech. Sci. Eng.* 35 (3): 257–264. https://doi.org/10.1007/s40430-013-0023-0.

7

Agro-Waste-Produced Biofuels

7.1 Introduction

Agricultural residues are valuable resources for the production of biofuels due to their high energy content. The agricultural waste is processed through the activities such as burning, dumping, land filling and many more [1, 2]. These approaches of handling the waste lead to pollution and waste of resources. Agricultural waste is a valuable resource that can be utilised in the form of energy with the suitable conversion processes. The agricultural waste can be reduced by utilising different forms such as fertilisers and biofuels production [1, 3, 4]. The details of some of the agro-waste converted into biomass using a suitable conversion technology, such as straw and husks of the rice, cob from the maize, fibrous bagasse from the sugar cane, coconut shell and fibre, peanuts leaf shells, are processed to convent into biofuel energy [5]. This chapter details the pre-treatment approach and process technology for converting the agricultural waste into production biofuels.

7.2 Agricultural Waste and Residues as Valuable Materials

Soil cultivation to grow the crops is the process that farmer practices on the agricultural land. Various stages followed in agricultural practices are as follows: (i) preparation of soil, (ii) seed selection and sowing, (iii) adding manures and fertilisers, (iv) irrigation, (v) weeding and crop protection and (vi) harvesting and storage [6, 7]. Some of the agro-waste and its by-product are shown in Figure 7.1, which shows that the materials in the form of waste generated during normal production activities are agricultural waste, such as paddy straws, sugar cane top, maize stalks, coconut shell, palm oil bunches, etc.

Agricultural waste is biodegradable at is broken down and returned to soil. Agricultural residues contain high energy that can be used to prepare the fertilizer, which can reduce the need of commercial fertilisers required for crop production [8]. However, the presence of fertilisers and pesticides in more amounts may cause harmful effect if returned to the soil. Management of agricultural waste is necessary to control the prevention of chemicals to the soil and water, along with the greenhouse gases to the environment. Agricultural manure and contaminated water are prevented from the surface and ground water through the waste management to

Sustainability in Biofuel Production Technology, First Edition. Pratibha S. Agrawal, Pramod N. Belkhode, and Samuel Lalthazuala Rokhum.

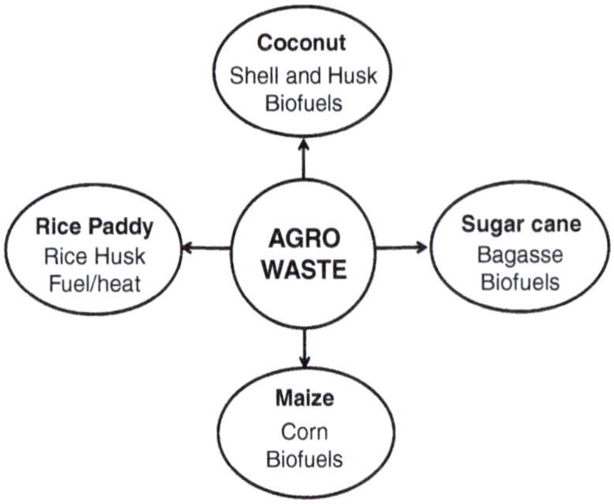

Figure 7.1 Some of the agro-waste and its by-product.

keep environment safe [9, 10]. The agricultural waste is processed through the activities, such as burning, dumping, land filling and many more. These approaches of handling the waste lead to pollution and waste of resources. Agricultural waste is a valuable resource that can be utilised in the form of energy with the suitable conversion processes. The agricultural waste can be minimised by proper treatment and converting it into a suitable form [11–13].

The residues are classified into two categories: primary residues and secondary residues [14, 15]. Primary residues are obtained during the harvesting such as paddy straw, sugar cane top, maize stalks, pal oil bunches, etc., whereas secondary residues or residues produced during the processing include husk obtained from the rice, bagasse, maize lob and parts obtained from the coconut, such as shell and husk, saw dust, etc. During harvesting, the agricultural residues, also known as agro-residues that include stalk, cane, seeds, leaves, roots, etc., that contain rich source of cellulose with lignin are obtained as by-products. The maximum residues are in the form of cereals crop (352 Mt), followed by fibres (66 Mt), oil seeds (29 Mt), pulses (13 Mt), and sugar cane (12 Mt). Cereal crops, such as rice, wheat and maize, are the major contributors to the residues of crops [16]. The types of agricultural waste and residues used for biofuel preparation are presented in the form of classification charts as shown in Figure 7.2, indicating the classification of agricultural waste and residues.

7.3 Pre-treatment of Agro-waste

The effect of greenhouse gases is more due to an increase in the consumption of fossil fuels, resulting in environmental pollution. The production of biodiesel from the food crops is not adopted due to the food and fuel issues. The utilisation of agricultural waste is the best alternative in terms of availability, lower cost and renewability for biofuel production. Biofuel production from the agro-waste using a suitable conversion process needs the pre-treatment of agro-waste [17]. Grasses, sawdust and wood chips are some of the lignocellulosic waste. The major agricultural waste is rice straw, wheat straw, corn straw and sugar cane bagasse. Agricultural residues in the form of cellulose, hemicelluloses and lignin are

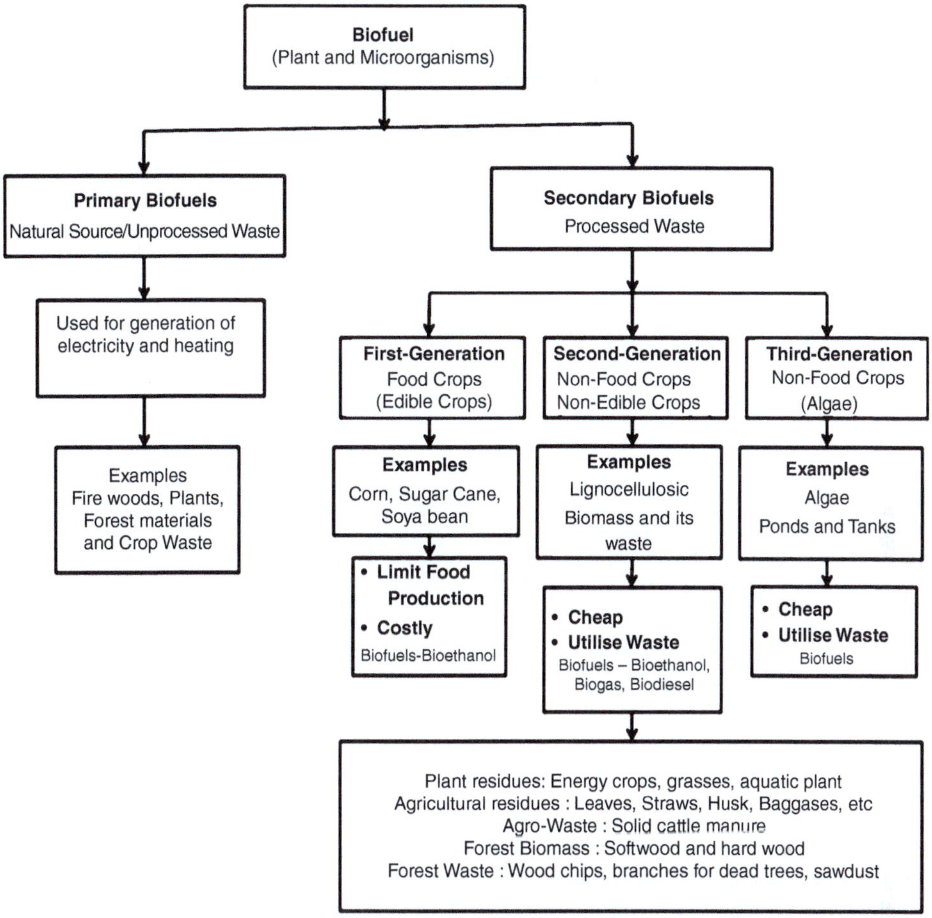

Figure 7.2 Classification of agricultural waste and residues. *Source:* Adapted from [14, 15].

pre-treated in which the structure of these residues are broken down and lignin is removed for high yield for the production of biodiesel. Table 7.1 shows pre-treatment for agricultural residues in which the pre-treatment process of breakdown is performed through the physical and chemical treatment. The pre-treatment processes can be controlled by variations in operating temperature and pressure, reaction timing, water availability, pH and the presence of chemicals.

Cellulose, hemicellulose and lignin are in the compact form with lignocellulosic biomass that is used for the biofuel production [21, 22]. The separation of these primary structures from the lignocellulosic is carried out in the pre-treatment. Pre-treatment of lignocellulosic is necessary for the production of biofuels. In the pre-treatment process, lignin is delignify and hemicellulose is decomposed to decrease the crystallinity of cellulosic. The selection of pre-treatment depends upon processing rate, cost involved and effect on the environment [23]. Various types of pre-treatment methods are as follows [24].

Table 7.1 Pre-treatment for agricultural residues.

Method	Pre-treatment	Methods
Residue pre-treatment method	Physical pre-treatment	Mechanical milling
		Microwave
		Ultrasonic
	Chemical pre-treatment	Acid
		Alkali
		Organosolvent
		Ozonolysis
	Physicochemical pre-treatment	Explosion of steam
		Subcritical water
		Supercritical CO_2
		Explosion of ammonia fibre
	Biological pre-treatment	Microbial delignification

Source: Adapted from [18–20].

7.3.1 Physical Pre-treatment

The biomass is broken down into fine particles for the active involvement of enzymes and microbes in the hydrolysis process. The production of fine particle of lignocellulosic is easily performed with milling and grinding. Reducing the size increases the product yield. Pre-treatment using microwave, electromechanical radiation and ultrasonic radiation is considered for the processing of the lignocellulosic biomass, which enhances the hydrolysis process [25–27].

These techniques can be followed by the conventional techniques, such as chemical treatment, to enhance the delignification process [28]. Cellulose can be treated with the pyrolysis at a temperature of around 300 °C to degrade cellulose and produce by-products that are further processed to produce bioethanol.

7.3.2 Chemical Pre-treatment

Acids and alkalis are used for the pre-treatment of lignocelluloses biomass [29–31]. Acids such as HCl, H_2SO_4 and peroxyacetic acid are used for breaking hemicellulose and cellulose with chemicals. Alkali pre-treatment with NaOH and KOH is used for mixing lignin [32]. Chemical pre-treatment is a good treatment for lignocellulosic but is not preferred due to environmental issues. New chemical pre-treatment such as organosolv and ozonolysis is used for the separation of lignin from the biomass mostly used in the pulp and paper industry [33]. Ozone gas is used to high lignin-containing agricultural waste for the solubilisation of lignin.

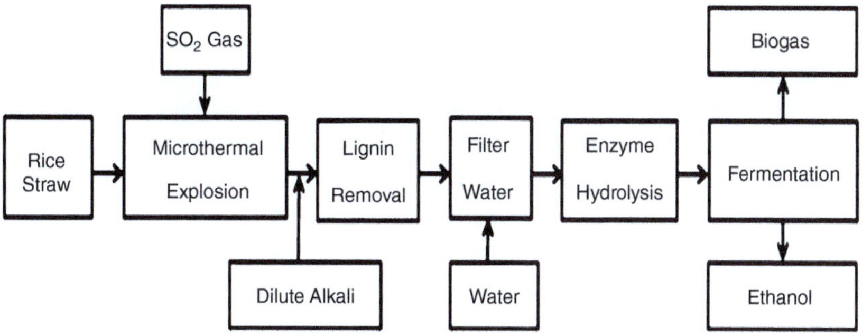

Figure 7.3 Pre-treatment process for rice straw. *Source:* Adapted from [38].

7.3.3 Physiochemical Treatment

Pre-treatment of lignocellulosic biomass is carried out using the physiochemical treatments, such as explosion of steam, subcritical water, supercritical CO_2 and explosion of ammonia fibre [34]. The fibres of the lignocellulosic are separated with superheated steam at high pressure in the steam explosion process that enhances the yield by breaking down the hemicelluloses. The presence of an acid or alkali solution with steam explosion enhanced the pre-treatment efficiency. Even hot water at temperatures of 175–225 °C and pressures of 5–6 bar shows the best performance for treating the agricultural waste such as lignocellulosic, bagasse, wheat straw and corncobs. Treatment of biomass in the presence of CO_2 medium or in the presence of NH_3 at the supercritical region for decomposing the hemicelluloses and lignin removal due to the presence of alkaline medium gives a high yield [35].

7.3.4 Biological Pre-treatment

Delignification and hemicellulose decomposition in the biological pre-treatment are performed with the help of microbial [36, 37]. Pre-treatment on agricultural residues using biological treatment improves the yield of hydrolysis for the production of biofuels. Figure 7.3 indicates the pre-treatment process for rice straw for the preparation of biogas and ethanol. The lignin from the corn is broken down with the fungi, and microorganisms reduce straw sugar, corn stalk, and rice husk by efficiently removing the lignin to improve the yield [38].

7.4 Process Technology – Agro-waste to Bioenergy

Agricultural residues are the source of the production of bioethanol using the fermentation process. High-yield ethanol is obtaining by removing the lignin and hemicelluloses that further improve the accessibility of cellulosic material to hydrolytic enzymes. Prior to fermentation, lignin and hemicellulose are separated from lignocelluloses biomass and then fermented to produce bioethanol [39]. Different methods of improving the yield by the pre-treatment of agro-waste are mentioned below.

7.4.1 Hydrolysis

The monomeric form is formed due to the degradation of cellulose and hemicellulose during the pre-treatment of lignocellulosic biomasses. Acids or enzymes are used for breaking down the lignocellulose biomass [40]. Dilute acid and concentrated acid with variations in temperature, pressure and duration are generally used to determine the hemicellulose degradation or the conversion of cellulose into glucose during the pre-treatment of cellulose. In concentrated acid hydrolysis, the complete conversion of cellulose and hemicellulose to sugar is performed at a lower cost compared to dilute acid hydrolysis but is restricted due to the limitation of corrosion of equipment and environmental issues. The enzymatic hydrolysis over the limitation occurs due to concentrated acid hydrolysis [41].

7.4.2 Anaerobic Digestion

Biological treatments of organic substances are carried out in the digester in the absence of oxygen in anaerobic digestion. The four main stages followed in the anaerobic digestion are (i) hydrolysis, (ii) acidogenesis, (iii) acetogenesis and (iv) methanogenesis. The hydrolysis phase involves the formation of simple manomeric forms, such as sugars, amino acids and fatty acids by breaking down the carbohydrates, proteins, and lipids [42, 43]. The breakdown is carried out by a hydrolase enzyme in an anaerobic digester.

7.4.3 Dark Fermentation

Dark fermentation is the process in which organic waste is utilised for the production of hydrogen. The technology is most suitable and efficient for hydrogen production from different agricultural residues, such as sugars, carbohydrates, proteins and lipids [44, 45]. Degradation of carbohydrate-rich substances forms hydrogen.

7.4.4 Transesterification

The process of esters and glycerol is obtained from the reaction of fat or oil with alcohol through transesterification. The reaction rate and the yield are improved with the help of catalysts. The conventional biodiesel is replaced with the biodiesel obtained from the transesterification of vegetable oils, animal fats or waste cooking oils [46]. Esterification is the process in which an ester is formed. Transesterification is the process in which an ester is modified. Transesterification is mostly preferred for the production of biodiesel due to its lower cost and high productivity. The pre-treatment process is performed using acids, bases or solid catalysts followed by a conversion process in which the resultant product is reacted with methanol or ethanol to produce biodiesel.

7.5 Creating Wealth from The Agricultural Waste

Agro-waste concerns agricultural and food processing waste. Agricultural waste is the residues obtained from the fruits, vegetables and crops. The pre-treatment of residues involved in the breakdown of waste requires more time, energy and expenses. The waste is a valuable resource that can be converted into a variety of useful products. The process of converting agricultural waste into a variety of products results in the creation of wealth. The approach of creation of wealth from different types of waste is detailed below [47].

Jute Residue

Jute residue is used for making a paper bag as a value-added product. The rural communities would get an opportunity for their livelihoods by utilising the jute waste; otherwise, waste is burnt or thrown away. Jute fibre is mixed with lignocelluloses fibres to produce low-cost pulp for handmade paper for making paper files, paper cardboards, paper bags, etc.

Corn Residue

Corn cob is used for the production of microbial proteins. Pre-treated corn cob is utilises as the carbon source during the fermentation process in the presence of nutrients and yeasts to produce proteins. The value-added products, such as mud cups and mud toys, are prepared by mixing waste corn cob with the red mud for biomass utilisation. Biochar is produced from the biomass in which agricultural residues are processed using the pyrolysis method. Biochar is used to enhance crop production, retention of water and soil amendment to improve the yield that reduces the requirement of fertilisation.

Groundnut Shell

Groundnut shell agricultural residues is used for the production of cellulose by microbial fermentation. Groundnut shell is used in composting wet materials and in fertiliser, pellets and plywood production. Around 2 Mt of groundnut shell is produced per year in India, which is available at a lower cost and is used in poultry feed and other purposes.

Cotton Stalk

Cotton stalk is agricultural residue used as lignocellulosic biomass for the production of mushrooms, which form the medium of income source in rural communities. It is also used in the manufacturing of particle boards, corrugated sheets and fuel billets.

Potato Residues

Potato residues are around 2 Mt per year in India, which causes disposal problem. Potato waste can be fed to animals. However, potato components as a raw ingredient can be used for pellet making.

Cabbage Waste

Cabbage waste in the form of leaf content contains large nutrients beneficial for health. Cabbage leaves are processed and converted into powder, which is mixed with wheat floor for baked production.

Stem Fibres

Stem fibres and sugar cane bagasse are the biomass utilised for the biofuel production. Stem fibres and sugar cane bagasse are used for the production of paper plates, which can easily be disposable. This value-added product replaced the conventional plastic paper plates. The lignin present in the fibres is processed, and the pulping raw material is compressed in the moulding machine to produce the paper plates. These plates hold good strength to sufficient food items. The concept is the start-up of small production units in the rural areas.

Cymbidium Orchids

Cymbidium orchid leaves are used for the development of value-added products, such as baskets, boards, furniture, etc. The basket strength is sufficient to hold the items and is easily degradable. This provides a solution to replace plastic components and enhance the rural activity in terms of earning.

Coconut Waste

Coconut waste, such as leafstalk, bunch waste, leaflets, etc., is used for oyster mushroom, which are nutritionally rich with a rich source of minerals. Coconut waste can be used in composting, particle boards, solid biofuels, biochar, etc. The production of many bioproducts, such as rope, brushes, floor mat, bowls, net and many more. The production of these items at a small scale gives the source of earning.

Crop Residues

Crop residue is used for the production of biofuels as it contains high carbohydrates. Organosolv pre-treated with rice straw is used to produce biohydrogen. It is also used for the production of particle boards biofertilisers, which are economically significant. Crop residues are suitable for the production of solid biofuels in the form of pellets, while the manure is suitable for the biogas production.

Vegetable Waste

Vegetable waste is rich in energy coconut with micronutrients. Vegetable waste is produced in larger quantities and causes environmental pollution due to its unpleasant odour. The material is processes to make silage used for feeding the animals.

Rice Husk

Rice husk is the waste generated from the rice, which contains around 30–50% of organic matter used for cattle feeding and fertilisation. Rice husk is used for improving the soil and source in the addition to cement. It is also used for the production of pellets, synthesis gas, building materials, fertilisers, etc. Suitable use of rice husk helps in boosting the farmer income and rural development.

Systematic processing of this waste residue provides a new approach to the economy. It is expected that the utilisation of the agricultural waste will become a potential resource to convert into usable low-cost products. Valuable products are processed from residues in practical applications.

7.6 Economic Valuation of Agro-waste

The cost–benefit analysis is highlighted in the context of cost with respect to agro-waste recycling and uses of waste as resources [48]. Bioeconomy is related to biomass organic waste of plants and animals. Bioeconomy is one of the economic models used to enhance the sustainable growth and development through the interaction between environment and society. Implementation of the policies is necessary to achieve the bioeconomy by using the sustainable natural resources [49, 50]. Environmental effect due to dependence on the non-renewable resources and increasing population demand is a challenge for government, resulting in failure to achieve bioeconomy. The best practices of the conversion of waste into value-added product and complete utilisation of goods as long as possible help to reduce the waste generation. The implementation of waste management would reduce the waste generation and create job opportunities, leading to overall development. An integrated system would be beneficial to achieve the waste management, resource utilisation and maximum uses of agricultural residues to promote the economy and reduce the environmental damage.

Analysis of agricultural waste is performed by recording the following observations related to agricultural residues [51]:

1) Volume of agricultural waste generated per year.
2) Volume of utilisation of agricultural waste in feeding per year.
3) Volume of agricultural waste left without benefit per year.
4) Volume of animal waste generated per year.
5) Volume of utilisation of animal waste as organic fertiliser per year.
6) Volume of animal waste left without benefit per year.
7) Total volume of agricultural (plant + animal) waste left without benefit per year.
8) Evaluation of total production and total crop price of each main product and secondary product.

Crop prices, particularly of maize, rice, cotton and sorghum along with their residues, should be analysed in terms of primary and secondary crops. Figure 7.4 shows the role model of bioeconomy used for the revenue generation. Cost analysis of secondary crops results in the development of bioeconomy, which will promote the set-up of the manufacturing unit to obtain the usable bioproduct at a lower cost, creating the source of revenue generation [52, 53].

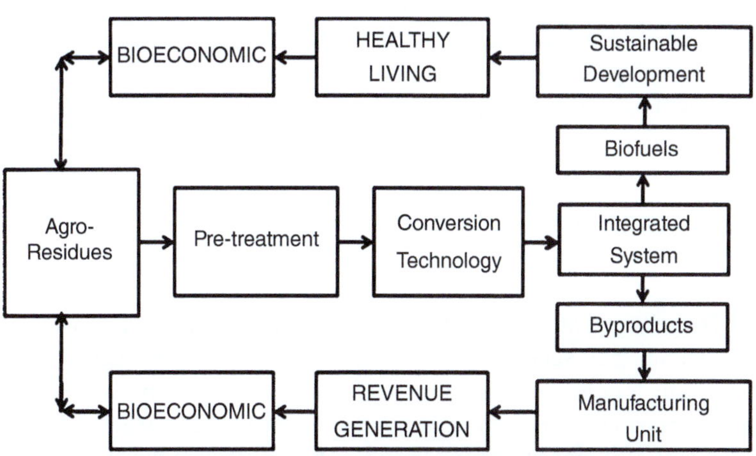

Figure 7.4 Role model of bioeconomy. *Source:* Adapted from [52, 53].

If the total volume of waste is not utilised without benefits, it may contaminate the environment. Awareness is necessary to utilise the large proportion of waste to convert it into usable material with an economic value, which would contribute to increase the productivity of agricultural crops, energy saving, environmental improvement and increase the self-sufficient rate [54]. Start-up grants should be made available to set up the manufacturing unit for processing agricultural residues in different valuable products.

7.7 Impact of Agricultural Waste

Management of agricultural waste reduces the environmental issues, promotes the practices of converting waste into usable by-product and creates the source of revenue generation [55]. It is an efficient method of achieving sustainable development in the agriculture treatment and processing the waste to enhance the bioeconomy. Improper handling of the agricultural waste in turn hampers food production. The agricultural waste contaminates water, water reservoir, ground water, surface water, soil erosion and sedimentation, livestock and organic contaminants. Agricultural waste can block the waterways that can lead to floods, resulting in losses of lives and property. Burning of the agricultural waste or improper dumping would create pollution that will affect the environment and human lives [56]. The solution to these problems is to manage the agricultural waste in efficient ways, such as recycling of the waste, which is the most promising and efficient method [57].

Agricultural waste could be used in the generation of biogas, electricity production, animal feeding, ethanol production, pulp and paper industry, cement industry, organic fertiliser and compost, silica production, soap making, mushroom and many valuable products. The prevention of wastage at all levels before they originate will restore some of these wastes, which will prevent from health and environmental issues. Awareness programmes would help agricultural activities at all levels and processing techniques to utilise the waste in a more beneficial way. Government would work on framing the policies that will ensure the efficient conversion of agricultural waste with financial support to achieve this conversion [58].

Figure 7.5 shows the impact of agricultural residues in the form of a chart that shows the classification of agro-residues and their impacts. The source of income can be enhanced by huge revenue generation by converting agricultural waste into useful products with the employment of people. Village people will get an opportunity to benefit the idle youth searching for employment by efficiently engaging the process of waste into wealth [59].

7.8 Current Challenges and Future Trends

The composition of the agricultural residues and procession techniques determine the production of biofuels. The selection of the correct pre-treatment for achieving high yield with suitable conversion technologies, such as biochemical, thermochemical, biological methods, is an environmentally friendly approach for the biofuel production with value-added products [60]. An integrated approach can be implemented for the large scale of biofuel production at a lower cost. Researchers around the world are engaged in the investigation

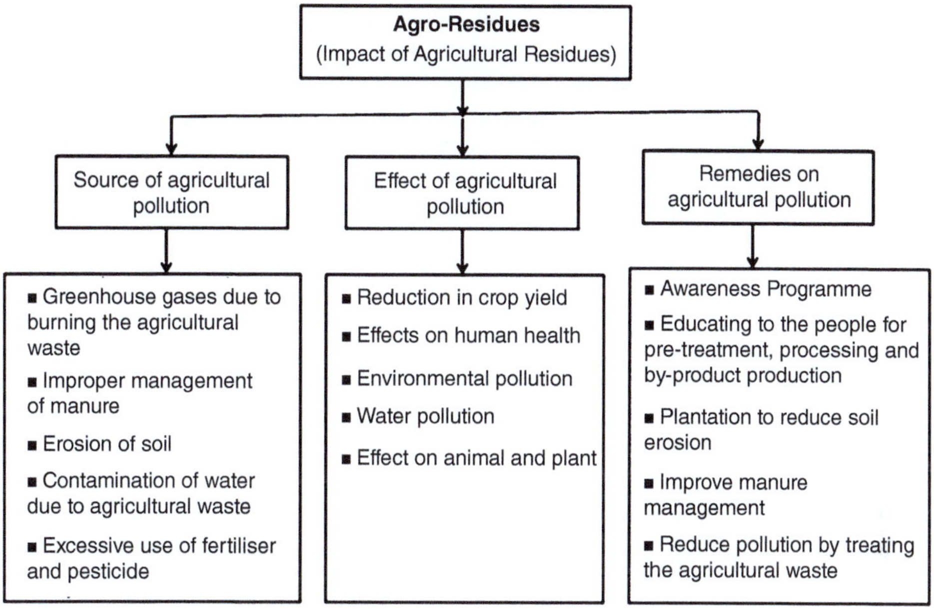

Figure 7.5 Impact of agricultural residues. *Source:* Adapted from [58, 59].

of the production of biofuels from the agricultural residues [61, 62]. The development of technological practices and policies ensure that the utilisation of agricultural residues across the world would benefit financially and environmentally.

7.9 Summary

Agriculture residues contain multiple nutrient elements that can be used in organic fertilizers, which reduce the dependency on commercial fertilisers. Agricultural waste is a valuable resource that can be utilised in the form of energy with the suitable conversion processes. The agricultural waste can be minimised by adopting the process of converting it into a usable product by recycling the production of biofuels. The main source of lignocellulose biomass, the conversion of these residues into biofuel production depending upon the pre-treatment and conversion technology adopted, would develop the economic model and enhance the source of income of the rural community.

References

1 Wasiak, A.L. (2017). Effect of biofuel production on sustainability of agriculture. *Procedia. Eng.* 182: 739–746. https://doi.org/10.1016/j.proeng.2017.03.192.

2 Viccaro, M., Cozzi, M., Rocchi, B., and Romano, S. (2019). Conservation agriculture to promote inland biofuel production in Italy: an economic assessment of rapeseed straight vegetable oil as a self-supply agricultural biofuel. *J. Cleaner Prod.* 217: 153–161. https://doi.org/10.1016/j.jclepro.2019.01.251.

3 Jaisamut, K., Paulová, L., Patáková, P. et al. (2013). Optimization of alkali pretreatment of wheat straw to be used as substrate for biofuels production. *Plant, Soil. Environ.* 59 (12): 537–542. https://doi.org/10.17221/7129-pse.

4 Srivastava, N., Mishra, P.K., and Upadhyay, S.N. (2020). Pectinases: significance in the digestion of pectin-rich agrowaste. *Ind. Enzym. Biofuels. Prod.* 183-204: https://doi.org/10.1016/b978-0-12-821010-9.00009-7.

5 Singh, S., Singh, B., Mishra, B.K. et al. (2013). Microbes in agrowaste management for sustainable agriculture. *Microorg. Sustain. Agric. Biotechnol.* 127-151: https://doi.org/10.1007/978-94-007-2214-9_8.

6 Awogbemi, O., Von, K.D.V., and Aigbodion, V.S. (2021). Trends in the development and utilization of agricultural wastes as heterogeneous catalyst for biodiesel production. *J. Energy Inst.* 98: 244–258. https://doi.org/10.1016/j.joei.2021.06.017.

7 Khan, H.M., Iqbal, T., Yasin, S. et al. (2021). Application of agricultural waste as heterogeneous catalysts for biodiesel production. *Catalys.* 11 (10): 1215. https://doi.org/10.3390/catal11101215.

8 Azadbakht, M., Safieddin Ardebili, S.M., and Rahmani, M. (2021). A study on biodiesel production using agricultural wastes and animal fats. *Biomass Convers. Biorefin.* 1–7. https://doi.org/10.1007/s13399-021-01393-1.

9 Toldrá-Reig, F., Mora, L., and Toldrá, F. (2020). Trends in biodiesel production from animal fatwaste. *Appl. Sci.* 10 (10): 3644. https://doi.org/10.3390/app10103644.

10 Pattanaik, L., Pattnaik, F., Saxena, D.K., and Naik, S.N. (2019). Biofuels from agricultural wastes. *Second Third Gener. Feed. Evol. Biofuel.* 103-142: https://doi.org/10.1016/B978-0-12-815162-4.00005-7.

11 Rao, P. and Rathod, V. (2019). Valorization of food and agricultural waste: a step towards greener future. *Chem. Rec.* 19 (9): 1858–1871. https://doi.org/10.1002/tcr.201800094.

12 Van, L.M., Chumpi, A., Belur, K. et al. (2022). Bioresource technology reports valorisation of food waste to sustainable energy and other value-added products : a review. *Bioresour. Technol. Rep.* 17: 100945. https://doi.org/10.1016/j.biteb.2022.100945.

13 Rajkumari, K., Das, D., Pathak, G., and Rokhum, L. (2019). Waste-to-useful: a biowaste-derived heterogeneous catalyst for a green and sustainable Henry reaction. *New J. Chem.* 43: 2134–2140. https://doi.org/10.1039/c8nj05029e.

14 Loulidi, I., Boukhlifi, F., Ouchabi, M. et al. (2020). Adsorption of crystal violet onto an agricultural waste residue: kinetics, isotherm, thermodynamics, and mechanism of adsorption. *Sci. World J.* 2020: https://doi.org/10.1155/2020/5873521.

15 Lubwama, M., Yiga, V.A., and Lubwama, H.N. (2020). Effects and interactions of the agricultural waste residues and binder type on physical properties and calorific values of carbonized briquettes. *Biomass Convers. Biorefin.* 1–21. https://doi.org/10.1007/s13399-020-01001-8.

16 Rocha-Meneses, L., Bergamo, T.F., and Kikas, T. (2019). Potential of cereal-based agricultural residues available for bioenergy production. *Data Br.* 23: 103829. https://doi.org/10.1016/j.dib.2019.103829.

17 Onu, P. and Mbohwa, C. (2021). Sustainable agricultural waste diversity: advances in green energy and materials production. *Agric. Waste Divers. Sustain. Issues* 55-73: https://doi.org/10.1016/b978-0-323-85402-3.00008-5.

18 Kushwaha, S., Soni, H., Ageetha, V., and Padmaja, P. (2013). An insight into the production, characterization, and mechanisms of action of low-cost adsorbents for removal

of organics from aqueous solution. *Crit. Rev. Env. Sci. Technol.* 43 (5): 443–549. https://doi.org/10.1080/10643389.2011.604263.

19 Jamaldheen, S.B., Sharma, K., Rani, A. et al. (2018). Comparative analysis of pretreatment methods on sorghum (Sorghum durra) stalk agrowaste for holocellulose content. *Prep. Biochem. Biotechnol.* 48 (6): 457–464. https://doi.org/10.1080/10826068.2018.1466148.

20 Thite, V.S. and Nerurkar, A.S. (2019). Valorization of sugarcane bagasse by chemical pretreatment and enzyme mediated deconstruction. *Sci. Rep.* 9 (1): 1–14. https://doi.org/10.1038/s41598-019-52347-7.

21 Tian, D., Shen, F., Hu, J. et al. (2022). Complete conversion of lignocellulosic biomass into three high-value nanomaterials through a versatile integrated technical platform. *Chem. Eng. J.* 428: 131373. https://doi.org/10.1016/j.cej.2021.131373.

22 Wan, J.Q., Wang, Y., and Xiao, Q. (2010). Effects of hemicellulose removal on cellulose fiber structure and recycling characteristics of eucalyptus pulp. *Bioresour. Technol.* 101 (12): 4577–4583. https://doi.org/10.1016/j.biortech.2010.01.026.

23 Hassan, S.S., Williams, G.A., and Jaiswal, A.K. (2018). Emerging technologies for the pretreatment of lignocellulosic biomass. *Bioresour. Technol.* 262: 310–318. https://doi.org/10.1016/j.biortech.2018.04.099.

24 Rezania, S., Oryani, B., Cho, J. et al. (2020). Different pretreatment technologies of lignocellulosic biomass for bioethanol production: an overview. *Energ.* 199: 117457. https://doi.org/10.1016/j.energy.2020.117457.

25 Nunes, L.J.R., De Oliveira Matias, J.C., and Da Silva Catalão, J.P. (2018). Physical Pretreatment of biomass. *Torrefact. Biomass. Energ. Appl.* 45-88: https://doi.org/10.1016/b978-0-12-809462-4.00002-x.

26 Piasecka, A., Krzemińska, I., and Tys, J. (2014). Physical methods of microalgal biomass pretreatment. *Int. Agrophys.* 28 (3): https://doi.org/10.2478/intag-2014-0024.

27 De Jong, W. (2014). Physical Pretreatment of biomass. Biomass as a sustain. Energy source Futur. *Fundam. Convers. Process.* 233: https://doi.org/10.1002/9781118916643.ch8.

28 Amriani, F., Salim, F.A., Iskandinata, I. et al. (2016). Physical and biophysical pretreatment of water hyacinth biomass for cellulase enzyme production. *Chem. Biochem. Eng. Q.* 30 (2): 237–244. 10.15255/CABEQ.2015.2284.

29 Badiei, M., Asim, N., Jahim, J.M., and Sopian, K. (2014). Comparison of chemical Pretreatment methods for cellulosic biomass. *APCBEE Procedia.* 9: 170–174. https://doi.org/10.1016/j.apcbee.2014.01.030.

30 Behera, S., Arora, R., Nandhagopal, N., and Kumar, S. (2014). Importance of chemical pretreatment for bioconversion of lignocellulosic biomass. *Renewable Sustainable Energy Rev.* 36: 91–106. https://doi.org/10.1016/j.rser.2014.04.047.

31 Faiz Norrrahim, M.N., Ilyas, R.A., Nurazzi, N.M. et al. (2021). Chemical Pretreatment of lignocellulosic biomass for the production of bioproducts: an overview. *Appl. Sci. Eng. Prog.* 14 (4): 588–605. 10.14416/j.asep.2021.07.004.

32 Lu, X., Han, T., Jiang, J. et al. (2020). Comprehensive insights into the influences of acid-base properties of chemical pretreatment reagents on biomass pyrolysis behavior and wood vinegar properties. *J. Anal. Appl. Pyrolysis* 151: 104907. https://doi.org/10.1016/j.jaap.2020.104907.

33 Ferreira, J.A. and Taherzadeh, M.J. (2020). Improving the economy of lignocellulose-based biorefineries with organosolv pretreatment. *Bioresour. Technol.* 299: 122695. https://doi.org/10.1016/j.biortech.2019.122695.

34 Halder, P., Kundu, S., Patel, S. et al. (2019). Progress on the pre-treatment of lignocellulosic biomass employing ionic liquids. *Renewable Sustainable Energy Rev.* 105: 268–292. https://doi.org/10.1016/j.rser.2019.01.052.

35 Tobita, H., Komatsu, M., Harayama, H. et al. (2019). Effects of combined CO2 and O3 exposures on net CO2 assimilation and biomass allocation in seedlings of the late-successional Fagus crenata. *Climate* 7 (10): 117. https://doi.org/10.3390/cli7100117.

36 Mishra, S., Singh, P.K., Dash, S., and Pattnaik, R. (2018). Microbial pretreatment of lignocellulosic biomass for enhanced biomethanation and waste management. 3. *Biotech.* 8 (11): 1–12. https://doi.org/10.1007/s13205-018-1480-z.

37 Keller, F.A., Hamilton, J.E., and Nguyen, Q.A. (2003). Microbial Pretreatment of biomass. *Biotechnol. Fuels Chem.* 27-41: https://doi.org/10.1007/978-1-4612-0057-4_3.

38 Sharma, H.K., Xu, C., and Qin, W. (2019). Biological Pretreatment of lignocellulosic biomass for biofuels and bioproducts: an overview. *Waste Biomass Valorization* 10 (2): 235–251. https://doi.org/10.1007/s12649-017-0059-y.

39 Zhao, R., Liu, Y., Zhang, Z. et al. (2018). Enhancing eco-efficiency of agro-products' closed-loop supply chain under the belt and road initiatives: a system dynamics approach. *Sustain.* 10 (3): 668. https://doi.org/10.3390/su10030668.

40 Rai, R., Bibra, M., Chadha, B.S., and Sani, R.K. (2019). Enhanced hydrolysis of lignocellulosic biomass with doping of a highly thermostable recombinant laccase. *Int. J. Biol. Macromol.* 137: 232–237. https://doi.org/10.1016/j.ijbiomac.2019.06.221.

41 Da Silva, A.S.A., Espinheira, R.P., Teixeira, R.S.S. et al. (2020). Constraints and advances in high-solids enzymatic hydrolysis of lignocellulosic biomass: a critical review. *Biotechnol. Biofuels* 13 (1): 1–28. https://doi.org/10.1186/s13068-020-01697-w.

42 Paul, S. and Dutta, A. (2018). Challenges and opportunities of lignocellulosic biomass for anaerobic digestion. *Resour. Conserv. Recycl.* 130: 164–174. https://doi.org/10.1016/j.resconrec.2017.12.005.

43 Sawatdeenarunat, C., Surendra, K.C., Takara, D. et al. (2015). Anaerobic digestion of lignocellulosic biomass: challenges and opportunities. *Bioresour. Technol.* 178: 178–186. https://doi.org/10.1016/j.biortech.2014.09.103.

44 Soares, J.F., Confortin, T.C., Todero, I. et al. (2020). Dark fermentative biohydrogen production from lignocellulosic biomass: technological challenges and future prospects. *Renewable Sustainable Energy Rev.* 117: 109484. https://doi.org/10.1016/j.rser.2019.109484.

45 Basak, B., Jeon, B.H., Kim, T.H. et al. (2020). Dark fermentative hydrogen production from pretreated lignocellulosic biomass: effects of inhibitory byproducts and recent trends in mitigation strategies. *Renewable Sustainable Energy Rev.* 133: 110338. https://doi.org/10.1016/j.rser.2020.110338.

46 Suzuki, S., Shibata, Y., Hirose, D. et al. (2018). Cellulose triacetate synthesis via one-pot organocatalytic transesterification and delignification of pretreated bagasse. *RSC Adv.* 8 (39): 21768–21776. https://doi.org/10.1039/c8ra03859g.

47 Anand, B., Sudha, S., and Keshaw, A. (2013). Value added products from agrowaste. *Rec. Res. Sci. Technol.* 5 (2): 7–12.

48 Koul, B., Yakoob, M., and Shah, M.P. (2022). Agricultural waste management strategies for environmental sustainability. *Environ. Res.* 206. https://doi.org/10.1016/j.envres.2021.112285.

49 Paul, S.L. and Sahni, H. (2019). Economic value of agro waste in developing countries. *Byprod. from Agric. Fish. Adding Value Food, Feed. Pharma Fuels* 581-595: https://doi.org/10.1002/9781119383956.ch25.

50 Atinkut, H.B., Yan, T., Arega, Y., and Raza, M.H. (2020). Farmers' willingness-to-pay for eco-friendly agricultural waste management in Ethiopia: a contingent valuation. *J. Cleaner Prod.* 261: 121211. https://doi.org/10.1016/j.jclepro.2020.121211.

51 Blonskaja, V. (2014). The analysis of biodegradable waste as renewable resource for alternative energy production in Estonia. *Ekologija* 60 (2): 27–37. https://doi.org/10.6001/ekologija.v60i2.2912.

52 Yamakawa, C.K., Qin, F., and Mussatto, S.I. (2018). Advances and opportunities in biomass conversion technologies and biorefineries for the development of a bio-based economy. *Biomass Bioenergy* 119: 54–60. https://doi.org/10.1016/j.biombioe.2018.09.007.

53 Mandegari, M.A., Farzad, S., and Görgens, J.F. (2017). Recent trends on techno-economic assessment (TEA) of sugarcane biorefineries. *Biofuel. Res. J.* 4 (3): 704–712. https://doi.org/10.18331/BRJ2017.4.3.7.

54 Bennich, T. and Belyazid, S. (2017). The route to sustainability-prospects and challenges of the bio-based economy. *Sustain.* 9 (6): 887. https://doi.org/10.3390/su9060887.

55 Chew, K.W., Chia, S.R., Chia, W.Y. et al. (2021). Abatement of hazardous materials and biomass waste via pyrolysis and co-pyrolysis for environmental sustainability and circular economy. *Environ. Pollut.* 278: 116836. https://doi.org/10.1016/j.envpol.2021.116836.

56 Voulvoulis, N. (2001). Environmental risk management for pharmaceutical compounds. *CIESM Work Monogr.* 26: 61–65.

57 Savon, D.Y., Kolotyri, K.P., and Romanov, A.V. (2019). Improving the ecological efficiency of the processing industry of agricultural security on the basis of economic instruments. *Russ. J. Ind. Econ.* 12 (3): 305–315. https://doi.org/10.17073/2072-1633-2019-2-305-315.

58 Bhatia, R.K., Ramadoss, G., Jain, A.K. et al. (2020). Conversion of waste biomass into gaseous fuel: present status and challenges in India. *Bioenergy Res.* 13 (4): 1046–1068. https://doi.org/10.1007/s12155-020-10137-4.

59 Razikordmahaleh, L. and Larijani, M. (2020). Identification and green grading of jobs in the renewable energy field of the biomass: a grounded theory study. *Iran. Occup. Heal.* 16 (6): 40–52.

60 Menyuka, N.N., Sibanda, M., and Bob, U. (2020). Perceptions of the challenges and opportunities of utilising organic waste through urban agriculture in the durban south basin. *Int. J. Environ. Res. Public Health* 17 (4): 1158. https://doi.org/10.3390/ijerph17041158.

61 Oluseun Adejumo, I. and Adebukola, A.O. (2021). Agricultural solid wastes: causes, effects, and effective management. *Strateg. Sustain. Solid. Waste. Manag.* 8: 139–150. https://doi.org/10.5772/intechopen.93601.

62 Zhai, Z., Martínez, J.F., Beltran, V., and Martínez, N.L. (2020). Decision support systems for agriculture 4.0: survey and challenges. *Comput. Electron. Agric.* 170: 105256. https://doi.org/10.1016/j.compag.2020.105256.

8

Biofuels for Aviation

8.1 Introduction

In the search of sustainable aviation fuel (SAF), eco-friendly feedstock is searched and grouped under the category of fourth-generation feedstock (4G) [1, 2]. All traditionally used aviation fuels contribute to global warming by its carbon emission. These carbon emissions should be reduced to protect the environment. IATA (2009) has set the following targets to reduce CO_2 emission from air transportation [3, 4]:

- Enhancing fuel efficiency
- Reducing aviation CO_2 emission
- Improving the aviation technologies
- Approving more efficient aircraft operations and engines
- Setting global standards and measures on SAF

Gasoline, diesel and jet fuels are typically used in aircraft. Gasoline fuels are hydrocarbons with C4–C12 and boiling points of about 35–200 °C [5, 6]. Diesel fuels are hydrocarbons with C8–C23 and boiling points of about 150–380 °C [7]. Jet fuels are hydrocarbons with C8–C16 and boiling points of about 125–290 °C [8–10]. Thus, jet fuel shows properties in between those of gasoline and diesel. The overlap of jet fuel boiling point with gasoline and diesel has implications during the distillation process, and due care is taken by fuel-manufacturing companies. It also has other side of consideration. If the market price of diesel increases, then the refinery process allows to produce diesel rather than jet fuel. Thus, biorefineries are producing desirable renewable fuel as per the market demand and price. The jet fuel market is smaller than gasoline [11] and diesel markets but is expected to grow in the coming future due to globalisation.

8.1.1 Types of Aviation Fuel

Aviation fuel is one of the most important requirements during flight operations. The quality of aviation fuel is more strictly maintained compared to transportation fuel. Though there are many worldwide available fuels, but the majority of the fuels used, which are commercially sustainable, are mainly two types of aircraft fuels, AVGAS (aviation gasoline) and jet fuels (Table 8.1). Depending on the type of engine, these fuels are used. AVGAS is

Sustainability in Biofuel Production Technology, First Edition. Pratibha S. Agrawal, Pramod N. Belkhode, and Samuel Lalthazuala Rokhum.

Table 8.1 Comparison of Jet and AVGAS.

Jet fuel	AVGAS
Jet fuel is easier to produce; it is obtained at earlier refining stage than AVGAS during the processing of crude oil.	It takes longer time to be distilled when crude oil is being refined. Additional reforming and combining processes are also required before its further use.
Simpler production and refining process makes jet fuel price cheaper.	Though used in small engines, it is still more expensive than jet fuel.
It is commonly used for a large number of aircraft.	It is only used in small designed pilot engines.
Sales volume is lower.	Sales volume is higher.

used for piston-engine aircraft [12–14], while jet fuel is used for aircraft with turbine-engine jets [15, 16]. AVGAS aircraft fly through the rotation of the propellers that generate the thrust and jet fuel fly with the thrust of expelled air.

Jet fuel is colourless, kerosene-based fuel used in aircraft with turbine engines. It can be Jet A and Jet A1. Both the fuels differ in their manufacturing and specification, and these can be still used in exchange for operating turbine engine aircraft. The main points of difference between the two are as follows:

- The freezing point of Jet A is −40 °C, while that of Jet A1 is −47 °C. The lower freezing point fuel that is Jet A1 is preferred for long distant international flights and also for the flights that have to fly via polar routes [17, 18].
- Inclusion of additives in Jet A is not in usual practice as it is responsible to decrease the static charges during the movement of fuel, whereas Jet A1 regularly works with static dissipater additives [19].
- Use of Jet A is preferred in the United States, while Jet A1 is commonly used in the rest of the world.

AVGAS is specifically used for small piston engines that operate in training jets or private jets. It contains traces of tetraethyl lead additive, which though toxic still prevents the engine knock. Research is going on to eliminate this substance from AVIGAS. It can be of two types: AVGAS 100 and AVGAS 100LL (100 stands for the octane rating) [20]. The main points of difference between the two are as follows:

- AVGAS 100 has a high tetraethyl lead additive content, while AVGAS 100LL (low lead) contains low amounts of tetraethyl lead.
- AVGAS 100 is dyed green in colour, whereas AVGAS 100LL is dyed blue in colour.

8.1.2 Comparison of Jet and AVGAS

The above-mentioned jet fuel variations and AVGAS variations are commonly used as aviation fuels. Under some extreme conditions or bad weather, other fuels that can still be used are follows:

TS-1 (used by Russia and CIS nations, a low freeze point of −50 °C, useful for flying in the cold areas, flashpoint is 28 °C and making it highly volatile) [21].

Jet B (it is 30% kerosene and 70% gasoline mixture used by Canada and Alaska, a low freeze point of −60 °C, useful for flying in the cold areas and a dangerously flammable kerosene–gasoline mixture makes it rare to use) [22].

JP-8 (specifically used in military aircraft and contains anti-icing and corrosion inhibitor additives to overcome the logistical and operational problems) [23–25].

JP-5 (kerosene-based military aircraft fuel and high flashpoint of 60 °C, thus avoids the risk of fire) [26–28].

8.2 Chemistry of Fuel Molecules

The main focus of jet fuel is reducing the aromatic contents and increasing the iso- and cycloalkanes to show improved properties. Aromatic molecules burn with tremendous amounts of soot, which is undesirable from environmental concerns. *n*-Alkanes with longer chains have high freezing point, whereas *n*-alkanes with smaller chains do not meet the flashpoint expectations [29]. As they are sourced from lipids, their acceptance in jet fuel can be observed [30]. Jet fuel with higher iso- and cycloalkanes showed higher energy content than Jet A fuel. It burns with cleaner emissions and also satisfies the freezing point and flashpoint demands. Thus, iso-alkanes, though expensive, are desirable. Substitution improves the freezing point and hence the fluidity of alkanes [31–33]. C12 has a freezing point of −10 °C, whereas iso-C12 alkane has a reduced freezing point of −46 °C [34]. Cycloalkanes with substitutions are always preferred in jet fuel. Whether lightly branched or heavily branched, researchers are trying to bring the production cost to minimum by correlating the molecular structures and fuel combustion–emission properties. Thus, two major conclusions can be drawn regarding fuel molecules in jet fuel – reduction in aromatics (increasing iso-alkanes) and replacing aromatics by cyclo-rings.

8.2.1 Iso-alkane

Iso-alkanes can be prepared by (i) cracking and isomerising larger HEFA or FT liquids, (ii) oligomerising and hydrotreating ATJ liquid or (iii) fermenting and hydrotreating sugar.

8.2.1.1 Cracking and Isomerising Larger HEFA and FT Liquids

Lipid content is the major source of renewable fuel production. Cleaving of larger molecules into smaller molecules is desirable for obtaining jet-range molecules. Commonly used methods for HEFA and FT processes are hydrotreating, hydrocracking and then isomerising *n*-alkanes into iso-alkanes [35, 36]. Hydrotreating helps to remove oxygen in carboxylate moieties, adding hydrogen to double bonds and saturating them. Hydrocracking and isomerisation result in the production of iso-alkane units. Another improved technology can include the following:

Hydrothermolysis is hydrothermal liquefication (HTL) using supercritical water followed by hydrotreating to produce *n*- and iso-alkanes [37].

Metathesis (reaction between unsaturated esters and ethylene) on fatty-acid-rich sources (C22) produces two C11 carbon fatty acid triglyceride compounds and C11 carbon alkene [38, 39]. This alkene is further saturated (for jet fuel demand) and esterified (for diesel fuel demand).

Thus, large molecules are cracked into smaller molecules using various processes to meet to the fuel standards.

8.2.1.2 Oligomerising and Hydrotreating ATJ Liquid

The ATJ liquid can be oligomerised to convert C8–C16 molecules into alkenes. Thus, the carbon length is controlled, which can be further cyclised or aromatised for the fuel use. Thus, the process is useful for the mixture of oxygenates, carboxylic acids and olefins. Various well-known chemical reactions can be used in this, such as alkene oligomerisation, Prins condensation (acid-catalysed electrophilic addition of aldehyde to alkene and), aldol condensation (reaction between two aldehydes giving aldol formation), ketonisation (converting two carboxylic acids to ketone) and Guebert reaction (base-catalysed coupling of two alcohols to obtain β-alkylated dimer alcohol) [40–42].

Thus, any functional group obtained from the feedstock can be converted to useful fuel components. Research is still going on for reducing the cost and meeting the fuel demands of today.

8.2.1.3 Fermenting and Hydrotreating Sugar

Fermentation of sugar produces fuel essential products along with many coproducts, which should be further hydrotreated carefully to increase the efficiency of the conversion. This method can produce larger molecules, and its efficiency can be balanced by the lightweight occurred due to oxygen loss [43, 44]. Currently, many biochemical technologies are integrated to reduce the separation cost.

8.2.2 Cycloalkane

Cyclohexane and its alkyl-substituted derivatives meet the fuel jet specifications of conventional fuels. Therefore, it can be successfully blended with a high percentage. Though its combustion property is not as good as that of iso-alkane, it is still much better than that of aromatics and is hence preferred from combustion and emission points of view [45]. Various routes of its production are available.

8.2.2.1 Aromatisation Followed by Hydrotreating

Different oxygenated alcohols, carbonyls, and carboxylic acids can be deoxygenated and aromatised using zeolite-type catalysts. The process is well established in the methanol-to-gasoline industry. Continuous efforts are still being made by scientists to improve the process for more complicated mixtures of carboxylic acids, where carbon is lost in coking. Hydropyrolysis is another method of incorporating hydrogen gas, followed by distillation and hydrotreatment of aromatics [46, 47]. Efforts have been made to stabilise the zeolite structure under hydrothermal conditions.

8.2.2.2 Phenol Hydrogenation

Phenols, the main building blocks of lignin, are produced by deconstructing lignin using various pyrolysis methods (producing mixtures of aromatics and phenols), liquefaction (producing mixtures of phenol and aromatics) or fermentation.

8.2.2.3 Ring Formation and Ring Closing Reactions

Five-membered rings can be produced from cyclohexanes using a catalytic route that causes a ring contraction (the process is used in petroleum refining), or various rings sizes are possible to produce from renewable biomass. Alkyl substituent in the cycloalkanes is also preferred. Thus, new ring modifications or the required structural modifications in the existing ring can be possible using chemical routes to improve the fuel properties.

The technique still has to face the following challenges [48]:

- Excess hydrogen use: For small oxygenated species, excess hydrogen is consumed, which results in significantly increased cost. Removing such small oxygenated species before hydrotreatment can be helpful.
- Pyrolytic techniques form the products of the gasoline range: Fractionation of gasoline- and jet-range material is required essentially. An alkylation of low-molecular-weight oxygenates with phenols will be helpful in increasing the carbon yield and also obtaining the fuel into the jet- and diesel-fuel ranges.
- When lignin is used as a feedstock for jet fuel, the pre-treatment technologies yield sugars in a highly usable form, which is not significant when environmental policies are considered.
- Some results show favourable swelling characteristics when using fused bicyclic alkanes and removing aromatics. This will lead to reduced emission and increased specific energies. More work is going on for improving the swelling characteristics of fused bicyclic alkanes.

8.2.3 Pathways for Producing Sustainable Aviation Fuel

SAF can reduce CO_2 emissions and thus is the most effective type of environmentally friendly fuels. It can be produced from any wasted material, such as agro-waste, used cooking oil or some solid waste. 'Farm to fly' initiatives taken by government help in producing biofuels, which can be used directly in engines without any modifications and hence accepted worldwide in the current stage of developing technology [49–51]. It is more expensive as compared to the above-mentioned traditionally used aviation fuels. From AVGAS price to JetA1, fuel prices are dropping worldwide to an average of \$4.60, with the lowest being \$2.20 and going maximum to \$11.40. With the current pandemic scenario, travel restrictions and government norms, the aviation fuel price has increased. Years will be required to fight back and return to the normal price line. However, the government supports that increasing research in improving the technology will be helpful in making it cost effective.

There are many pathways to produce SAF. Major four techniques are as follows:

- Alcohol to Jet (ATJ)
- Oil to Jet (OTJ)
- Gas to Jet (GTJ)
- Sugar to Jet (STJ)

Many process technologies are available for converting biomass to jet fuel out of which some are commercially set and some are in premature state of development. These technologies basically depend upon the feedstock used. Accordingly, the process technology will also vary. The product yield obtained, environmental emissions and the cost of production technology are the prime factors in commercialising the process. Aviation fuel has contributed approximately 2–6% to global carbon emission, and thereby impacting greenhouse gases. Life cycle assessment will analyse the emissions from harvesting the feedstock to the fuel application engine emission. Thus, it is important to review each technology before judging the more sustainable pathway.

8.3 Alcohol to Jet (ATJ)

Fatty alcohols are converted to jet fuel. Ethanol is already used as a B10–B15 blend in transportation. To increase the blend beyond this percentage has not yet been achieved [52, 53]. Therefore, another way is to use these alcohols in the aviation fuel blend for upgrading their properties.

8.3.1 Feedstock Used

There are many pathways for alcohol productions, but the basic production uses biomass only. Alcohol can be produced mainly from corn (hydrolysis of starch), commercially from lignocellulosic biomass, and also from sugar cane, sugar beet (fermentation of sugar), wheat and wood (thermochemical methods). This alcohol produced is further processed to convert it into aviation fuel. This conversion can either be biochemical (cellulose can be hydrolysed to soluble sugar that is again fermented by bacteria or yeast to provide ethanol) or thermochemical (ethanol is produced by gasification) [54]. The major drawback of using ethanol in an aviation fuel blend is its high volatile nature, tendency to corrode, low flashpoint and low density compared to jet fuel, making it less compatible for use (Figure 8.1) [55]. However, its use with upgradation in the properties is still favoured due to the easy accessibility of ethanol. Other alcohols that can be used are iso-butanol or *n*-butanol, which is less soluble in water and hence proved to be a better candidate, and isopropanol, long-chain fatty alcohols or sometimes methanol [54]. γ-Valerolactone obtained from biomass can also be catalytically converted into bio-1-butene. Lighter olefins (ethylene and propylene) can also be catalytically converted to gasoline blend.

8.3.2 Process Analysis

The ATJ process that converts alcohol into jet fuel basically involves a three-step process of alcohol dehydration, oligomerisation, and hydrogenation followed by the fractionation of paraffin products as shown in Figure 8.2 for ethanol and butanol. Oligomerisation is useful to generate a variety of fuels satisfying the jet fuel range, and the residues are used in the diesel range [56, 57].

The ATJ process for ethanol has been analysed by earlier researchers using a variety of catalysts, such as alumina and transition metal oxides, zeolites, heteropolyacid catalysts, Ziegler–Natta catalysts, silico-aluminium phosphates and homogeneous nickel/phosphorous ligand catalysts. Syndol, a heterogeneous catalyst, showed high selectivity and efficient conversion of ethanol to ethylene. Dehydration of higher alcohols has been less explored. Efficient

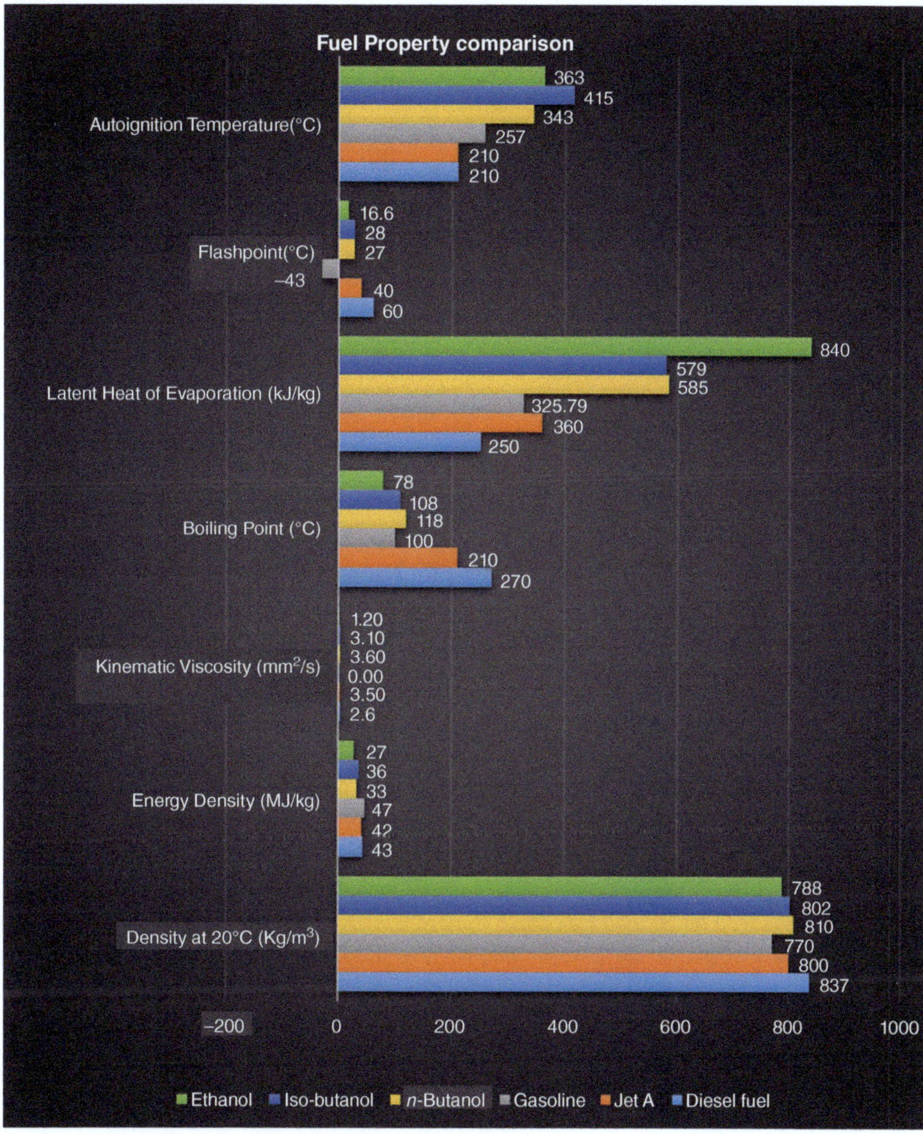

Figure 8.1 Fuel properties compared to various feedstock.

dehydration of iso-butanol is reported using zeolites, alumina, and Amberlyst acid catalysts [58]. Amongst a mixture of isomers of butane, iso-butane is selectively converted into iso-butanol as dictated by the catalyst choice. Iso-butanol can be converted to iso-butane, *n*-butane and 2-butane (cis and trans) after dehydration. Water molecule is removed during this process so that the dehydrating agent should be able to withstand the water content. However, the next oligomerisation reaction may not be tolerant to water content [59]. Therefore, due care must be taken to remove any traces of water before proceeding to the next step by the distillation/sieve method. Unreacted alcohol is again recycled many times to increase the output of the dehydration unit. Moreover, strong acidic catalysts (ZSM-5 and Amberlyst-35)

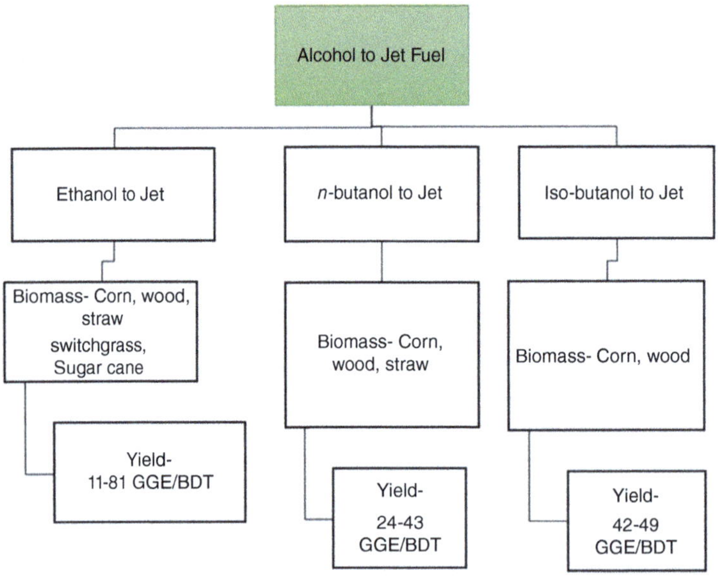

Figure 8.2 The ATJ process flow chart.

can also initiate butene oligomerisation reaction [60]. However, in such a one-step reaction, a product with poor fuel properties may be obtained. Hence, two separate step reactions are preferred.

The oligomerisation of alkenes is a well-established technique in the presence of a variety of homogeneous or heterogeneous catalysts in a single or multiple step reactor. Alkene produced after dehydrations is oligomerised now to the desired length C8–C16, making it compatible in the kerosene range. The step is used to achieve a typical carbon length distribution. The process was well developed by earlier researchers using different catalyst systems. All these processes are by this time commercialised, as the blend is already used in transport fuel. Researchers are now focusing on the development of integrated biomass-based cost optimisation technology. Primary intention of the process is to minimise the difference in physical and chemical properties of alcohols and jet fuels [61]. In the United States, 99.9% pure anhydrous ethanol is essentially used as a blend to avoid phase separation (Figure 8.3).

Iso- and *n*-butene oligomerisation is less developed commercially than ethylene. Usually, C12 trimers and C16 tetramers are preferred for the jet range. Using an Amberlyst-35 catalyst, a mixture of C8, C12 (major) and C16 is obtained [62]. Some additional side reactions may form C9 or C11 chains, which improves the product composition. Branched alcohols like iso-butanol will result in a higher degree of carbon branching in the oligomerised product and final product. Branching has unpredictable complex effects on fuel properties, such as cold flow properties, freezing points and cetane numbers [63]. As the branching increases, the cetane number is lowered, which affects cold flow properties and is favourable. Another preferred process is hydrogenation, that is, the addition of hydrogen to the unsaturated site of the oligomerised molecule making it saturated. It is carried out on a solid catalyst surface in excess of hydrogen gas for the complete saturation of olefins to paraffins. The remaining excess of hydrogen can be recycled and reused. The unreacted olefins can also be distilled and sent back to the dimerisation process or recycled back to the oligomerisation reactor.

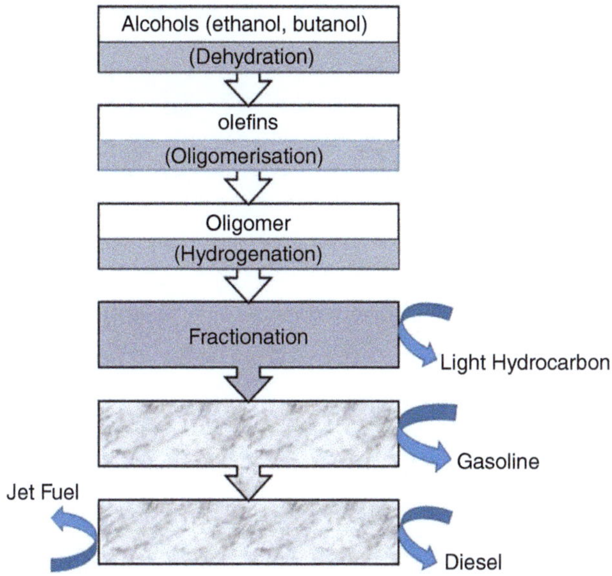

Figure 8.3 ATJ flow chart for ethanol and butanol.

When *n*-butanol is used, acetone and ethanol are also formed during the fermentation process. These formed co-products either to be upgraded catalytically to use as fuel or separated and sold as an individual product [64–66]. *n*-Butanol can be dehydrated to 1-butene using an alumina catalyst. Butane is then subjected to the oligomerisation process using Ziegler–Natta catalyst and Amberlyst-70 catalyst to obtain mixed unsaturated oligomers that should be further hydrogenated.

In addition to above-mentioned alcohol, methanol can also be converted into gasoline or olefin [67]. Methanol from syngas is turned into gasoline or olefin in a fluidise bed reactor. After fractionation, the olefins are sent to a fixed bed reactor and converted into 82% distillate, 15% gasoline, and 3% light gases over a ZSM-5 catalyst. The Jet-fuel-range product is recovered from the gasoline and light gases. Figure 8.4 shows the general flow chart of the methanol-to-olefin process.

8.3.3 Economic and Life-cycle Analysis

Economic analysis of biojet fuel depends on the cost of producing alcohols and the biochemical and thermochemical process cost, which should be studied extensively. Most of the researchers compared the minimum selling prices. The minimum ethanol selling price for the biochemical conversion was reported to be $2.8/gallons or $4.2/gallon of gasoline equivalent (GGE) (in 2011 US dollars), while that for the thermochemical route was reported to be $2.5/gallons or $3.9/GGE (in 2011 US dollars) [68–70]. Similarly, the minimum butanol selling price for the fermentation using corn grain was reported to be $2.9/gal and that using cellulosic biomass was reported to be $4.1/gal. The selling price of isobutanol from cellulosic biomass is estimated at $3.7 per gallon of butanol [71]. All these prices depend on the conversion efficiency, feedstock availability and productivity, feedstock market cost, and the cost of coproducts formed. To evaluate the overall commercial

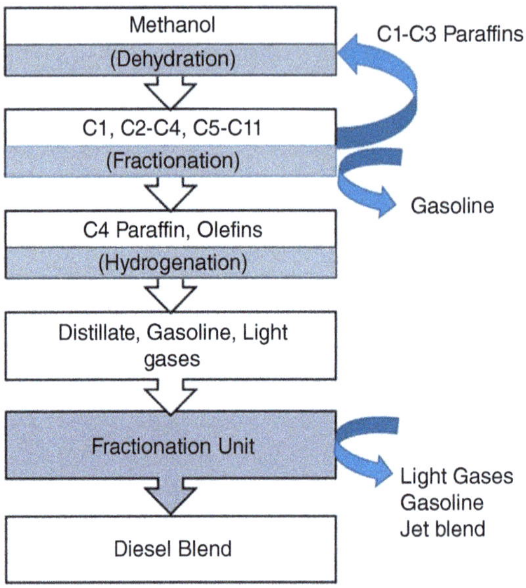

Figure 8.4 ATJ flow chart for methanol.

feasibility of the ATJ pathway, the economics of various involved processes, such as dehydration, oligomerisation, dimerisation and hydrogenation, should also be considered [72]. As these processes are still in progress, more data will be required to accomplish this goal.

LCA studies for the ATJ process depend on the production of alcohols via different pathways. Studies have also focused on emissions, water use and global warming. It has been observed that *n*-butanol releases more direct emissions, such as CO_2, nitrogen dioxide, and sulphur dioxide, than the iso-butanol, while ethanol releases the lowest emissions [73]. CO_2 is produced during the combustion process; nitrogen dioxide is formed through high-temperature oxidation of the diatomic nitrogen in the combustion air; and sulphur dioxide emissions are observed when using sulphuric acid in the pre-treatment process. The biomass feedstock is mainly responsible for the carbon footprint and fossil energy consumption. The LCA of the ATJ fuel upgrading processes is still unknown and requires more attention in future studies. Various companies are working for a more economical and renewable jet-fuel production solution from waste gas resources.

8.4 Oil to Jet (OTJ)

OTJ conversion is classified into mainly three types as shown below:

- Hydroprocessed renewable jet (HRJ or hydroprocessed esters and fatty acids or HEFA)
- Catalytic hydrothermolysis (CH or hydrothermal liquefaction)
- Pyrolysis (hydrotreated depolymerised cellulosic jet [HDCJ])

Presently, only the HRJ pathway has been approved for blending as per the ASTM specification.

8.4.1 Feedstock Used

OTJ fuel should fulfil the norms of the biodiesel [74–76]. The feedstock used in this process are plant oils, waste oils, algal oils and pyrolysis oils.

Plant oil: Oil from plants, such as canola, soya bean, rapeseed, palm and corn, can be used as feedstock in the process. Depending upon the production and availability of the plants, the respective plant oil is used in the concerned country, such as soya bean oil has been used extensively in the United States, rapeseed oil is used in Europe and palm oil in Europe is imported from Indonesia [77]. However, the use of such plant oil leads to GHG emissions due to changes in the forestry and land use. The search for non-edible oil and easy-to-grow plants stops at the Camelina crop. Jatropha, algae oils and waste cooking oils are also supposed to be good options in this case. Feedstock produces high-quality diesels of required specifications and also helps in reducing harmful emissions. Jatropha produces more oil than many other feedstock (35–55% of oil of the dry seed weight) [78].Various coproducts can also be obtained from its husk and seed shells. Thus, Jatropha proves to be a desirable feedstock for biodiesel production.

Algal oil: As discussed in the previous chapters, algal biofuel is of great interest in biodiesel production as described below:

1) algae have high productivity all over the year,
2) algal cultivation requires less pure water than other plants as it can also grow on industrial wastewater,
3) algae are cultivated on water bodies and hence fight for the use of agricultural land stopped,
4) algae have high oil content (20–50% of dry cell weight),
5) algae can take nutrients from wastewater for growth,
6) coproducts formed, and residual biomass can be used as fertilisers and
7) less GHG compared to other plant oils [79].

The algal biomass harvesting includes flocculation, filtration, flotation, and centrifugal sedimentation steps, which are costly ones [17, 80]. After harvesting, the drying can be performed in open sunlight, or various drying processes can be implemented, which again contribute to increasing the cost of the technology. Freeze drying is expensive; nut oil extraction is very easy.

Pyrolysis oil: Heating the biomass in the absence of oxygen produces pyrolysis gas, biochar and pyrolysis oil (bio-oil). Bio-oil is different from plant oil as it contains a mixture of oxygenated organic compounds from C1 to C21 or even more [81]. It can be refined in the jet fuel, diesel or gasoline. Market value-added biochar helps to reduce the production cost of bio-oil and thus can be used for jet fuel [82]. Due to the presence of unsaturated fatty acids, more hydrogen will be needed during the processing of bio-oils.

Thus, plant oil, algae oil and the pyrolysis oil are successfully used in biofuel production. The cost of production is greatly affected by the value-added coproducts if formed.

8.4.2 Process Analysis

HRJ and CH give TG feedstock FFA, and in pyrolysis, bio-oil is produced from biomass feedstock. The general outlined OTJ process is presented in Figure 8.5.

All the pathways are discussed in detail as follows:

Hydroprocessed renewable jet (HRJ or hydroprocessed esters and fatty acids or HEFA): This is a highly and commercially developed technique. In the past decade, this process was also used for jet fuel. The fuel shows an improved cetane number with comparatively fewer aromatic compounds, which results in lower emissions. The process is also developed catalytically. Due care must be taken for unsaturation units or glycerides by hydrogenating these units [83].

First, the crude oil is extracted from the biomass. The unsaturated portion or glycerides present in this crude oil will be hydrogenated to provide saturation. The glycerol portion will be converted into propane after hydrogenation, which will be cleaved to remove propane and yield FFA. Another route of producing FFA is the conversion of TG into MG with the value-added side product, glycerol [84].

The FFA produced will undergo decarboxylation and hydrodeoxygenation either sequentially or simultaneously to produce C17 and C18 paraffin, respectively. This paraffin will undergo isomerisation and cracking, producing paraffin, iso-paraffin and finally will be converted into jet fuel and diesel (HEFA). The process is presented in Figure 8.6.

All the products obtained will be fractionated to separate the kerosene, diesel, naphtha and lighter gases. The maximum permissible blend is 50%. A suitable catalyst is used for selective action in each stage.

Catalytic hydrothermolysis (CH, or hydrothermal liquefaction): It turns triglycerides into a mixture of straight chain, branched and cyclic hydrocarbons. First, TG

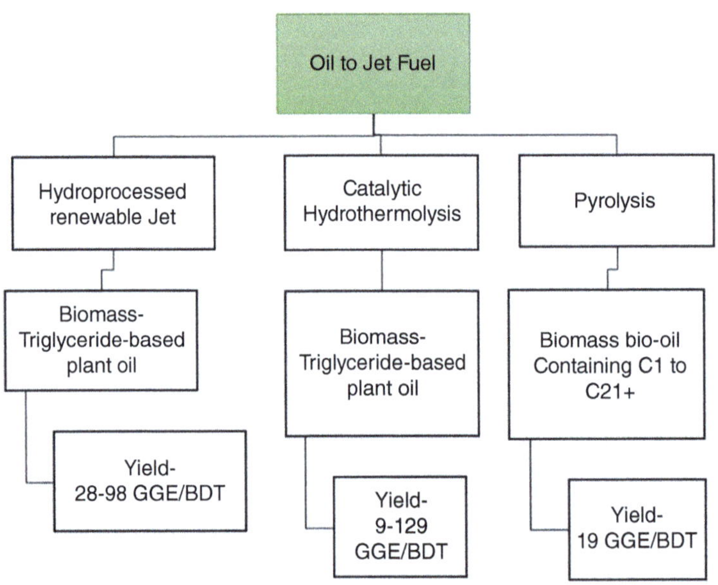

Figure 8.5 OTJ process flow chart.

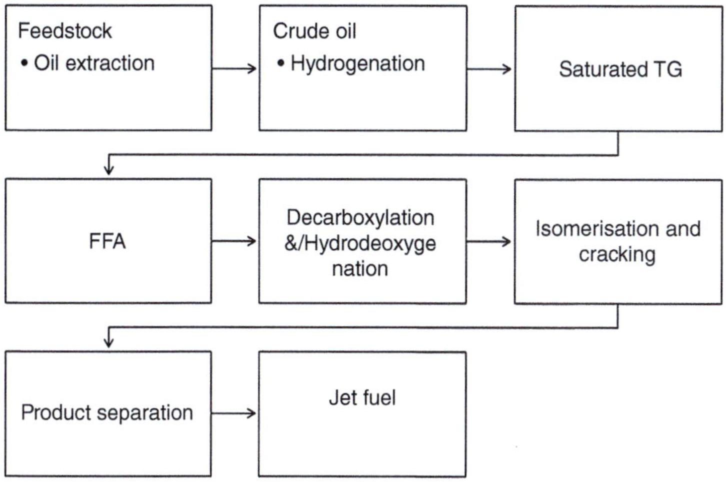

Figure 8.6 HRJ/HEFA pathway for jet fuel production.

undergoes several pre-treatment steps, such as catalytic conjugation, cyclisation and hydrothermolysis. The organic phase obtained was then separated and exposed to hydrogenation, decarboxylation and hydrotreatment to produce a mixture of paraffin, which was fractioned to yield naphtha, jet fuel and diesel separately.

Pyrolysis (HDCJ): It is the recently developed method of converting bio-oil from pyrolysis into jet fuel. Biomass is subjected to pyrolysis to give biogas, biochar and bio-oil, which are collected for further biofuel production. This oil undergoes hydrotreatment to produce hydrocarbons [85, 86]. The obtained hydrocarbon is fractioned to obtain the desired jet fuel yield.

8.4.3 Economic and Life-cycle Analysis

As HRJ requires hydrogen supply during jet fuel production, its processing cost is 20% higher than other two methods. However, at the same time, the side products, such as naphtha, propane and diesel, have more market value than glycerol produced in the transesterification reaction. The CH process when applied for algae ($3.3–$4.4/GGE) or dairy waste ($4.8/gal), the economically feasible jet fuel is obtained. PNNL and National Renewable Energy Laboratory (NREL) investigated the fast pyrolysis of biomass. Based on corn stover feedstock, the product value was estimated to be $3.1/GGE.

Another important factor to be considered is GHG emissions. The GHG emissions for soya bean oil (31–68%), low soya bean (26–44%), rapeseed oil (45–87%), jatropha oil (36–52%) algae oil (16–21%) were compared to those of conventional jet fuel. It is found that the GHG emission biojet fuel can be reduced by improvements in the techniques. When hydrogen, which is required in the process, is generated from natural gas, the emissions are reduced by 45% relative to conventional fuels. Jet fuel producing companies are trying to obtain sustainable fuel that minimises emissions and produces compatible jet fuel. Due to the variation in processes and coproducts formations, the biorefinery concept should be used to apply the technology cost effectively.

8.5 Gas to Jet (GTJ)

The GTJ process describes the conversion of biogas, natural gas or syngas into biojet fuel. The following two pathways (Figure 8.7) are used in this process:

Fischer–Tropsch Synthesis
Gas Fermentation

8.5.1 Feedstock Used

To obtain an efficient conversion process, hydrogen and carbon contents are preferred in the feedstock used. The most common feedstock used to produce gas are coal, natural gas and biomass. The carbon content depends on the biomass used. Woody biomass has a high carbon content. The choice of feedstock and pre-treatment steps are dependent on the economic results [87, 88]. A wide variety of woody biomass and agricultural waste feedstock are usually used for thermochemical routines. Biogas produced from the anaerobic digestion of animal manure can also be used as feedstock.

8.5.2 Process Analysis

8.5.2.1 Fischer–Tropsch Process

This process is used for producing fuels from syngas. The advantages of this technique are it uses very low sulphur and carbon contents, which result in reduced emissions when used as jet fuel. This can provide a good environmentally friendly alternative to the conventional fuels used.

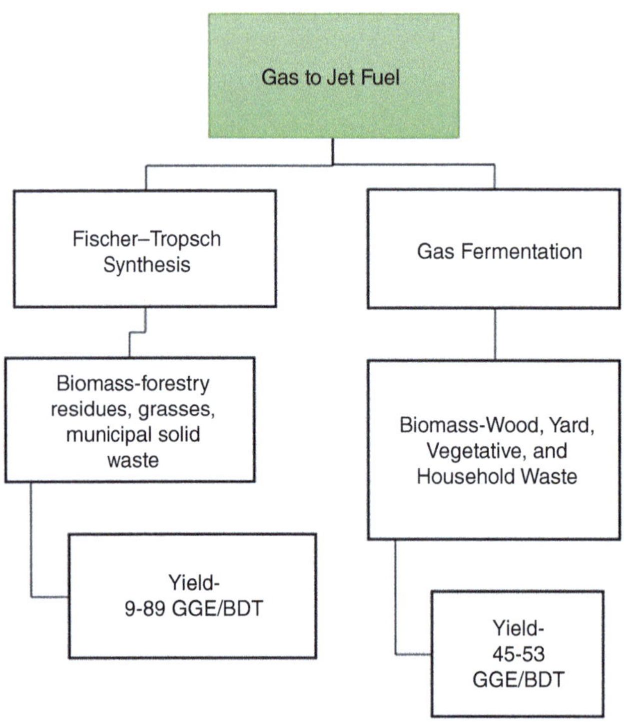

Figure 8.7 The generalised GTJ path.

In this process, the biomass is first dried and undergoes pre-processing. It is then exposed to steam for gasification. Various techniques are available for gasification. It can be performed via partial oxidation or direct gasifiers and indirect steam-blown gasifiers [89]. A direct gasifier requires a purest oxygen reaction, which breaks organic compounds into gases. The process requires a high capital investment. Indirect steam gasifiers are heated by the heat produced from a hot solid surface and have comparatively cheaper paths. It deconstructs the biomass into a mixture of syngas, tars and solid char.

The syngas is further cleaned and conditioned in a fluid catalytic cracker before being sent to FT synthesis. It is washed and scrubbed with water to remove any impurities. FT is a set of catalytic processes for converting syngas into liquid hydrocarbons. It can be operated at high temperatures (at 340 °C with iron-based catalysts) or low temperatures mode (at 230 °C with either iron or cobalt catalysts). The products formed are from methane to higher alkanes, alkenes and oxygenated compounds such as alcohols, acids or aldehydes. The heat produced in the exothermic FT process should be removed continuously to avoid excess heating and inhibit catalytic activities [90]. Supported metal CO/Fe) catalysts are preferred for the FT synthesis in the GTL process. The FT process further continues into polymerisation and hydrogenation. The process is presented in Figure 8.8.

FT reactors can be of three types: multitubular fixed bed, fluidised catalyst bed and slurry reactor. In a multitubular fixed bed reactor, the catalyst is embedded in narrow long tubes with a water-circulating outer jacket typically used for wax-based catalyst processes. Fluidised catalyst (typically iron based) beds operate at 320–350 °C and are used to produce alkenes, gasoline, diesels and jet fuels. In slurry reactors, finely divided catalysts are used, where the gas is bubbled continuously.

After the FT process, the unconverted syngas is sent back to the FT reactor. Various refinery processes (hydrocracking, isomerisation, hydrogenation and fractionation) are then used to improve the quality of the product. This resulting fuel is found to be superior to any other processes and alternating fuels.

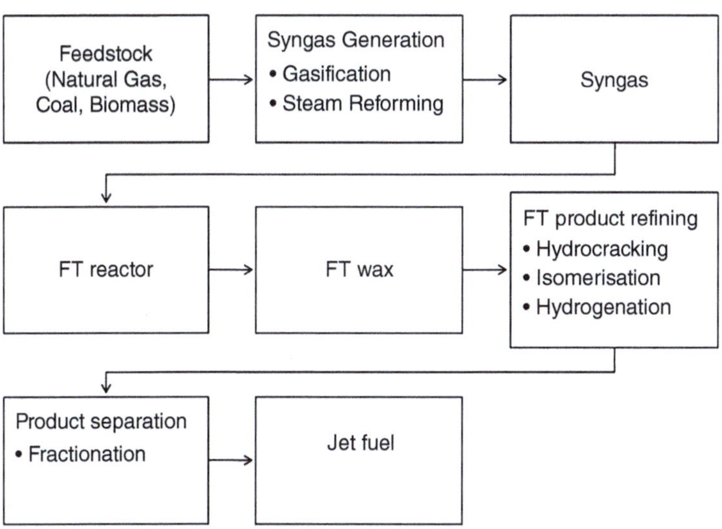

Figure 8.8 FT reaction pathway for jet fuel production.

8.5.2.2 Gas Fermentation

An alternative to the catalytic upgradation of syngas is its fermentation.

The syngas can be fermented to alcohol (ethanol or butanol) by acetogenic bacteria *Clostridium*. It can consume CO and hydrogen to produce acetate, ketones and alcohols. This mixed product can be exposed to ATJ technology as discussed in an earlier section. The process is shown in Figure 8.9.

The fermentation process has the advantage of producing a greater variety of products and increasing the yield at a lower cost. It has a higher tolerance limit of impurities, and it is also comfortable with the feedstock type. The process is observed to be more efficient than the FT process [91, 92]. Low operating temperature, pressure and low-cost enzymes decrease the operating cost of the process. Usually, a bubble column reactor is used for this process. A trickle-bed bioreactor with methanotrophic bacteria can also be used. This is one of the best methods as far as the utilisation of CH_4 in the bioprocess development is concerned [93].

8.5.3 Economic and Life-cycle Analysis

Low-temperature gasification is good for decreasing the cost, but the other difficulties encountered are the formation/flow of slag [94]. The FT process has a lower GHG contribution as woody feedstock is used. These reactors were built on a commercial level during World War II. The first commercial operation was started in the United States in 2010 [95]. Such produced fuels are highly recommended.

8.6 Sugar-to-Jet (STJ) Fuel

There are basically the following three pathways to produce jet fuel from sugar-containing feedstock:

Catalytic APR
Catalytic HMF and DMF
Biological conversion

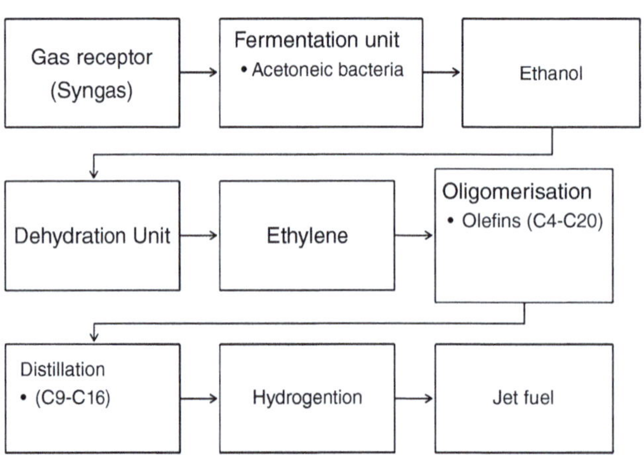

Figure 8.9 Gas fermentation process.

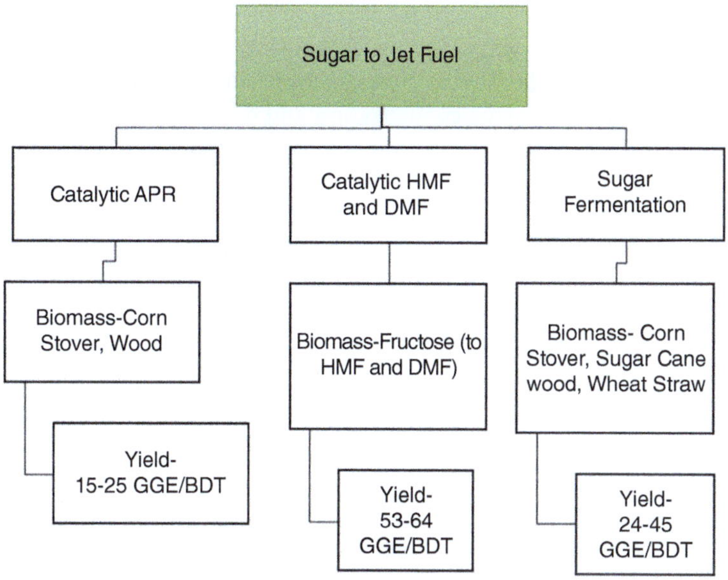

Figure 8.10 Pathway for the STJ process.

The required feedstock and the yield details are shown in Figure 8.10.

8.6.1 Feedstock Used

The feedstock for catalytic upgrading of sugars to hydrocarbons can be a variety of sugar crops, such as sugar cane or sugar beets, corn sugar from corn starch and lignocellulosic sugars from the hydrolysis of hemicellulose and cellulose. Derived sugar feedstock can also be used for aerobic biological conversion to hydrocarbons.

8.6.2 Process Analysis

In the process, biomass feedstock is first converted into solubilised sugars either by catalytic or biological methods.

The biomass feedstock is first converted into solubilised sugar. This sugar is then purified and concentrated using a microvapour recompression evaporator, microfiltration and ion exchange filtration to obtain better efficiency of the process. Unconverted sugar and impurities are completely removed in this process [96]. It is sent back to the combustor. Such a closed loop helps to improve the process cost.

The purified hydrolysate is sent now to reforming reactors for converting the carbohydrates into polyhydric alcohols in the presence of hydrogen (hydrogenation) or short-chain oxygenates (hydrogenolysis). The produced polyhydric alcohols are now sent to the APR reactor at temperatures of 450–575 K and pressures of 10–90 bar in the presence of a heterogeneous catalyst.

Various reactions occurring in the APR reactor are reforming (to produce hydrogen), dehydrogenation of alcohols, deoxygenation, hydrogenolysis and cyclisation. The products formed in APR are hydrogen, CO_2, alkenes and oxygenates [97]. Oxygenates produced are converted into jet fuel using condensation and dehydration. The lighter alkanes (C1–C4) from APR are resent to the combustor to provide additional process heat.

Another catalytic pathway is the 5-hydroxymethylfurfural (HMF) route. Sugars are transformed to HMF via fructose or glucose dehydration, which is then turned into dihydroxymethylfuran (DMF) through hydrogenolysis over a copper-ruthenium (CuRu) catalyst. This DMF has shown better fuel properties than traditional ones.

The general process for biological conversion of biomass into jet fuel starts with the pretreatment of the biomass, enzymatic hydrolysis and separation of C5–C6 sugars. It is then concentrated and sent to biological conversion (either aerobic or anaerobic) to form the hydrocarbon intermediate product [98]. Many intermediates, such as alkanes, aromatics and cycloparaffins, are formed during the reaction, which should be removed properly using the separation techniques. This product undergoes oligomerisation and hydrotreating to yield jet fuel. LS9 is a global sugar-to-jet-fuel-producing company using biological conversion.

8.6.3 Economic and Life-cycle Analysis

The minimum selling price of jet fuel formed by this process would be higher, as not all hydrocarbons produced are usable as jet fuel [99]. The cost of the copper-ruthenium-carbon catalyst used in the DMF process may play an important role in the final production cost. The search for a low-cost route to produce jet fuel that is compatible with the current state of the art is essential. The use of food crops keeps this technology on the last preference when sustainability is considered.

8.7 Overview of Blending Sustainable Aviation Fuel

Table 8.2 shows different internationally approved processes for SAF. It is not necessary that the approved technology is also accepted on commercial and global levels. Currently, HEFA-based fuel is used globally.

Once the fuel is developed in the laboratory, it is tested in the laboratory for its properties. It is compared to the properties of conventional jet fuels. Engine performance is tested well. Once the ground level testing is finished, it is then tested on aircraft on test flights. Once the operating conditions are ensured, they are further tested from various angles for real-time use.

8.8 Summary

Biological as well as non-biological natural resources are sunlight, water and air, which are supposed to be used in the generation of aviation fuel. Some researchers have worked on the generation of jet fuel from CO_2, water and solar energy. If power to gas and power to

Table 8.2 Various pathways approval and blending limits.

Pathways and processes	Date of approval	Current blending limit
Fischer–Tropsch for synthetic paraffinic kerosene [100, 101]	2009	Up to 50%
Hydroprocessed esters and fatty acids [100, 102]	2011	Up to 50%
Hydroprocessed fermented sugars to synthetic iso-paraffins [103]	2014	Up to 10%
FT-SPK with aromatics [104]	2015	Up to 50%
Alcohol-to-jet synthetic paraffinic kerosene (iso-butanol) [61, 105]	2016	Up to 30%
Alcohol-to-jet synthetic paraffinic kerosene (ethanol)	2018	Up to 50%
Catalytic hydrothermolysis synthetic jet fuel [106, 107]	2020	Up to 50%

liquid are integrated, which can capture a large amount of CO_2 and utilise it in successful fuel production, it can be more sustainable with less carbon emission. More work is being carried out in improving technology, process operations, value-added by-products, and market value. Improved advanced technologies must be carefully integrated into new designs to reduce any environmental impacts. Effective coordination between all stakeholders will surely bring up the improved technology in the long term.

References

1 Halder, P., Azad, K., Shah, S., and Sarker, E. (2019). Prospects and technological advancement of cellulosic bioethanol ecofuel production. *Adv. Eco-Fuels Sustain. Environ.* 211–236. https://doi.org/10.1016/b978-0-08-102728-8.00008-5.

2 Bajpai, P. (2021). Production of bioethanol. *Green Energy Technol.* 41–110. https://doi.org/10.1007/978-981-15-8779-5_5.

3 Gutiérrez-Antonio, C., Gómez-Castro, F.I., de Lira-Flores, J.A., and Hernández, S. (2017). A review on the production processes of renewable jet fuel. *Renew. Sustain. Energy Rev.* 79: 709–729. https://doi.org/10.1016/j.rser.2017.05.108.

4 Cho, K.S., Li, G., and Bardell, N. (2020). Towards meeting the iata-agreed 1.5% average annual fuel efficiency improvements between 2010 and 2020: the current progress being made by u.s. air carriers. *Aviation* 23 (4): 123–132. https://doi.org/10.3846/aviation.2019.12019.

5 Ibarra, Á., Rodríguez, E., Sedran, U. et al. (2016). Synergy in the cracking of a blend of bio-oil and vacuum gasoil under fluid catalytic cracking conditions. *Ind. Eng. Chem. Res.* 55 (7): 1872–1880. https://doi.org/10.1021/acs.iecr.5b04502.

6 Pradelle, F., Braga, S.L., Martins, A.R.F.A. et al. (2015). Gum formation in gasoline and its blends: a review. *Energy Fuel* 29 (12): 7753–7770. https://doi.org/10.1021/acs.energyfuels.5b01894.

7 Striebich, R.C., Smart, C.E., Gunasekera, T.S. et al. (2014). Characterization of the F-76 diesel and jet-a aviation fuel hydrocarbon degradation profiles of Pseudomonas aeruginosa and Marinobacter hydrocarbonoclasticus. *Int. Biodeter. Biodegr.* 93: 33–43. https://doi.org/10.1016/j.ibiod.2014.04.024.

8 Duan, D., Zhang, Y., Wang, Y. et al. (2020). Production of renewable jet fuel and gasoline range hydrocarbons from catalytic pyrolysis of soapstock over corn cob-derived activated carbons. *Energy* 209: 118454. https://doi.org/10.1016/j.energy.2020.118454.

9 Li, J., Yang, G., Yoneyama, Y. et al. (2016). Jet fuel synthesis via Fischer-Tropsch synthesis with varied 1-olefins as additives using Co/ZrO_2-SiO_2 bimodal catalyst. *Fuel* 171: 159–166. https://doi.org/10.1016/j.fuel.2015.12.062.

10 Kim, H., Kim, D., Park, Y.K., and Jeon, J.K. (2018). Synthesis of jet fuel through the oligomerization of butenes on zeolite catalysts. *Res. Chem. Intermed.* 44 (6): 3823–3833. https://doi.org/10.1007/s11164-018-3385-1.

11 Gaudenzi, B. and Bucciol, A. (2016). Jet fuel price variations and market value: a focus on low-cost and regular airline companies. *J. Bus Econ. Manage.* 17 (6): 977–991. https://doi.org/10.3846/16111699.2016.1209784.

12 Kumar, T., Mohsin, R., Majid, Z.A. et al. (2019). Response surface methodology application in optimization of performance and exhaust emissions of RON 98, aviation gasoline 100LL and the blends in Lycoming O-320 engine. *Fuel* 256: 115909. https://doi.org/10.1016/j.fuel.2019.115909.

13 Altuntas, O. (2020). Lead emissions from the use of leaded avgas in Turkey. *Aircr. Eng. Aerosp. Technol.* 93 (3): 493–501. https://doi.org/10.1108/AEAT-05-2020-0108.

14 Kumar, T., Mohsin, R., Ghafir, M.F.A. et al. (2018). Concerns over use of leaded aviation gasoline (AVGAS) fuel. *Chem. Eng. Trans.* 63: 181–186. https://doi.org/10.3303/CET1863031.

15 Yang, J., Xin, Z., He, Q. et al. (2019). An overview on performance characteristics of bio-jet fuels. *Fuel* 237: 916–936. https://doi.org/10.1016/j.fuel.2018.10.079.

16 Wei, H., Liu, W., Chen, X. et al. (2019). Renewable bio-jet fuel production for aviation: a review. *Fuel* 254: 115599. https://doi.org/10.1016/j.fuel.2019.06.007.

17 Bwapwa, J.K., Akash, A., and Trois, C. (2019). Jet fuel blend from algal jet fuel and jet A1 in 50/50 volume ratio. *Int. J. Low-Carbon Technol.* 14 (2): 234–240. https://doi.org/10.1093/ijlct/ctz014.

18 Padwal, M.B. and Mishra, D.P. (2016). Experimental characterization of gelled jet A1 spray flames. *Flow Turbul. Combust.* 295–337: https://doi.org/10.1007/s10494-015-9676-3.

19 Maghrabie, H.M. (2021). Heat transfer intensification of jet impingement using exciting jets – a comprehensive review. *Renew. Sustain. Energy Rev.* 139: 110684. https://doi.org/10.1016/j.rser.2020.110684.

20 Kumar, T., Mohsin, R., Majid, Z.A. et al. (2019). Response surface methodology (RSM) in optimization of performance and exhaust emissions of RON 97, RON 98, and RON 100 (motor gasoline) and AVGAS 100LL (aviation gasoline) in Lycoming O-320 engine. *SAE Int. J. Engines* 12 (4): 427–454. https://doi.org/10.4271/03-12-04-0029.

21 Fedorov, E.P., Ivanov, V.F., and Yanovskii, L.S. (2004). The "iodine number" of TS-1 fuel. On elimination of the index from GOST 10227-86. *Chem. Technol. Fuels Oils* 40 (5): 298–302. https://doi.org/10.1023/B:CAFO.0000046262.08920.50.

22 ASTM. Standard Specification for Jet B Wide-Cut Aviation Turbine Fuel 1. ASTM 2015, 05.03.

23 Tran, D.T., Palomino, J.M., and Oliver, S.R.J. (2018). De of JP-8 jet fuel: challenges and adsorptive materials. *RSC Adv.* 8 (13): 7301–7314. https://doi.org/10.1039/c7ra12784g.

24 Violi, A., Yan, S., Eddings, E.G. et al. (2002). Experimental formulation and kinetic model for JP-8 surrogate mixtures. *Combust. Sci. Technol.* 174 (11–12): 399–417. https://doi.org/10.1080/00102200215080.

25 Pleil, J.D., Smith, L.B., and Zelnick, S.D. (2000). Personal exposure to JP-8 jet fuel vapors and exhaust at air force bases. *Environ. Health Perspect.* 108 (3): 183–192. https://doi.org/10.1289/ehp.00108183.

26 Cowart, J., Foley, M.P., and Luning, P.D. (2019). The development and testing of navy jet fuel (JP-5) surrogates. *Fuel* 249: 80–88. https://doi.org/10.1016/j.fuel.2019.03.096.

27 Lee, H. (2021). Spray, combustion, and air pollutant characteristics of JP-5 for naval aircraft from experimental single-cylinder CRDI diesel engine. *Energies* 14 (9): 2362. https://doi.org/10.3390/en14092362.

28 Korres, D.M., Karonis, D., Lois, E. et al. (2008). Aviation fuel JP-5 and biodiesel on a diesel engine. *Fuel* 87 (1): 70–78. https://doi.org/10.1016/j.fuel.2007.04.004.

29 Liu, Y., Deng, D., and Bao, X. (2020). Catalysis for selected C1 chemistry. *Chem* 6 (10): 2497–2514. https://doi.org/10.1016/j.chempr.2020.08.026.

30 Anaizi, N. (2019). Fat facts: an overview of adipose tissue and lipids. *Ibnosina J. Med. Biomed. Sci.* 11 (1): 5. https://doi.org/10.4103/ijmbs.ijmbs_6_19.

31 Chime, A.H., Kalia, A., Kramlich, J. et al. (2018). Alternative aviation fuels evaluated for lean-flame NOx and blowout, and rich-flame soot threshold. *Proc. ASME Turbo Expo* 4A–2018A. https://doi.org/10.1115/GT201875564.

32 Hoseyni Chime A, Kalia A, Kramlich J, Novosselov I, Malte PC, Vijlee S. Alternative aviation fuels evaluated for lean-flame NOx and blowout, and rich-flame soot threshold, *Proceedings of the ASME Turbo Expo 2018: Turbomachinery Technical Conference and Exposition. Volume 4A: Combustion, Fuels, and Emissions.* Oslo, Norway (June 11–15, 2018). V04AT04A042. ASME, http://dx.doi.org/10.1115/gt2018-75564.

33 Chen, F., Li, N., Li, S. et al. (2015). Solvent-free synthesis of C9 and C10 branched alkanes with furfural and 3-pentanone from lignocellulose. *Catal. Commun.* 59: 229–232. https://doi.org/10.1016/j.catcom.2014.11.003.

34 Wang, T., Zhang, Q., Ding, M. et al. (2017). Bio-gasoline production by coupling of biomass catalytic pyrolysis and oligomerization process. *Energy Proc.* 105: 858–863. https://doi.org/10.1016/j.egypro.2017.03.401.

35 Brandão, R.D., de Freitas Júnior, A.M., Oliveira, S.C. et al. (2021). The conversion of coconut oil into hydrocarbons within the chain length range of jet fuel. *Biomass Convers. Biorefin.* 11 (3): 837–847. https://doi.org/10.1007/s13399-020-01046-9.

36 Zech, K.M., Dietrich, S., Reichmuth, M. et al. (2018). Techno-economic assessment of a renewable bio-jet-fuel production using power-to-gas. *Appl. Energy* 231: 997–1006. https://doi.org/10.1016/j.apenergy.2018.09.169.

37 de Klerk, A. (2016). Aviation turbine fuels through the Fischer–Tropsch process. *Biofuels Aviat.* 241–259: https://doi.org/10.1016/b978-0-12-804568-8.00010-x.

38 Boelhouwer, C. and Mol, J.C. (1984). Metathesis of fatty acid esters. *J. Am. Oil Chem. Soc.* 61 (2): 425–430. https://doi.org/10.1007/BF02678808.

39 Bonin, H., Keraani, A., Dubois, J.L. et al. (2015). Cross-metathesis of fatty acid methyl esters with acrolein: an entry to a variety of bifunctional compounds. *Eur. J. Lipid Sci. Technol.* 117 (2): 209–216. https://doi.org/10.1002/ejlt.201400362.

40 Paxton, B.T., Fugger, C.A., Tomlin, A.S., and Caswell, A.W. (2020). Experimental investigation of fuel type on combustion instabilities in a premixed bluff-body combustor. *AIAA Scitech 2020 Forum* 0174. https://doi.org/10.2514/6.2020-0174.

41 Kang, D., Kim, D., Kalaskar, V. et al. (2019). Experimental characterization of jet fuels under engine relevant conditions – part 1: effect of chemical composition on autoignition of conventional and alternative jet fuels. *Fuel* 239: 1388–1404. https://doi.org/10.1016/j.fuel.2018.10.005.

42 Pechstein, J., Neuling, U., Gebauer, J., and Kaltschmitt, M. (2017). Alcohol-to-Jet (AtJ). *Biokerosene Status Prospect.* 543–574. https://doi.org/10.1007/978-3-662-53065-8_21.

43 Adegboye, M.F., Ojuederie, O.B., Talia, P.M., and Babalola, O.O. (2021). Bioprospecting of microbial strains for biofuel production: metabolic engineering, applications, and challenges. *Biotechnol. Biofuels* 14 (1): 1–21. https://doi.org/10.1186/s13068-020-01853-2.

44 Berłowska, J., Pielech-Przybylska, K., Balcerek, M. et al. (2016). Simultaneous saccharification and fermentation of sugar beet pulp for efficient bioethanol production. *Biomed. Res. Int.* https://doi.org/10.1155/2016/3154929.

45 Kosir, S.T., Behnke, L., Heyne, J.S. et al. (2019). Improvement in jet aircraft operation with the use of high-performance drop-in fuels. *AIAA Scitech 2019 Forum* 0993. https://doi.org/10.2514/6.2019-0993.

46 Zhou, M., Chen, C., Liu, P. et al. (2020). Catalytic hydrotreatment of β-O-4 ether in lignin: cleavage of the C-O bond and hydrodeoxygenation of lignin-derived phenols in one pot. *ACS Sustain. Chem. Eng.* 8 (38): 14511–14523. https://doi.org/10.1021/acssuschemeng.0c04941.

47 Wu, S., Huang, K., Liu, W. et al. (2019). Test on influence of aromatic hydrocarbons on smoke emission characteristics in turboshaft engine. *Hangkong Dongli Xuebao J. Aerosp. Power* https://doi.org/10.13224/j.cnki.jasp.2019.05.015.

48 Zhang, C., Hui, X., Lin, Y., and Sung, C.J. (2016). Recent development in studies of alternative jet fuel combustion: progress, challenges, and opportunities. *Renew. Sustain. Energy Rev.* 54: 120–138. https://doi.org/10.1016/j.rser.2015.09.056.

49 Heyne, J., Rauch, B., Le Clercq, P., and Colket, M. (2021). Sustainable aviation fuel prescreening tools and procedures. *Fuel* 290: 120004. https://doi.org/10.1016/j.fuel.2020.120004.

50 Chiaramonti, D. (2019). Sustainable aviation fuels: the challenge of decarbonization. *Energy Proc.* 158: 1202–1207. https://doi.org/10.1016/j.egypro.2019.01.308.

51 Pechstein, J., Bullerdiek, N., and Kaltschmitt, M. (2020). A "book and claim"-approach to account for sustainable aviation fuels in the EU-ETS – development of a basic concept. *Energy Policy* 136: 111014. https://doi.org/10.1016/j.enpol.2019.111014.

52 Guzman, J., Kukkadapu, G., Brezinsky, K., and Westbrook, C.K. (2021). Oxidation of an iso-paraffinic alcohol-to-jet fuel and n-heptane mixture: an experimental and modeling study. *Int. J. Chem. Kinet.* 53 (9): 1014–1035. https://doi.org/10.1002/kin.21501.

53 Guo, X., Guo, L., Zeng, Y. et al. (2021). Catalytic oligomerization of isobutyl alcohol to jet fuels over dealuminated zeolite Beta. *Catal. Today* 368: 196–203. https://doi.org/10.1016/j.cattod.2020.04.047.

54 Jagtap, S.S. (2019). Assessment of feedstocks for blended alcohol-to-jet fuel manufacturing from standalone and distributed scheme for sustainable aviation. *AIAA Propuls. Energy Forum Expo.* 2019: 3887. https://doi.org/10.2514/6.2019-3887.

55 Romero-Izquierdo, A.G., Gómez-Castro, F.I., Gutiérrez-Antonio, C. et al. (2021). Intensification of the alcohol-to-jet process to produce renewable aviation fuel. *Chem. Eng. Process. Process Intensif.* 160: 108270. https://doi.org/10.1016/j.cep.2020.108270.

56 Geleynse, S., Brandt, K., Garcia-Perez, M. et al. (2018). The alcohol-to-jet conversion pathway for drop-in biofuels: techno-economic evaluation. *ChemSusChem* 11 (21): 3728–3741. https://doi.org/10.1002/cssc.201801690.

57 Han, G.B., Jang, J.H., Ahn, M.H., and Jung, B.H. (2019). Recent application of bio-alcohol: bio-jet fuel. *Alcohol Fuels Curr. Technol. Future Prospect.* 109–121.

58 Pelucchi, M., Namysl, S., Ranzi, E. et al. (2020). Combustion of n-C3-C6 linear alcohols: an experimental and kinetic modeling study. Part II: speciation measurements in a jet-stirred reactor, ignition delay time measurements in a rapid compression machine, model validation, and kinetic analysis. *Energy Fuel* 34 (11): 14708–14725. http://dx.doi.org/10.1021/acs.energyfuels.0c02252.

59 Ryu, J.I., Motily, A.H., Lee, T. et al. (2021). Effect of hot probe temperature on ignition of alcohol-to-jet (Atj) fuel spray under aircraft propulsion system conditions. *AIAA Scitech 2021 Forum* 0985: https://doi.org/10.2514/6.2021-0985.

60 Dagle, V.L., Winkelman, A.D., Jaegers, N.R. et al. (2020). Single-step conversion of ethanol to n-butene over Ag-ZrO_2/SiO_2 catalysts. *ACS Catal.* 10 (18): 10602–10613. https://doi.org/10.1021/acscatal.0c02235.

61 Brooks, K.P., Snowden-Swan, L.J., Jones, S.B. et al. (2016). Low-carbon aviation fuel through the alcohol to jet pathway. *Biofuels Aviat.* 109–150: https://doi.org/10.1016/b978-0-12-804568-8.00006-8.

62 Luning Prak, D.J., Jones, M.H., Trulove, P. et al. (2015). Physical and chemical analysis of alcohol-to-jet (ATJ) fuel and development of surrogate fuel mixtures. *Energy Fuel* 29 (6): 3760–3769. https://doi.org/10.1021/acs.energyfuels.5b00668.

63 He, M., Wang, M., Tang, G. et al. (2018). From medium chain fatty alcohol to jet fuel: rational integration of selective dehydration and hydro-processing. *Appl. Catal. A. Gen.* 550: 160–167. https://doi.org/10.1016/j.apcata.2017.11.009.

64 Harvey, B.G. and Meylemans, H.A. (2011). The role of butanol in the development of sustainable fuel technologies. *J. Chem. Technol. Biotechnol.* 86 (1): 2–9. https://doi.org/10.1002/jctb.2540.

65 Pelucchi, M., Namysl, S., Ranzi, E. et al. (2020). Combustion of n-C3-C6 linear alcohols: an experimental and kinetic modeling study. Part I: reaction classes, rate rules, model lumping, and validation. *Energy Fuel* 34 (11): 14708–14725. http://dx.doi.org/10.1021/acs.energyfuels.0c02251.

66 Wang, X., Liu, H., Zheng, Z., and Yao, M. (2015). Development of a reduced n-butanol/biodiesel mechanism for a dual fuel engine. *Fuel* 157: 87–96. https://doi.org/10.1016/j.fuel.2015.04.053.

67 Wassermann, T., Schnuelle, C., Kenkel, P., and Zondervan, E. (2020). Power-to-methanol at refineries as a precursor to green jet fuel production: a simulation and assessment study. *Comput. Aided Chem. Eng.* 48: 1453–1458. https://doi.org/10.1016/B978-0-12-823377-1.50243-3.

68 Martinez-Hernandez, E., Ramírez-Verduzco, L.F., Amezcua-Allieri, M.A., and Aburto, J. (2019). Process simulation and techno-economic analysis of bio-jet fuel and green diesel production – minimum selling prices. *Chem. Eng. Res. Des.* 146: 60–70. https://doi.org/10.1016/j.cherd.2019.03.042.

69 Baral, N.R., Kavvada, O., Mendez-Perez, D. et al. (2019). Techno-economic analysis and life-cycle greenhouse gas mitigation cost of five routes to bio-jet fuel blendstocks. *Energ. Environ. Sci.* 12 (3): 807–824. https://doi.org/10.1039/c8ee03266a.

70 Liu, H., Zhang, C., Tian, H. et al. (2021). Environmental and techno-economic analyses of bio-jet fuel produced from jatropha and castor oilseeds in China. *Int. J. Life Cycle Assess.* 26 (6): 1071–1084. https://doi.org/10.1007/s11367-021-01914-0.

71 Julio, A.A.V., Batlle, E.A.O., Rodriguez, C.J.C., and Palacio, J.C.E. (2021). Exergoeconomic and environmental analysis of a palm oil biorefinery for the production of bio-jet fuel. *Waste Biomass Valorization* 12 (10): 5611–5637. https://doi.org/10.1007/s12649-021-01404-2.

72 Tongpun, P., Wang, W.C., and Srinophakun, P. (2019). Techno-economic analysis of renewable aviation fuel production: from farming to refinery processes. *J. Clean. Prod.* 226: 6–17. https://doi.org/10.1016/j.jclepro.2019.04.014.

73 Di Gruttola, F. and Borello, D. (2021). Analysis of the eu secondary biomass availability and conversion processes to produce advanced biofuels: use of existing databases for assessing a metric evaluation for the 2025 perspective. *Sustainability* 13 (14): 7882. https://doi.org/10.3390/su13147882.

74 Itthibenchapong, V., Srifa, A., Kaewmeesri, R. et al. (2017). Deoxygenation of palm kernel oil to jet fuel-like hydrocarbons using Ni-MoS_2/γ-Al_2O_3 catalysts. *Energ. Conver. Manage.* 134: 188–196. https://doi.org/10.1016/j.enconman.2016.12.034.

75 Mäki-Arvela, P., Martínez-Klimov, M., and Murzin, D.Y. (2021). Hydroconversion of fatty acids and vegetable oils for production of jet fuels. *Fuel* 306: 121673. https://doi.org/10.1016/j.fuel.2021.121673.

76 Du, X., Li, D., Xin, H. et al. (2019). The conversion of jatropha oil into jet fuel on NiMo/Al-MCM-41 catalyst: intrinsic synergic effects between Ni and Mo. *Energ. Technol.* 7 (5): 1800809. https://doi.org/10.1002/ente.201800809.

77 Jung, S., Jung, J.M., Lee, K.H., and Kwon, E.E. (2021). Biodiesels from non-catalytic transesterification of plant oils and their performances as aviation fuels. *Energ. Conver. Manage.* 244: 114479. https://doi.org/10.1016/j.enconman.2021.114479.

78 Groendyk, M. and Rothamer, D.A. (2019). Optical investigation of mixing-controlled combustion using a novel transgenic plant oil. *Fuel* 252: 675–698. https://doi.org/10.1016/j.fuel.2019.03.094.

79 Carter, N.A. (2012). Environmental and economic assessment of microalgae-derived jet fuel. *Massachusetts Inst. Technol.* 141–149.

80 Bwapwa, J.K., Anandraj, A., and Trois, C. (2017). Possibilities for conversion of microalgae oil into aviation fuel: a review. *Renew. Sustain. Energy Rev.* 80: 1345–1354. https://doi.org/10.1016/j.rser.2017.05.224.

81 Lee, N., Joo, J., Lin, K.Y.A., and Lee, J. (2021). Waste-to-fuels: pyrolysis of low-density polyethylene waste in the presence of H-ZSM-11. *Polymers (Basel)* 13 (8): 1198. https://doi.org/10.3390/polym13081198.

82 Park, C. and Lee, J. (2021). Pyrolysis of polypropylene for production of fuel-range products: effect of molecular weight of polypropylene. *Int. J. Energy Res.* 45 (9): 13088–13097. https://doi.org/10.1002/er.6635.

83 Chiaramonti, D., Prussi, M., Buffi, M. et al. (2017). Review and experimental study on pyrolysis and hydrothermal liquefaction of microalgae for biofuel production. *Appl. Energy* 185: 963–972. https://doi.org/10.1016/j.apenergy.2015.12.001.

84 Hsu, H.W., Chang, Y.H., and Wang, W.C. (2021). Techno-economic analysis of used cooking oil to jet fuel production under uncertainty through three-, two-, and one-step conversion processes. *J. Clean. Prod.* 289: https://doi.org/10.1016/j.jclepro.2020.125778.

85 Graham, M. (1997). A test of magnitude: does the strength of predictors explain differences in drug use among adolescents? *J. Drug Educ.* 27 (1): 83–104. https://doi.org/10.2190/11L4-HDCJ-TTK1-4DV9.

86 Neuling, U. and Kaltschmitt, M. (2017). Conversion routes from biomass to biokerosene. *Biokerosene Status Prospect.* 435–473. https://doi.org/10.1007/978-3-662-53065-8_18.

87 De Jong, S., Antonissen, K., Hoefnagels, R. et al. (2017). Life-cycle analysis of greenhouse gas emissions from renewable jet fuel production. *Biotechnol. Biofuels* 10 (1): 1–8. https://doi.org/10.1186/s13068-017-0739-7.

88 Shen, B.X. and Liu, W.Q. (2020). Effect of hot fuel gas on a combinational opposing jet and platelet transpiration thermal protection system. *Appl. Therm. Eng.* 164: 114513. https://doi.org/10.1016/j.applthermaleng.2019.114513.

89 Safarian, S., Unnthorsson, R., and Richter, C. (2021). Performance analysis of power generation by wood and woody biomass gasification in a downdraft gasifier. *Int. J. Appl. Power Eng.* 10 (1): 80. https://doi.org/10.11591/ijape.v10.i1.pp80-88.

90 Falgout, Z., Rahm, M., Sedarsky, D., and Linne, M. (2016). Gas/fuel jet interfaces under high pressures and temperatures. *Fuel* 168: 14–21. https://doi.org/10.1016/j.fuel.2015.11.061.

91 Smith, C.H., Pineda, D.I., Zak, C.D., and Ellzey, J.L. (2013). Conversion of jet fuel and butanol to syngas by filtration combustion. *Int. J. Hydrogen Energy* 38 (2): 879–889. https://doi.org/10.1016/j.ijhydene.2012.10.102.

92 Minamoto, Y., Kolla, H., Grout, R.W. et al. (2015). Effect of fuel composition and differential diffusion on flame stabilization in reacting syngas jets in turbulent cross-flow. *Combust. Flame* 162 (10): 3569–3579. https://doi.org/10.1016/j.combustflame.2015.06.013.

93 Stolecka, K. and Rusin, A. (2020). Analysis of hazards related to syngas production and transport. *Renew. Energy* 146: 2535–2555. https://doi.org/10.1016/j.renene.2019.08.102.

94 Kim, J.-K., Park, J.Y., Yim, E.S. et al. (2015). Bio-jet fuel production technologies for GHG reduction in aviation sector. *Trans. Korean Hydrogen New Energy Soc.* 26 (6): 609–628. https://doi.org/10.7316/khnes.2015.26.6.609.

95 Walde, T.W. (2008). US foreign oil policy since world war I – for profits and security. *J. World Energy Law Bus* 1 (1): 113–117. https://doi.org/10.1093/jwelb/jwn002.

96 Pamula, A.S.P., Lampert, D.J., and Atiyeh, H.K. (2021). Well-to-wake analysis of switchgrass to jet fuel via a novel co-fermentation of sugars and CO_2. *Sci. Total Environ.* 782: 146770. https://doi.org/10.1016/j.scitotenv.2021.146770.

97 Michailos, S. (2018). Process design, economic evaluation and life cycle assessment of jet fuel production from sugar cane residue. *Environ. Prog. Sustain. Energy* 37 (3): 1227–1235. https://doi.org/10.1002/ep.12840.

98 Han, J., Tao, L., and Wang, M. (2017). Well-to-wake analysis of ethanol-to-jet and sugar-to-jet pathways. *Biotechnol. Biofuels* 10 (1): 1–5. https://doi.org/10.1186/s13068-017-0698-z.

99 Ganguly, I., Pierobon, F., Bowers, T.C. et al. (2018). "Woods-to-wake" life cycle assessment of residual woody biomass based jet-fuel using mild bisulfite pretreatment. *Biomass Bioenergy* 108: 207–216. https://doi.org/10.1016/j.biombioe.2017.10.041.

100 Jürgens, S., Oßwald, P., Selinsek, M. et al. (2019). Assessment of combustion properties of non-hydroprocessed Fischer-Tropsch fuels for aviation. *Fuel Process. Technol.* 193: 232–243. https://doi.org/10.1016/j.fuproc.2019.05.015.

101 Sahir, A.H., Zhang, Y., Tan, E.C.D., and Tao, L. (2019). Understanding the role of Fischer–Tropsch reaction kinetics in techno-economic analysis for co-conversion of natural gas and biomass to liquid transportation fuels. *Biofuels Bioprod. Biorefin.* 13 (5): 1306–1320. https://doi.org/10.1002/bbb.2035.

102 Starck, L., Pidol, L., Jeuland, N. et al. (2016). Production of hydroprocessed esters and fatty acids (HEFA) – optimisation of process yield. *Oil Gas Sci. Technol. Rev d'IFP Energies Nouv* 71 (10): 1–13. https://doi.org/10.2516/ogst/2014007.

103 Delfino, J.R., da Silva, J.L., Marques, A.L.B., and Stradiotto, N.R. (2020). Antioxidants detection in aviation biokerosene by high-performance liquid chromatography using gold nanoparticles anchored in reduced graphene oxide. *Fuel* 260: 1–2. https://doi.org/10.1016/j.fuel.2019.116315.

104 Jürgens, S., Selinsek, M., Bauder, U. et al. (2020). Potential of decentralized container-scale PtL plants for aviation: from crude to post- processed FT-SPK. *Proc. ASME Turbo Expo.* 84126: V04AT04A011. https://doi.org/10.1115/GT2020-14306.

105 Donnelly, J., Horton, R., Gopalan, K. et al. (2016). Branched ketone biofuels as blending agents for jet-A1 aviation kerosene. *Energy Fuel* 30 (1): 294–301. https://doi.org/10.1021/acs.energyfuels.5b01629.

106 Gawron, B., Białecki, T., Janicka, A., and Suchocki, T. (2020). Combustion and emissions characteristics of the turbine engine fueled with HeFA blends from different feedstocks. *Energies* 211–236. https://doi.org/10.3390/en13051277.

107 Geiselman, G.M., Kirby, J., Landera, A. et al. (2020). Conversion of poplar biomass into high-energy density tricyclic sesquiterpene jet fuel blendstocks. *Microb. Cell Fact.* 19 (1): 1–6. https://doi.org/10.1186/s12934-020-01456-4.

9

State of the Art Design and Fabrication of a Reactor in Biofuel Production

9.1 Introduction

A bioreactor is a vessel for carrying chemical changes of the substance involving the formation or destruction of chemical bonds, which involve transformation with chemical reaction; with nature friendly, the process carried out in the bioreactor may be aerobic or anaerobic type [1, 2]. The selection of the bioreactor of the process depends upon the shapes, capacity and material used for manufacturing. The bioreactor consists of a reactor tank, stirred mechanism, aeration tubes, agitator, baffles, stands, head plates, condensers, thermowells, heat exchangers and stoppers with different biocatalysts [3–4].

The design of bioreactors is a difficult task as it is controlled in a suitable environment to achieve the growth or formation of the product under optimum condition; the production process in the bioreactor at control temperature, pH range, pressure, microorganisms and cell cultures works within certain temperatures and pH ranges as the prescribe limits process is slow and it depends on the performance of highly dependent catalysts; on the other hand, unfavourable environmental condition may even be destroyed [5, 6]. Therefore, the selection of measurable sensors plays an important role, as pH changes alter growth conditions of the system in addition to culture mixing and feeding nutrients, which can easily be measured and controlled facilitating various biochemical reactions [7–8].

The most common types of the bioreactor are simple stirred-type, column, airlift, fluidised bed and packed bed bioreactors [9, 10]. The bioreactor design includes agitation for mixing, aeration for aerobic fermentation, process temperature, potential of hydrogen, pressure, mixing and transfer of heat to maintain uniform environment and capacity [11]. Glass material is used for smaller vessels, while steel is used as the material for larger vessels depending upon the capacity [12].

As the bioreactors of various types are used for wastewater treatment in environmental protection, cell culture and tissue engineering in the health care sector, the production of high-value pharmaceuticals, bulk chemicals, cultivation of algae and strong demand upon application in the field, the bioreactor is simulated and developed with novel design with bioreactor geometries, control parameters and control systems that are efficient, robust and economical [13, 14].

Sustainability in Biofuel Production Technology, First Edition. Pratibha S. Agrawal, Pramod N. Belkhode, and Samuel Lalthazuala Rokhum.

9.2 Limitation of Conventional Production Technology

The conventional bioreactors are cylindrical vessels with dome-shaped structure at the bottom and top surrounded by a jacket with a sparger located at the base for the introduction of air. Power for the motor is transferred to the agitator shaft for mixing of the microbial culture, ensuring optimum conditions for the supply of nutrients and oxygen [15, 16], so that good products are formed under favourable environmental condition. Foam is formed in bioreactors as a result of agitated aeration, and minerals oil or vegetable oil is used as antifoam agents to lower surface tension or break the foam bubbles in the fermentation process [17, 18]. Provision of ports is used for the measurement of pH, temperature and oxygen sensor.

The aeration system is designed to supply oxygen and remove carbon dioxide from the bioreactor [19]. Oxygen is introduced from the lower level of the base through a sparger. This allows an upwards flow of air bubbles. Figure 9.1 indicates the conventional bioreactor that shows different components of the conventional bioreactor, such as inoculation pipe, stirrer, baffle plate, impeller, etc. The aeration capacity depends on dissolved oxygen in the medium that can be enhanced by stirring. The aeration capacity of the stirred fermentation is proportional to the stirring speed, rate of air flow and internal pressure [20].

The performance of the fermentation process can be obtained by regularly monitoring the variables, such as temperature, amount of oxygen dissolved, proportionate mixing, concentration of nutrient and pH. Optimum values of pH 5.5 and 8.5 for fermentation as microorganism grow release metabolites, resulting in changes in pH necessary to continuously monitor and maintain optimum levels [21]. Furthermore, the optimum level can be achieved with the addition of acid or alkali base while mixing the fermentation contents.

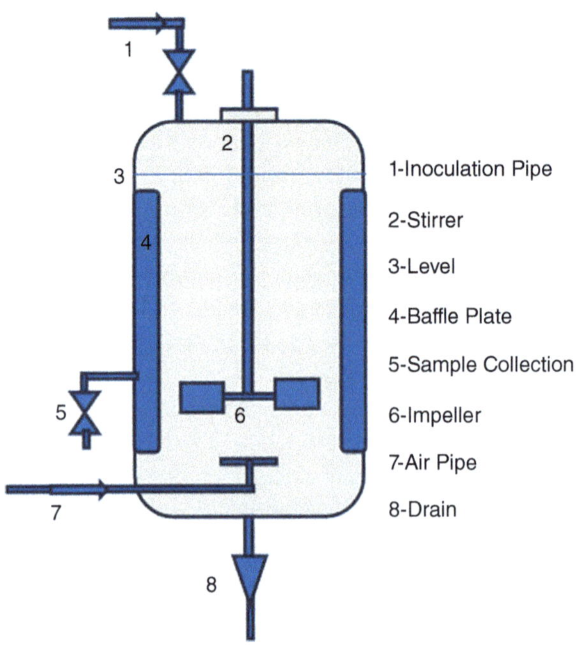

Figure 9.1 Conventional bioreactor. *Source:* Adapted from [20].

A good formation process also depends on the temperature; a lower temperature results in a reduction in product formation, while a higher temperature affects the growth of microorganisms. Therefore, it is provided with heating and cooling systems that can maintain the optimum temperature.

Oxygen is sparingly soluble in water culture medium due to the availability of oxygen during the sterilisation [22]. Therefore, air is introduced if the oxygen concentration is maintained for optimum product formation.

The contents are removed for processing after the fermentation process is completed. The bioreactor is cleaned for the next fermentation process. Time required for cleaning depends upon the size of the bioreactor as the cleaning of the bioreactor is carried out by using high-pressure water jets from the nozzle fitted at the top of the bioreactor [23].

Biofuels are promising renewable sources of energy as alternative fuels over fossil fuels due to a number of advantages. Production of biodiesel using conventional and novel technologies can be used for achieving the efficiency and safety of the biodiesel production process [24, 25]. Selection of feedstock used for the production of biodiesel has been more challenging due to various issues related to the processing of converting biomass into biofuels, which consists of pre-treatments, hydrolysis microbial fermentation and fuel separation [26, 27]. The cost of feedstock processing was higher due to lower advance technology processing costs.

Biorefineries will be successful by properly utilising biomass feedstock and transforming all into valuable products by selecting a suitable bioreactor for processing [28]. Designing an economical and efficient bioreactor for biofuel production has a great potential [29, 30].

Aeration and agitation were modified in the conventional bioreactor for improving the mixing process by preventing setting and saving power consumption. The conventional technologies for producing biofuels can be categorised based on the processing techniques [31]. The challenges for each technology include feedstock availability process design and process economic and life-cycle assessment of greenhouse gas emission. Conventional bioreactors include a smooth plate, stirred tank and horizontal- or vertical-type tubular and airlift column with improvement in illumination, aeration and agitation, which improve the biomass yield and productivity [32].

9.3 Ideal Reactors

An ideal reactor is one in which stirring is so efficient that the contents are always in composition and temperature throughout the process. In an ideal reactor, the transport and mixing processes can be described in such a way that energy molecules of the reactant spend an equal amount of time in the reaction [33]. An ideal reactor can be considered a basic reactor that does not exist and is affected by many factors, such as temperature, pressure, nature of feed, nature of reaction, etc. A reactor in which the temperature effect is not accounted for only by axial flow mixing is an isothermal ideal reactor [34]. An ideal continuously stirred tank reactor (CSTR) will exhibit well-defined flow behaviour that can be characterised by the reactor residence time distribution or exit age distribution. In an ideal plug flow reactor (PFR), the reaction mixture moves with the same speed along the direction of the flow, thus preventing mixing or back flow. The batch reaction consists of a simple vessel into which stirring ensures that at any instant the composition and temperature of the mixture are the same [33].

The reagent travels through the PFR in the chemical reaction. For the same volume, the efficiency of PFR has a higher theoretical efficiency than a CSTR. The completion rate of reaction is higher in a PFR than in a CSTR for the same residence time. The reactors in which molecules have different space times are called non-ideal reactors. The reasons for non-ideal may be dead time, bypass, etc.; for a non-ideal reactor, there are different models to explain non-ideal characteristics.

9.4 Reaction Designing from an Engineering Aspect

Reactor designing refers to reaction engineering for the output production with minimum wastage, low energy conservation and water conservation [35]. Table 9.1 shows the important aspects of chemical reactor design in which important terms are involved in the design of the reactor. The aspect ratio is the length:diameter ratio that produces high gas velocity and high pressure drops across the catalyst bed; the space velocity is expressed as $SV = V_0/V$, that is, the ratio of the volumetric flow rate of the reactants entering the reactor to the volume of the reactor that plays a significant role for reactor designing. The reactor space time is proportional to V_0/V, and the space velocity is determined based on the reactor inlet condition, catalyst type and fractional conversion [36].

One of the important aspects of reactor design is the fermentation used to make the product successfully using mechanically agitated fermenters. They control mass transfer rates and viscosities. The optimum aspect ratio depends on every aspect, including mass transfer, electric power consumption, agitator cost, vessel cost, heat transfer, mass transfer and pressure.

9.5 Process Parameters in Reactor Designing

Good reactor design refers to reaction design for the production of chemicals at minimum wastage with conservation of energy and water and minimum utilisation of fossil fuels [37]. The design of the reactor means the start of the reaction with maximum efficiency towards

Table 9.1 Important aspects of chemical reactor design.

Important aspects of chemical reactor design	
Reactor design	Flow mode: Batch/semi-batch/continuous Transport: Mass transfer/heat transfer/diffusion Reaction: Homogenous/heterogeneous Mixing

Source: Adapted from [35, 36].

the desired output product, which is able to produce the highest product yield at the lowest cost and operation. Different factors are required for designing.

The reactor design consists of the following factors:

1) Reactor size
2) Reactor type
3) Reactor time
4) Material used for reaction and temperature
5) Heat involved
6) Fluid flow pattern

General steps to be followed for designing the reactor are as follows:

1) Collection or required data
2) Selection of reactor conditions
3) Selection of material for construction
4) Sizing parameters
5) Sizing, layout and costing
6) Reactor performance

The important aspect of reactor design is the heat transfer, which controls industrial processes that maintain safety. Temperature is a dominant variable that plays an important role in finding the economics, worth value and safety of the reactor [38]. The ratio of diameter to height affects both the heat transfer area and the level of mixing. Some of the important process parameters are as follows.

9.5.1 Kinetics and Reaction Equilibrium

The chemical reaction rate depends on the catalyst concentration, initiator concentration, pH, temperature and the concentration of the reactants and products. As in multiphase systems, the chemical reaction rate is complex due to chemical kinetics, mass or heat transfer, which often play a significant role. The transfer of mass and heat with chemical kinematics are important parameters to consider for the design of reactors [39].

9.5.2 Collection of Required Data

Enthalpies of the reaction and rate at which the reaction is performed with the mass and heat transfer coefficient are the process input data find out by a simulation model or experimentally [40]. Equilibrium constants can be found useful for the evaluation of the rates of the forward and reverse reaction, process condition, temperature, pressure and concentration to find the time required for the conversion process.

9.5.3 Reaction Condition

The operation condition depends on the type of reaction, catalyst and temperature, which further affects the type of exothermic or endothermic reaction and the cost of reaction [41]. Material selection for the construction of the reaction depends on the specified reaction condition, process and material use for the construction [42]. Mixing involves the mass transfer rate, heat transfer rate and mixing time for the fast reaction or reaction condition, which depends on the sizing of the reactions [43]. The reaction estimation and process volume and the size of the reaction are related to the reaction depending on the flow pattern and mixing requirement.

9.6 Safety Consideration of Reaction Design

Safety plays a significant role in the plants. Safety should be considered in all aspects of manufacturing units and operations. The process should be designed to minimise the storage and discharge of waste materials [44]. The reactor should be designed to be safe so that its equipment, control and operational system can be easily and safely handled.

Reaction design should meet the demand of enhanced safety, resource utilisation and reliability. In the reactor design, the type of reaction is one of the challenges; while designing the reactor, a suitable flow rate should be selected, which can prevent the chance of runaway to have good instrumentation that feeds into a reaction to give advance warning with an emergency rapid cooling system to lower the reaction rate [45].

Most of the time, while designing, the reactor unwillingness to invest in the safety measures is the main reason for all safety issues [45]. Temperature measurement of the reaction could be forwarded to various interlock, and even the valves would open up in the case of uncontrolled reaction that would stop the reaction altogether.

The key consideration for an inherently safe reactor with regard to reaction is one that has no potential for hazardous consequences even under difficult situations as it consists of the release of energy, which includes the following cases:

1) Reactor size: Smaller reactors have better heat removal capability.
2) Reaction shape: Reactor shapes affect the area/volume return.
3) Heat transfer: Heat transfer coefficients can be enhanced by having fins on the beds or designing for natural cooling.
4) Cooling media: Cooling media that utilised air cooled condenser may enhance safety.
5) Reactor feed: Maximum reactor feed rates can be controlled by limiting the feed piping size.
6) Concentration: Concentrations of the reacting stream act as a catalyst in the bed.

Consideration of such factors can be an inherently safer design. Adequate operator training as a trained operator followed the procedure available in the control room at all times. Setting of pre-alarms on the safe operating limits and reliable operating limits in 30–40 minutes advance gives enough time for the operation to take place. Alarms, controls, training and many other factors should be considered for reactor design safety.

Reactor design starts with respect to customers used with the development of a profitable solution. Design consists of modification in the existing reactor, improvement in the production capacity and the design of a new process. The chemical engineer plays an important role in the rector design as it involves the calculation of mass and energy balance, sizing of equipment and cost associated with components and equipment materials used in the manufacturing of the reactor.

The design process begins with desired production rate, system of units, design codes, raw materials and utilities and follows the design concept, analysis and evaluation in computer simulation before final selection. All the factors are related to process design, which undergoes procurement and construction and operations of the reactor [46].

The existing device or system is improved by the process of optimisation so that the best possible chemical process can be achieved. The design variables involved can be classified as continuous and discrete. The temperature and pressure of the reactor are considered continuous variables as they affect the speed and feed the discrete variable.

9.7 Reactors for Biodiesel Production

A bioreactor is the device that controls the biological activity during the process of the production of biofuels. Different types of reactors such as chemical reactors, fusion reactors and nuclear reactors are selected based on the process [47]. Biodiesel production is generally carried out with batch reactors, semicontinuous flow reactors and continuous flow reactors.

Batch reactors are generally preferred for inexpensive processes in which less investment and capital are required with flexibility in operations for varieties in feedstock type, composition and productivity with the variation of product quality at intensive labour and energy requirements. In laboratory, a batch reactor is considered suitable for low-quantity materials as it involves time for charging and discharging [48]. It gives high-value products at low production quantities. It is considered suitable for a variety of products, such as in the case of pharmaceuticals. Generally, for high production quantities with less product, a continuous type of reactor is used. Literature review suggests to use the static reactor mixer reactor for the processing of biodiesel.

It consists of a simple tank equipped with an agitator that is operated when the tank is filled with the reactant for the process for a period of time. The content of the reactor is drained out for further processing. The important characteristic is that it start with unreacted material, which reacts and converts into the reacted material [49]. It is considered suitable for small biodiesel production plants. Figure 9.2 shows the batch reactor used for the production of bioreactor. For large production, the batch reactor is considered unsuitable due to the physical size of the reactor.

A semi-continuous flow reactor is almost similar to a batch process in which production begins at a small volume as compared to the size of the vessel with the addition of ingredients until the vessel becomes full; therefore, this process is costly due to time consumption and high labour cost, so it is commonly not used in the production of biodiesel [50]. Figure 9.3 indicates the semi-continuous flow reactor used for the production of biodiesel. Mixing of the reactant leads to uncontrollable temperature rises as in the batch reactor, which is controlled by slowly adding the reactant that controls the reaction rate and the corresponding temperature maintains good heat transfer.

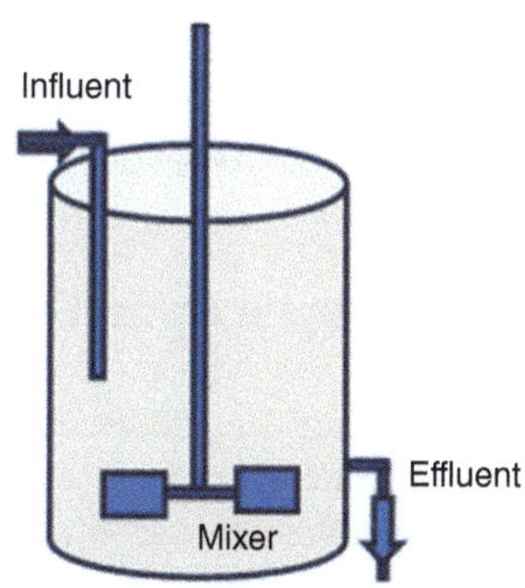

Figure 9.2 Batch reactor.
Source: Adapted from [49].

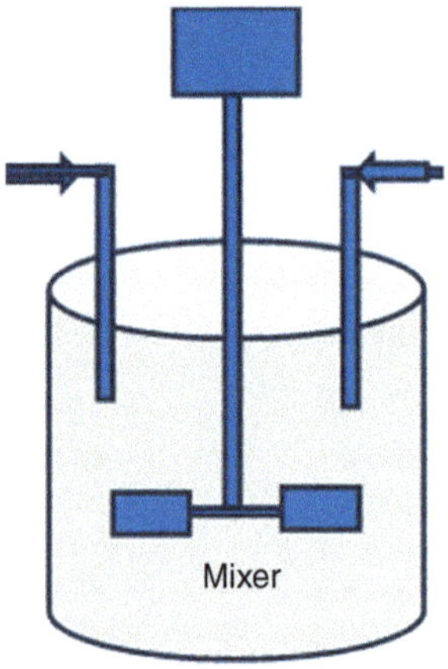

Figure 9.3 Semi-continuous flow reactor. *Source:* Adapted from [50].

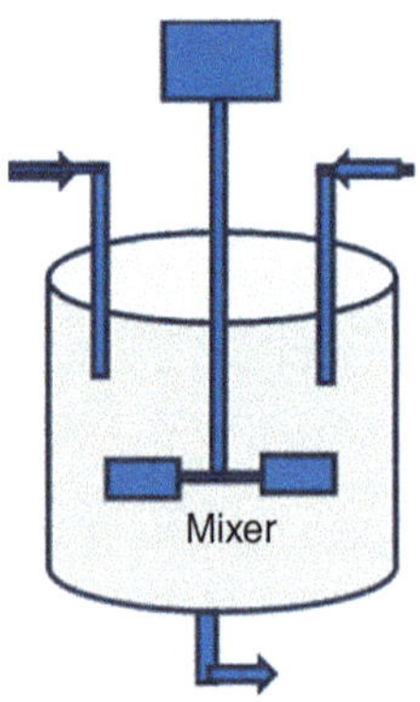

Figure 9.4 Continuous flow reactor. *Source:* Adapted from [49].

Continuous transesterification processes provide best results compared to batch processes as they require larger scale production due to constant product quality at lower operating costs [48]. The most suitable is the continuous flow reactor in which the continuous stirred tank reactor (CSTR) is commonly used for the production of biofuels. It is used for commercial purpose along with the ultrasonic and supercritical reactors.

Biodiesel production is mostly performed with the help of CSTR. It is identical to the batch reactor but equipped with necessary monitoring devices to control the reactor setup in the case of a continuous flow system. As shown in the figure, the continuous flow reactor reactant is continuously added and the product that consists of the mixture of different chemicals along with unreacted reactants is continuously discharged from the bottom [49]. Figure 9.4 indicates the continuous flow reactor used for the production of biodiesel. Suitable agitation is required for uniform chemical composition, and maintaining the temperature for process control and monitoring of product quality are performed during the process.

9.8 Ultrasonic Biodiesel Reactors

The importance of ultrasound is to form a mixture of liquids that tend to separate. Sufficient mixing is required for the production of biodiesel so that enough mixing between the biomass and alcohol occurs more effectively at the start of the reaction. Ultrasonic waves promote for more mixing so that the reaction rate can reach a high value [47]. The ultrasound waves transfer the energy to the mixture and create the vibration, which causes cavitation bubbles. When these bubbles burst quickly, fluid contracts and the constituents are mixed in the region of bubbles, resulting in an increase in the reactivity without increasing the temperature and energy.

9.9 Supercritical Reactors

The supercritical process is the catalyst-free process in the transesterification as in the normal production of biodiesel, catalysts, such as sodium or potassium hydroxide, are used, which are removed after reaction to improve the quality of fuel. The supercritical state is expensive as it demands very high temperature and pressure [48]. The reaction take place quickly due to a large reactor size, which is beneficial to the large biodiesel producers as it is cost effective.

9.10 Static Mixers as Biodiesel Reactors

It is the simple device in which spiral-shaped internal parts in the shape of a pipe create the turbulent flow. It is easy to use due to the absence of moving parts and effective mixing of liquids. As in the production of biodiesel, the initial mixing is difficult due to the solubility of alcohol in biomass [48]. The static mixers help to mix the reactant before processing [49].

9.11 Reactive Distillation

Chemical reaction and product separation occur concurrently in one unit. It is effective due to the combination of both the reactor and separation units. Techniques remove the reacted product from the reaction zone, which prevents the reaction from reversing and improves the overall conversion rate. It consists of many chambers arranged in an alternative manner, which opens from first to the second, second to the third and so on. The mixture with the ingredients is transferred from one chamber to the next, and the reaction progresses until the last chamber where the reaction is completed. It facilitates the large hold-up of liquid, which increases the retention time [50]. Due to the complex process, generally, it is not preferred for the commercial production of biodiesel; however, this complexity can be reduced with certain modifications [51].

9.12 Capital Cost and Performance Analysis of Reactors

The overall cost of the reactor is analysed with the consideration of the combined effects of different parameters, such as the medium in which the reaction occurs and temperatures depend on conditions of bioreactor cultivation [52]. The evaluation of the direct costs is performed on the cost of raw materials involved and their uses in the bioproducts. The cost evaluation is used to compare the bioreactor operational conditions.

9.13 Summary

High production cost, low product efficiency and long reaction time have been the major challenges in biodiesel synthesis. Recently, at a larger scale, various reactors and low-cost catalysts have been designed to obtain maximum benefits. Batch-type, continuous-type stirred, plug flow and packed bed bioreactors are some of the examples of reactors. Depending on the type of catalyst, reactor used, energy consumption and feedstock used, a right combination of bioreactors is chosen for obtaining a higher biodiesel yield at a lower cost.

References

1 Kumari, A., Mahapatra, P., Garlapati, V.K., and Banerjee, R. (2009). Enzymatic transesterification of Jatropha oil. *Biotechnol. Biofuels* 2009: 1. https://doi.org/10.1186/1754-6834-2-1.

2 Jeevan Kumar, S.P., Vijay Kumar, G., Dash, A. et al. (2017). Sustainable green solvents and techniques for lipid extraction from microalgae: a review. *Algal Res.* 21: 138–147. https://doi.org/10.1016/j.algal.2016.11.014.

3 Beebe, D.J., Mensing, G.A., and Walker, G.M. (2002). Physics and applications of microfluidics in biology. *Annu. Rev. Biomed. Eng.* 4: 261–286. https://doi.org/10.1146/annurev.bioeng.4.112601.125916.

4 Jeevan Kumar, S.P. and Banerjee, R. (2019). Enhanced lipid extraction from oleaginous yeast biomass using ultrasound assisted extraction: a greener and scalable process. *Ultrason. Sonochem.* 52: 25–32. https://doi.org/10.1016/j.ultsonch.2018.08.003.

5 Dhutekar, P., Mehta, G., Modak, J. et al. (2021). Establishment of mathematical model for minimization of human energy in a plastic moulding operation. *Mater. Today Proc.* 47: 4502–4507. https://doi.org/10.1016/j.matpr.2021.05.330.

6 Garlapati, V.K., Kant, R., Kumari, A. et al. (2013). Lipase mediated transesterification of Simarouba glauca oil: a new feedstock for biodiesel production. *Sustain. Chem. Process.* 1 (1): 1–6. https://doi.org/10.1186/2043-7129-1-11.

7 Funke, A. and Ziegler, F. (2010). Hydrothermal carbonization of biomass: a summary and discussion of chemical mechanisms for process engineering. *Biofuels Bioprod. Biorefin.* 4 (2): 160–177. https://doi.org/10.1002/bbb.198.

8 Lu, X., Pellechia, P.J., Flora, J.R.V., and Berge, N.D. (2013). Influence of reaction time and temperature on product formation and characteristics associated with the hydrothermal carbonization of cellulose. *Bioresour. Technol.* 138: 180–190. https://doi.org/10.1016/j.biortech.2013.03.163.

9 Dunne, P.W., Munn, A.S., Starkey, C.L. et al. (2015). Continuous-flow hydrothermal synthesis for the production of inorganic nanomaterials. *Philos. Trans. R Soc. A Math. Phys. Eng. Sci.* 373 (2057): 20150015. https://doi.org/10.1098/rsta.2015.0015.

10 Belkhode, P.N., Shelare, S.D., Sakhale, C.N. et al. (2021). Performance analysis of roof collector used in the solar updraft tower. *Sustain. Energy Technol. Assess.* 48: 101619. https://doi.org/10.1016/j.seta.2021.101619.

11 Pant, D., Misra, S., Nizami, A.S. et al. (2019). Towards the development of a biobased economy in Europe and India. *Crit. Rev. Biotechnol.* 39 (6): 779–799. https://doi.org/10.1080/07388551.2019.1618787.

12 Minowa, T., Kondo, T., and Sudirjo, S.T. (1998). Thermochemical liquefaction of Indonesian biomass residues. *Biomass Bioenergy* 14 (5–6): 517–524. https://doi.org/10.1016/S0961-9534(98)00006-3.

13 Demirbaş, A. (1997). Calculation of higher heating values of biomass fuels. *Fuel* 77 (9–10): 1117–1120. https://doi.org/10.1016/S0016-2361(97)85520-2.

14 Parshetti, G.K., Chowdhury, S., and Balasubramanian, R. (2014). Hydrothermal conversion of urban food waste to chars for removal of textile dyes from contaminated waters. *Bioresour. Technol.* 161: 310–319. https://doi.org/10.1016/j.biortech.2014.03.087.

15 Miedema, J.H., Benders, R.M.J., Moll, H.C., and Pierie, F. (2017). Renew, reduce or become more efficient? The climate contribution of biomass co-combustion in a coal-fired power plant. *Appl. Energy* 187: 873–885. https://doi.org/10.1016/j.apenergy.2016.11.033.

16 Huang, Y.F., Kuan, W.H., Lo, S.L., and Lin, C.F. (2010). Hydrogen-rich fuel gas from rice straw via microwave-induced pyrolysis. *Bioresour. Technol.* 101 (6): 1968–1973. https://doi.org/10.1016/j.biortech.2009.09.073.

17 Wu, T.N. (2008). Environmental perspectives of microwave applications as remedial alternatives: review. *Pract. Period Hazardous Toxic Radioact. Waste Manage.* 12 (2): 102–115. https://doi.org/10.1061/(ASCE)1090-025X(2008)12:2(102).

18 Berge, N.D., Li, L., Flora, J.R.V., and Ro, K.S. (2015). Assessing the environmental impact of energy production from hydrochar generated via hydrothermal carbonization of food wastes. *Waste Manag.* 43: 203–217. https://doi.org/10.1016/j.wasman.2015.04.029.

19 Dillen, S.Y., Djomo, S.N., Al Afas, N. et al. (2013). Biomass yield and energy balance of a short-rotation poplar coppice with multiple clones on degraded land during 16 years. *Biomass Bioenergy* 56: 157–165. https://doi.org/10.1016/j.biombioe.2013.04.019.

20 Demir-Cakan, R., Baccile, N., Antonietti, M., and Titirici, M.M. (2009). Carboxylate-rich carbonaceous materials via one-step hydrothermal carbonization of glucose in the presence of acrylic acid. *Chem. Mater.* 21 (3): 484–490. https://doi.org/10.1021/cm802141h.

21 Dai, L., He, C., Wang, Y. et al. (2017). Comparative study on microwave and conventional hydrothermal pretreatment of bamboo sawdust: Hydrochar properties and its pyrolysis behaviors. *Energ. Conver. Manage.* 146: 1–7. https://doi.org/10.1016/j.enconman.2017.05.007.

22 Zhang, J., An, Y., Borrion, A. et al. (2018). Process characteristics for microwave assisted hydrothermal carbonization of cellulose. *Bioresour. Technol.* 259: 91–98. https://doi.org/10.1016/j.biortech.2018.03.010.

23 Shelare, S., Kumar, R., and Khope, P. (2021). Assessment of physical, frictional and aerodynamic properties of Charoli (buchanania Lanzan Spreng) nut as potentials for development of processing machines. *Carpathian J. Food Sci. Technol.* 13 (2): 174–191. https://doi.org/10.34302/crpjfst/2021.13.2.16.

24 Sparks, D., Smith, R., Riley, D. et al. (2010). Monitoring and blending biofuels using a microfluidic sensor. *J. ASTM Int.* 7 (8): 1–9. https://doi.org/10.1520/JAI102473.

25 Saxenaa, K., Jaina, S., Sharmab, D. et al. (2014). Applications of integrated microfluidic devices in environmental monitoring: a review. *J. Energy Environ. Sci. Phot.* 2014 (128): 521–530.

26 Christoffersson, J., Bergstrom, G., Schwanke, K. et al. (2016). A microfluidic bioreactor for toxicity testing of stem cell derived 3D cardiac bodies. *Methods Mol. Biol.* 159–168. https://doi.org/10.1007/7651_2016_340.

27 Chintagunta, A.D., Ray, S., and Banerjee, R. (2017). An integrated bioprocess for bioethanol and biomanure production from pineapple leaf waste. *J. Clean. Prod.* 165: 1508–1516. https://doi.org/10.1016/j.jclepro.2017.07.179.

28 Waghmare, S.N., Shelare, S.D., Tembhurkar, C.K., and Jawalekar, S.B. (2020). Pyrolysis system for environment-friendly conversion of plastic waste into fuel. *Lect. Notes Mech. Eng.* 131–138. https://doi.org/10.1007/978-981-15-4748-5_13.

29 Pasirayi, G., Auger, V., Scott, S.M. et al. (2012). Microfluidic bioreactors for cell culturing: a review. *Micro Nanosyst.* 3: 137–160. https://doi.org/10.2174/1876402911103020137.

30 Malahat, S., Iovenitti, P.G., and Sbarski, I. (2010). Influence of tool fabrication process on characteristics of hot embossed polymer microfluidic chips for electrospray. *Microsyst. Technol.* 16 (12): 2075–2085. https://doi.org/10.1007/s00542-010-1135-4.

31 Das, T. and Chakraborty, S. (2009). Biomicrofluidics: recent trends and future challenges. *Sadhana Acad. Proc. Eng. Sci.* 34 (4): 573–590. https://doi.org/10.1007/s12046-009-0035-8.

32 Han, A., Hou, H., Li, L. et al. (2013). Microfabricated devices in microbial bioenergy sciences. *Trends Biotechnol.* 31 (4): 225–232. https://doi.org/10.1016/j.tibtech.2012.12.002.

33 Won, W., Motagamwala, A.H., Dumesic, J.A., and Maravelias, C.T. (2017). A co-solvent hydrolysis strategy for the production of biofuels: process synthesis and technoeconomic analysis. *React. Chem. Eng.* 2 (3): 397–405. https://doi.org/10.1039/c6re00227g.

34 Gautam, R. and Vinu, R. (2020). Reaction engineering and kinetics of algae conversion to biofuels and chemicals: via pyrolysis and hydrothermal liquefaction. *React. Chem. Eng.* 5: 1320–1373. https://doi.org/10.1039/d0re00084a.

35 Thatikayala, D., Pant, D., and Min, B. (2021). A mesoporous silica-supported CeO_2/cellulose cathode catalyst for efficient bioelectrochemical reduction of inorganic carbon to biofuels. *React. Chem. Eng.* 6 (10): 1993–2001. https://doi.org/10.1039/d1re00166c.

36 Lukić, M. and Vrsaljko, D. (2021). Effect of channel dimension on biodiesel yield in millireactors produced by stereolithography. *Int. J. Green Energy* 18 (2): 156–165. https://doi.org/10.1080/15435075.2020.1831513.

37 Lee, H.L.T., Boccazzi, P., Ram, R.J., and Sinskey, A.J. (2006). Microbioreactor arrays with integrated mixers and fluid injectors for high-throughput experimentation with pH and dissolved oxygen control. *Lab Chip* 6 (9): 1229–1235. https://doi.org/10.1039/b608014f.

38 Sasaki, K., Sasaki, D., Tsuge, Y. et al. (2021). Enhanced methane production from cellulose using a two-stage process involving a bioelectrochemical system and a fixed film reactor. *Biotechnol. Biofuels* 14 (1): 1–2. https://doi.org/10.1186/s13068-020-01866-x.

39 Kadi, M.A., Akkouche, N., Awad, S. et al. (2019). Kinetic study of transesterification using particle swarm optimization method. *Heliyon* 5 (8): https://doi.org/10.1016/j.heliyon.2019.e02146.

40 Belkhode, P.N., Ganvir, V.N., Shende, A.C., and Shelare, S.D. (2021). Utilization of waste transformer oil as a fuel in diesel engine. *Mater. Today Proc.* 49: 262–268. https://doi.org/10.1016/j.matpr.2021.02.008.

41 Losey, M.W., Schmnidt, M.A., and Jensen, K.F. (2001). Microfabricated multiphase packed-bed reactors: characterization of mass transfer and reactions. *Ind. Eng. Chem. Res.* 40 (12): 2555–2562. https://doi.org/10.1021/ie000523f.

42 Patil, A.D., Baral, S.S., Dhanke, P.B. et al. (2018). Parametric studies of methyl esters synthesis from Thumba seed oil using heterogeneous catalyst under conventional stirring and ultrasonic cavitation. *Mater. Sci. Energy Technol.* 1: 106–116. https://doi.org/10.1016/j.mset.2018.06.004.

43 Lee, H., Yang, W., Wei, X. et al. (2015). A microsized microbial fuel cell based biosensor for fast and sensitive detection of toxic substances in water. *Proc. IEEE Int. Conf. Micro Electro Mech. Syst.* 573–576. https://doi.org/10.1109/MEMSYS.2015.7051020.

44 Lutz, W.K. (1995). Take chemistry and physics into consideration in all phases of chemical plant design. *Process Saf. Prog.* 14 (3): 153–160. https://doi.org/10.1002/prs.680140304.

45 Ackrill, R. and Kay, A. (2015). Biofuels policy design and external implementation challenges. *Growth Biofuels 21st Century* 14 (3): 153–160. https://doi.org/10.1057/9781137307897.0014.

46 Wijakmatee, T., Hemra, N., Wongsakulphasatch, S. et al. (2021). Process intensification of biodiesel production with integrated microscale reactor and separator. *Chem. Eng. Process Process Intensif.* 164: https://doi.org/10.1016/j.cep.2021.108422.

47 Müller TE. Biodiesel Prod. Syst. Reactor Technol., 2019, 15–25, https://doi.org/10.1007/978-3-030-00985-4_2 Biodiesel Production Systems: Reactor Technologies. In: Tabatabaei, M., Aghbashlo, M. (eds) Biodiesel. Biofuel and Biorefinery Technologies, vol 8. Springer, Cham. https://doi.org/10.1007/978-3-030-00985-4_2.

48 Rahmat, B., Setiasih, I.S., and Kastaman, R. (2013). Biodiesel reactor design with glycerol separation to increase biodiesel production yield. *MAKARA J. Technol. Ser.* 17 (1): 11–16. https://doi.org/10.7454/mst.v17i1.1921.

49 Okoye, E.K., Edeh, C.P.C., Ezumezu, C.O., and Ejiogu, E.C. (2013). Biodiesel production from used cooking oil using controlled reactor plant. *IEEE AFRICON Conf.* 1–4. https://doi.org/10.1109/AFRCON.2013.6757725.

50 Benavides, P.T. and Diwekar, U. (2013). Studying various optimal control problems in biodiesel production in a batch reactor under uncertainty. *Fuel* 103: 585–592. https://doi.org/10.1016/j.fuel.2012.06.089.

51 Nugroho, A.T., Antariksawan, A.R., and Biyanto, T.R. (2019). Optimization to review capital cost of operation heat exchanger of kartini reactor. *AIP Conf. Proc.* 2088 (1): 020002. https://doi.org/10.1063/1.5095254.

52 Mata, T.M., Martins, A.A., and Caetano, N.S. (2010). Microalgae for biodiesel production and other applications: a review. *Renew. Sustain. Energy Rev.* 14: 217–232. https://doi.org/10.1016/j.rser.2009.07.020.

10

Modelling and Simulation to Predict the Performance of the Diesel Blends

10.1 Introduction

The chemical stability of oil is influenced by three factors: temperature, oxygen availability and the existence of catalysts [1]. Oil deterioration could be induced by the breakdown of hydrocarbons in oil at high temperatures. The oxygen concentration of insulating oil may cause an increase in acidity and the production of sludge. Catalysts such as iron and copper dissolve in oil while ageing and may hasten the procedure [2].

Because of the probable exhaustion and rising price of petroleum, as well as environmental problems generated by the burning of fossil fuels, the quest for alternative fuels has drawn a lot of attention [3]. The alternative fuel not only avoids the petroleum crises but also reduces pollutant gases emitted by an engine.

The waste transformer oil is efficient for operating at high temperature stresses. It is highly viscous and contains sludge with colour containing compound and oxidation product. First, the adsorption of these impurities from waste transformer oil is performed by using silica gel, and the product obtained after the adsorption is colourless and named as treated transformer oil [4], which shows the same properties as that of diesel fuel. Hence, the blends of various proportions, such as a 10% blend that is 10% treated transformer oil and 90% diesel fuel, likewise 20%, 25%, 30% and 40%, are made and then compared to the properties of blends with the pure diesel fuel [5]. The performance of engine that is fuelled with blends is evaluated in terms of the following values:

1) Brake thermal efficiency
2) Brake-specific fuel consumption

Various factors of the operations are identified so as to optimise the brake thermal efficiency and brake-specific fuel consumption. There are many approaches to develop/upgrade engine performance, such as method study, time study, etc. [6]. The experimental-data-based modelling approach is proposed to correlate the fuel and engine characteristic with the engine observation, such as brake thermal efficiency and brake-specific fuel consumption system [7].

Sustainability in Biofuel Production Technology, First Edition. Pratibha S. Agrawal, Pramod N. Belkhode, and Samuel Lalthazuala Rokhum.

10.2 Cause and Effect Relationships

The formulation of logic-based model correlating causes and effects is not possible for these types of complex phenomena. The only approach appropriate for this type of study of phenomenon is the experimental-data-based modelling [8]. The experimental-data-based model correlates the inputs or causes, in other words outputs of such activity by formulating the quantitative mathematical modelling. The indices of the causes of model, i.e. mathematical model, indicate the most influencing inputs [9]. Such correlation indicates the deficiency and the strength of the blende diesel system that helps to improve the performance of the brake thermal efficiency and brake-specific fuel consumption. Hence, for improving system/activity performance it is necessary to procedure such analytical cause–effect relationships conceptualised as experimental-data-based models.

10.3 Approach to Formulate

The main objective is to explain the approach to formulate the mathematical model for the man–machine system [10]. In order to form the mathematical model, the most critical activity – performance of diesel engine – is identified and studied. Diesel engine is tested with the pure diesel, and diesel blends in various proportions, such as a 10% blend that is 10% treated transformer oil and 90% diesel fuel, likewise 20%, 25%, 30% and 40%, are made and then compared to the properties of blends with pure diesel fuel. The formulation of mathematical model for brake thermal efficiency and brake-specific fuel consumption is selected as the case study. In this case, the approximate generalised mathematical models have been established applying the concepts of theories of experimentation for the brake thermal efficiency and brake-specific fuel consumption. The general procedure adopted is as follows.

1) Possibility of the formulation of model for improving engine performance. The engine performance is very difficult to analyse. Therefore, there is a need for developing a model that helps to improve brake thermal efficiency and brake-specific fuel consumption. Data is collected based on the sequence of engine performance by direct measurement. From this measurement, data input and output variables are decided, and the model is formed by forming dimensionless equation using regression analysis.
2) Possibility to validate the output and the model of the system – the aim is to find out the utility and effectiveness of the model. The effectiveness of the model is decided by artificial neural network (ANN) simulation, sensitivity analysis and optimisation technique.

10.4 Concept of Man–Machine System

The concept of experimental/field-data-based modelling with the help of examples comes across in our life. A detailed concept of the man–machine system is discussed with effects and causes [11]. Activity is mainly dominated by human beings, and its influence is explained. The accuracy of models depends on the magnitude of the curve-fitting constant, and the importance of indices has been explained.

In our life, we come across very many activities. These activities have some environmental system in which these activities take place [12]. The environment or system can be specified in terms of its parameters; some of which are always constant in their magnitudes, whereas some are variable. The activities are set in action by some parameters that are considered as causes. These causes interact with parameters of the system as a result of this interaction, and some effects are produced. The above-discussed matter in a diagrammatic form can be presented as shown in Figure 10.1.

Figure 10.1 shows one rectangular block within which is written activity along with its nature, i.e. activity may be totally physical or it may be a combination of human directed/operated physical activity stated as man–machine system or it may be an activity mainly dominated by human beings stated in the block as 'Totally Human System'. The functioning of an activity is influenced by two sets of parameters: one set is characterising the features of environment of an activity and the other set is planned parameters or causes that influence the functioning of the system [13].

Accordingly, Figure 10.1 shows these parameters. The parameters characterising the features of an environment are E1, E2, E3, E4, etc. Some of which are permanently fixed, such as E1 and E2, whereas remaining are time variant, such as E3 and E4, on which one has no control. The planned parameters are known as causes shown as A, B, C, D, E, etc., and effects of an activity are shown as Y1, Y2, Y3, Y4, etc.

If one analyses any activity of society, one would be able to identify the causes A to E, etc., system parameters E1 to E4, etc., and the effects Y1 to Y4, etc. This may be treated as the qualitative analysis of the societal activity. This can be demonstrated by one example of everyday life of man-machine system of our life.

Let us consider a gardener preparing flower beds in a kitchen garden of a conventional house of a slightly upper middle-class family.

Suppose the house owner has to prepare around six to eight flower beds. Each is 3 m in length, 1 m wide and 0.5 m in depth. The house owner instructs his gardener accordingly. Let us assume that gardener decides to start the work from a specific day along with his team of two to three helpers. The tools necessary for this operation are (i) a kudali (axe), (ii) a phawada (a spade) and (iii) a soil collector a ghamela. The work would take place in

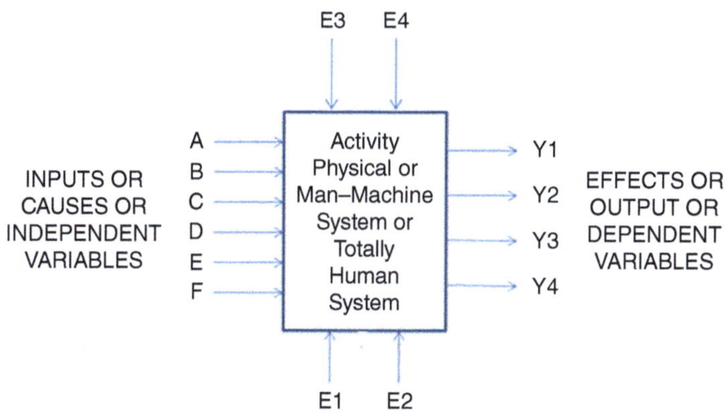

Figure 10.1 Block diagrammatic representation of an activity.

a shift of eight hours, i.e. since 8.00 a.m. in the morning till 5.00 p.m. in the late afternoon with a rest cum lunch break from 12.30 p.m. to 1.30 p.m. Let us assume that it is a team of three members with one supervisor performing this task. One member, e.g. A, digs the soil, B collects the dry soil with spade and puts it in ghamela, and the third member, e.g. C, carries the soil to the place where heap of soil is made.

The planned sequence of working is A dugs for 10 minutes with rest pause for three minutes at the end of every seven minutes. During this rest pause of three minutes, B collects this dug soil in ghamela and C carries this soil to the heap. Like this, A, B, C workers will work for every 30 minutes in a sequence. During these three minutes, the leader of the team takes down the observations of activity performed in seven minutes of digging and three minutes of carrying dug soil to heap and measuring the dug quantity in kgf. The measurements taken are as follows: (i) the initial pulse rate of A, B and C, (ii) the pulse rate at the end of seven minutes of digging for A, (iii) the pulse rates of B and C at the end of the 10th minute, (iv) the measurement of the rise in body temperature, (v) the measurement of soil dug in kgf at the end of every 10 minutes and (vi) the surface finish of sides of rectangular space created by A.

The complete set of observations of the above-listed parameters at the end of every 30 minutes would be recorded. Along with this record, the records of attitude and enthusiasm of workers would also be maintained during the beginning and end of 30 minutes. The specifications of the tools are recorded in terms of their geometry, weight, sharpening of digging point and edge of the spade. At the end of the shift on every day, a total of all causes and effects will be made. Referring to Figure 10.1 for this operational cause, A indicates anthropometric dimensions of operators including their numbers, B indicates experience and qualifications of operators, E indicates enthusiasm and attitude of operators, C indicates geometric dimensions of the tools, their weights and their conditions (i.e. sharpens of edges and tips), D indicates soil condition at the spot before digging, i.e. at the beginning of every 10 minutes and F indicates the time of operation, whereas E1 and E2 indicate the general features of kitchen garden, E3 indicates ambient temperature, pressure and humidity and E4 indicates noise level. E1 and E2 indicate constant parameters and E3 and E4 indicate extraneous variables of the system, i.e. activity under consideration in this case. Obviously, the responses of the activity could be total soil dug out in every 10 minutes, say Y1, human energy input in terms of the pulse rate or blood pressure rise, say Y2, and the quality of operation performed, say Y3.

Thus, in this case, one can say that E1 and E2 are constant parameters of the system, whereas E3 and E4 are system extraneous variables A, B, C, D, E and F, which are planned and/or actually unplanned but measured causes or inputs, and Y1, Y2, Y3 are responses/outputs of the activity [14].

In the above list of variables, there are some variables that are very difficult to measure, e.g. enthusiasm and attitude of the operator. These are categorised as abstract inputs or causes and sometimes effects, i.e. how the worker feels psychologically at the end of every 10 minutes. However, these abstract quantities can be measured using the concept of weightages. In short, all causes, constant parameters of the system, some of extraneous variables and effects can be quantified. In this case, one can arrive at the complete observation table of this activity. In the context of a completely physical system, it could have been specified as phenomenal or experimental observations.

Once these observations are ready, one can form the mathematical correlationships amongst (i) causes, (ii) effects and to some extent (iii) extraneous variables of the form as far as this concerned activity:

$$Y1 = K1(A)^{a1}(B)^{b1}(C)^{c1}(D)^{d1}(E1)^{e11}(E2)^{e21} \quad (10.1)$$

$$Y2 = K2(A)^{a2}(B)^{b2}(C)^{c2}(D)^{d2}(E1)^{e12}(E2)^{e22} \quad (10.2)$$

$$Y3 = K3(A)^{a3}(B)^{b3}(C)^{c3}(D)^{d3}(E1)^{e13}(E2)^{e23} \quad (10.3)$$

Equations (10.1)–(10.3) can be formed based on information presented and using the already available mathematical treatments, especially matrix algebra. All exponents of Eqs. (10.1)–(10.3) can be obtained. The quantities K1, K2 and K3 are known as curve-fitting constants. They collectively represent what is not identified as causes logically and/or as a result of action of extraneous variables quantitatively but collectively.

Once these Eqs. (10.1)–(10.3) are formed, they can then be considered as a design tool for planning similar activity in the future and not necessarily with the same conditions but could be applicable for variations of all causes and constant parameters of the system within the 80–120% range of their variation. Eqs. (10.1)–(10.3) can be considered as field-data-based models, the concept first launched by one of the authors of this book, Dr. J. P. Modak, in the body of knowledge. It is a field-data-based model because it is formed based on actual field studies. The approach is strongly advocated for any man–machine system of our life. The concept of field-data-based modelling is applied in several man–machine systems, such as (i) human assembly operation, (ii) sewing machine operations and (iii) agricultural weeding operation, street noise, face drilling in coal mines, pharmaceutical industry, heavy-duty press operation, cotton-ginning operation, civil construction activities, factory maintenance jobs, floor mill operation and plastic industry operation [15–24].

The accuracy of models is dependent on the magnitude of curve-fitting constants K1, K2 and K3. Ideally, if these are numerically 1, then the model very rightly simulates the man–machine system. If it is too low, the causes are overestimated; if it is too high, the causes are underestimated. This would decide when to repeat the investigation again or to refine the approach in subsequent attempts. The magnitude of exponents of the causes, i.e. the quantities on the right-hand side of Eqs. (10.1)–(10.3), indicates the degree of influence of these causes on the specific response. The ratio of indices indicates the relative influences of two causes, i.e. if the exponent of B is higher than that of C, then the interpretation would be that the qualification and experience of the operator are more influential than that of the quality of tools used.

The approach is extremely important for technologists working in industries. Presently, for some industry while designing a new plant for the product similar to the one being manufactured, the design approach is normally based on the opinion of individual group members. It goes on like this at the first meeting of seniors some members may opine saying, ‘I think it is sophisticated equipment and is more influential than the qualities of human resource'. A model is formulated for functioning of their present setup, if available, and then that model itself will indicate what is more influential for productivity, product quality, production turnover or getting low maintenance cost.

This approach is similar to Dr. Modak's approach, who has a very rich experience of applying an approach of 'Theories of Engineering Experimentation', by H. Schenck Jr., in the formulation of experimental-data-based models for many manufacturing process machines energised by Human Powered Flywheel Motor HPFM. These process machines are (i) rectangular sectioned brick extruders, (ii) keyed sectioned brick extruders, (iii) wood turning, (iv) low head water lifting, (v) algal formation machine, (vi) chaff cutting, (vii) fertiliser ingredient mixing, (viii) ballets making, (ix) food grain crushing, (x) concrete ingredient mixing, (xi) electricity generation and (xii) ground nut crushing for manufacturing ground nut oil. Dr. Modak and his research scholars applied this approach of experimental-data-based modelling to these process machines energised by HPFM and also two subsystems, i.e. energy source HPFM itself and its mechanical power transmission [15–24].

HPFM stands for human powered flywheel motor. This comprises (i) a bicycle-like peddling system, (ii) speed-increasing gear pair with a speed rise ratio $G = 2.5$–30 and (iii) a fairly big size flywheel of about 1 m rim diameter, 10.0 cm rim width and 2.0 cm rim thickness [25, 26].

A young boy in the age group of 20–25 years, with slim stature and a height of about 160–165 cm, spins the flywheel at a speed of 800–1000 RPM in a minute's time. Thereby, around 40 000 N m energy is stored in the flywheel. Then, a specially designed torsionally flexible clutch (TFC) is engaged, and this flywheel is connected to the process units mentioned earlier.

Upon engaging the clutch, instantaneous energy and momentum transfer takes place from the high-speed spinning flywheel to the input shaft of the processor, which results in the manufacturing of the product meant by that processor, i.e. bricks in the case of clay extruder so on and so forth.

The energy utilisation time is in the range of 5–15 seconds depending on the average process resistance of the process units. This means that the process units need processes around 11.0–4.0 hp., and when the product quality is not getting affected by continuously changing the rotating speed of the input shaft of the processor, they can be energised by such a non-conventional energy source HPFM.

The functional feasibility and economic viability of this novel machine system are established. All the three subsystems of this machine system, i.e. (i) HPFM of the energy source, (ii) process units and (iii) hence the intermediate system, i.e. TFC and torque amplification gear pair G, operate all the while in a highly transient state.

It becomes very difficult to evolve logic-based design data for such a machine system. Hence, the only approach leftover was to generate experimental-data-based design data. In this approach, all the independent quantities should be experimentally varied over a wide possible range to collect the response data, and based on this collected data 'experimental-data-based mathematical models' are formulated for all the subsystems. Obviously, these mathematical models can now work as design data for such systems [27–30].

The approach of experimental-data-based modelling is deduced based on the application of the approach proposed in the book *Theories of Engineering Experimentation* [21].

The stepwise approach is as follows:

i) Identify all causes, responses and extraneous variables of the phenomena. This can be performed by the qualitative analysis of phenomena.
ii) Treat causes as inputs/independent variables.
iii) Treat responses as outputs/dependent variables.
iv) Combine independent variables by dimensional analysis to reduce them in size. This gives independent Π terms.
v) Decide test envelopes, test points and test sequence for all independent Π terms.
vi) Design and fabricate the experimental set-up.
vii) Execute experimentation as per its planning.
viii) Collect the experimental data. Upon obtaining experimental data regarding causes/ inputs, independent variables and similar data for responses, formulate the mathematical models. These models can work as design data.

Based on the above discussion, a totally physically system is operated in a highly transient state, and logic-based modelling is highly importable for such systems. However, what about man–machine systems?

For all man–machine systems, the approach of preparing flower beds is only applicable as mentioned earlier. This approach is similar to the experimental-data-based modelling approach. This is true, but the design of the experimentation phase of this approach is not possible for man–machine systems. The field activities should be allowed as planned by their administrators.

Appropriate sites, where activities take place for a specific objective should be carefully selected, but with variations of all inputs/causes/independent variables and hence the corresponding variations of effects/outputs/dependent variables [31, 32].

Upon obtaining this information from mathematical models, this approach is conceptualised as field-data-based modelling for all man–machine systems. Regarding the formulation of the model, there are no other substitutes but to adopt the methodology of experimentation, more suitably the one suggested by Hilbert. This method is applied to a complex totally physical phenomenon for which logic-based modelling is highly improbable. The inputs to the phenomenon, the outputs and the extraneous variables are identified. The inputs are experimentally varied over a broad yet practically possible range, and response data is collected. The curve-fitting constant of the model collectively represents extraneous variables that affect the phenomena but cannot be measured. On the basis of the gathered data, the models are formed.

10.5 Formulation of the Mathematical Model

Evaluation of engine performance activity using the pure diesel and various proportions of a diesel blend is a complex phenomenon. There are many factors, such as load on engine, blend, flashpoint, aniline point, kinematic viscosity, density, API gravity, diesel index, cetane number, calorific value, time, mass of fuel, bore diameter, stroke length, cubic capacity, fuel tank capacity, engine speed, that affect the performance of diesel engine [4]. To study the interaction of these independent variables on the dependent variables, such as brake thermal efficiency and brake-specific fuel consumption, in engine performance, an experimental-data-based modelling approach is proposed.

The experimental studies to be carried out for such investigations need proper planning. Normally, a large number of variables are involved in such experimental studies. It is expected that the influence of all the variables and parameters should be studied economically without sacrificing the accuracy by reducing the number of variables to a few dimensionless terms through the technique of dimensional analysis [33, 34]. Theory of engineering experimentation is a study of scientific phenomenon that includes the analysis and synthesis of the scientific phenomenon. The study may include a theoretical approach and an experimental approach.

In the theoretical approach, the laws of mechanics and physics are applied, which include (i) force balance, (ii) momentum balance, (iii) energy balance, and (iv) quantity balance. Effects of several independent variables on complex process are studied to formulate the process. In the experimental approach, various steps are involved in formulating the model for such a complex phenomenon as discussed below.

10.5.1 Identify the Causes and Effects

Performing qualitative analysis of the process to identify various physical quantities. These are the causes (inputs). Experiments take more time and become complex if a large number of independent variables are involved. By deducing dimensional equation from the phenomenon, numbers of variables are reduced.

10.5.2 Perform Test Planning

This involves the selection of test envelope, test points, test sequence, and plan of experimentation [35–39].

i) **Test envelope**: To decide the range of variation of individual independent π terms.
ii) **Test points**: To decide and specify values of independent π terms at which the experimental set-up is set during experimentation.
iii) **Test sequence**: To decide the sequence in which the test points are set during experimentation. Sequence may be in ascending order, descending order or random order. Usually, ascending or descending order, depending on the nature of the phenomenon, is adopted for irreversible experiments, while random order is adopted for reversible experiments.
iv) **Plan of experimentation**: In the planning, the decision is made regarding how to vary the independent variable. Planning may be the classical plan or factorial plan. In the classical plan, only one variable is varied at a time maintaining all other variables at a constant level. In the factorial plan, more than one or all independent variables are varied at a time.

10.5.3 Physical Design of an Experimental Set-up

Here, it is necessary to figure out the physical design of an experimental set-up, including deciding specifications and procurement of instrumentation and experimentation. The next step is to execute experimentation as per test planning. This generates experimental data regarding causes (inputs) and effects (responses).

10.5.4 Checking and Rejection of Test Data

Based on the experimental results, it is necessary to check the set-up for its reliability. The erroneous data is identified and removed from the gathered data; for this purpose, some statistics-based rules are adopted.

10.5.5 Formulation of the Model

A quantitative relationship is formulated in terms of the dimensional equation between the dependent and independent π terms. This establishes the relationship between outputs (effects) and inputs (causes).

10.6 Limitations of Adopting the Experimental Database Model

For experimental systems in complex activities, unfortunately in many such systems, the test planning component of the experimental technique is not possible to implement [40–42]. For a variety of reasons, it is necessary to enable the activity to proceed as planned. This happens when one wishes to formulate a model for any activity in engine performance, civil construction activities, human assembly operations, industry manufacturing activities, etc.

However, the links between the dependent and independent factors are defined qualitatively on the basis of the existing published studies, and the generalised quantitative correlations are not always understood. As a result, establishing a quantitative relationship is impossible. Since there is no potential of developing a scientific model (logic-based), the only option is to develop a fact-based or, to be more exact, experimental-data-based model. As a result, it is recommended that such a model should be developed in the current study.

The technique given by Schenck H. Jr. for building generalised experimental-data-based models has been presented in the current investigation, which entails the following stages [9, 27, 43]:

- Considering a phenomenon
- Identification of variables or parameters affecting the phenomenon
- Reduction of variables through dimensional analysis
- Selecting a sufficient number of cities with variations in causes and extraneous variables
- Executing experimental work for data collection
- Rejection of absurd data
- Formulation of the model

Depending upon that filtered data, a quantitative link between dependent and independent components of the dimensional equation must be developed.

10.7 Identification of Causes and Effects of an Activity

Identification of causes and effects is the first step. These causes and effects vary as time elapses. Selection of variables such as dependent and independent, as the phenomena take place, should be performed based on the observed qualitative analysis of the process.

These variables are of four types as listed below:

1) Independent variables
   Independent variables can be defined as causes of an activity that influences the activity. This can be changed independently of other variables of the activity.
2) Dependent variables
   The phenomenal quantity or parameters that change due to the change in the values of the independent variables are called the response or the dependent variables.
3) Extraneous variables
   Any parameter that influences the process, but its magnitude cannot be changed or altered at our wish, such as ambient pressure, humidity, temperature, operator-related parameters such as enthusiasm, attitude, etc.
4) Controlled variables
   Controlled variables are the phenomenal quantities that remain constant all through the duration of the activity; basically, these are independent variables, but due to practical reasons they are not alterable, for example, acceleration due to gravity.

10.8 Dimensional Analysis

The dimensional analysis was primarily used as an experimental tool to combine any experimental variable into one (or to reduce the number of experimental variables); this technique was then mainly applied in the fluid mechanics and heat transfer for almost all experiments [44–46].

a) Dimensions/quantities
   There are two types of quantities or dimensions.
   i) Fundamental quantities or fundamental units
      Mass (M), length (L) and time (T) are three constant dimensions. If heat is involved, then temperature (θ) is also taken as a fundamental quantity.
   ii) Derived dimensions (secondary quantities)
      If the physical quantity is designed based on some fundamental aspects, then it is known as a derived quantity. The derived quantities are classified as shown in Tables 10.1 and 10.2.

10.8.1 Dimensional Equation

If each quantity is represented by its dimensional formula in an equation with a physical quantity, the resulting equation is known as the dimensional equation [47].

$$\text{Kinetic energy} = \frac{1}{2} mv^2$$

Here, m is the mass of the body and v is the velocity. Writing the formula for kinetic energy in the dimensional equation form, we have

$$[\mathrm{M}] \times [\mathrm{MLT}^{-1}]^2 = [\mathrm{ML}^2\mathrm{T}^{-2}]$$

The above equation is known as the dimensional equation.

Table 10.1 Types of the derived quantities.

S.N.	Variables/constants	Distinguishing feature of the physical quantity	Example
1	Dimensional variables	Have dimension but no physical value	Force, velocity, power
2	Dimensional variables	Neither have dimension nor fixed value	Specific gravity
3	Dimensional Constants	Have fixed dimensions and fixed values	Gravitational constant
4	Dimensional Constants	Have no dimensions but have fixed values	1, 2, 3

The connection between the dependent and independent variables is evaluated in terms of its fundamental dimensions in the dimensional analysis of a physical phenomenon to obtain insights into the underlying link between the dimensionless factors that influence the event. There are numerous methods to reduce the complexity of dimensionless parameters. The most commonly used methods are Rayleigh's method and the Buckingham pi theorem method. Sometimes even by observations and preliminary qualitative analysis, a dimensional equation can be formed.

10.8.2 Rayleigh's Method

Let *Y* be an independent variable that depends on x_1, x_2, x_3, x_4, etc. According to Rayleigh's method, *Y* is a function of x_1, x_2, x_3, x_4, . . ., etc., and mathematically it can be written as

$$Y = f\left(x_1, x_2, x_3, x_4\right)$$

The equation can be written as follows:

$$\mathrm{Y} = \mathrm{k}\,\mathrm{x}_1^{\mathrm{a}}, \mathrm{x}_2^{\mathrm{b}}, \mathrm{x}_3^{\mathrm{c}}, \mathrm{x}_4^{\mathrm{d}}$$

where *k* is a constant, and *a*, *b*, *c* and *d* are arbitrary indices. The values of *a*, *b*, *c* and *d* are obtained by comparing the powers of the fundamental dimension on both sides.

10.8.2.1 Buckingham Pi theorem method

The Buckingham pi theorem asserts that if the factors determining a physical event have *m* main dimensions, then the phenomenon may be represented by $(n - m)$ independent dimensionless groups [25, 30, 48–52]. This theorem can be used for reducing the number of variables affecting the process. According to the theorem, if an equation is dimensionally homogeneous, it may be reduced to a connection between a full set of dimensionless products. In this procedure, *m* numbers of repeating factors are chosen, and dimensional lower groups are generated at a time by each of the remaining variables. Rayleigh's approach is also known as the repeating variable method. The recurring

Table 10.2 Dimensional formulae of the derived quantities.

Sl. No	Physical quantity	Relation with other pi quantities	Dimensional formula	SI unit
1	Area	Length × breadth	$[L]\times[L] = [L^2]$	m^2
2	Volume	Length × breadth × Height	$[L]\times[L]\times[L] = [L^3]$	m^3
3	Density	Mass/volume	$[M]/[L^3] = [M^1L^{-3}\,T^0]$	$kg\ m^{-3}$
4	Velocity	Distance/time	$[L]/[T] = [M^0L^1T^{-1}]$	$m\ s^{-1}$
5	Acceleration	Velocity/time	$[M^0L^1T^{-1}]/[T] = [M^0L^1T^{-2}]$	$m\ s^{-2}$
6	Force	Mass × acceleration	$[M]x[M^1L^1T^{-2}] = [M^1L^1L^{-2}]$	N
7	Momentum	Mass × velocity	$[M]x[M^0L^1T^{-1}] = [M^1L^1T^{-1}]$	$kg\ m\ s^{-1}$
8	Work	Force × distance	$[MLT^{-2}]\times[L] = [ML^2T^{-2}]$	N m
9	Power	Work/time	$[ML^2T^{-2}]/[T] = [ML^2T^{-3}]$	W
10	Pressure	Force/area	$[MLT^{-2}]/[L^2] = [ML^{-1}\,T^{-2}]$	$N\ m^{-2}$
11	Kinetic energy	½ × Mass × (velocity)^2	$[M]x[MLT^{-1}]^2 = [ML^2T^{-2}]$	N m
12	Potential energy	Mass × g × distance	$[M]x[M^0LT^{-2}]\times[L] = [ML^2T^{-2}]$	N m
13	Impulse	Force × time	$[MLT^{-2}]\times[T] = [MLT^{-1}]$	N s
14	Torque	Force × distance	$[MLT^{-2}]\times[L] = [ML^2T^{-2}]$	N m
15	Stress	Force/area	$[MLT^{-2}]/[L^2] = [ML^{-1}\,T^{-2}]$	$N\ m^{-2}$
16	Strain	Extension in length/ original length	$[L]/[L] = [M^0L^0T^0]$	Number
17	Elasticity	Stress/strain	$[ML^{-1}\,T^{-2}]/[M^0L^0T^0] = [ML^{-1}\,T^{-2}]$	$N\ m^{-2}$
18	Surface tension	Force/length	$[MLT^{-2}]/[L^1] = [ML^0T^{-2}]$	$N\ m^{-1}$
19	Force constant of spring	Applied force/extension in length	$[MLT^{-2}]/[L^1] = [ML^0T^{-2}]$	$N\ m^{-1}$
20	Gravitational constant	Force × (distance)^2/ (mass)^2	$[MLT^{-2}]\times[L^2]/[M^2] = [M^{-1}L^3T^{-2}]$	$N\ m^2\ kg^{-2}$
21	Frequency	1/time period	$1/[T] = [M^0L^0T^{-1}]$	s^{-1}
22	Angle	Arc/radius	$[L]/[L] = [M^0L^0T^0]$	rad
23	Angular velocity	Angle/time	$[M^0L^0T^0]/[T] = [M^0L^0T^{-1}]$	$rad\ s^{-1}$
24	Angular acceleration	Angular velocity/time	$[M^0L^0T^{-1}]/[T] = [M^0L^0T^{-2}]$	$rad\ s^{-2}$
25	Moment of inertia	Mass × (distance)^2	$[M]/[L^2] = [M^1L^{-2}\,T^0]$	$kg\ m^2$
26	Angular momentum	Moment of inertia × angular velocity	$[M^1L^{-2}T^0]\times[M^0L^0T^{-1}] = [ML^2T^{-1}]$	$kg\ m^2\ s^{-1}$
27	Heat	Energy	$[ML^2T^{-2}]$	J
28	Planck's constant	Energy/frequency	$[ML^2T^{-2}]/[M^0L^0T^{-1}] = [ML^2T^{-1}]$	J s
29	Velocity gradient	Change in velocity/ distance	$[M^0LT^{-1}]\times[L] = [M^0L^0T^{-1}]$	s^{-1}
30	Radius of gyration	Distance	$[M^0LT^0]$	m

variables must be chosen with care. They must have all of the essential dimensions involved in the situation.

1) The dependent variable must not be chosen as a repeating variable.
2) The repeating variables should be chosen in such a way that one variable contains geometric property, the other variable contains flow properly and the third variable contains fluid property.
3) Usually, a length parameter (*D* or *H*); a typical velocity *V* and the fluid density are convenient sets of repeating variables.

When the number of variables exceed the number of fundamental dimensions, Rayleigh's approach of dimensional analysis becomes more arduous (M.L.T.). Buckingham's pi theorem is used to solve this problem.

Using this idea, the current tests can significantly enhance their working procedures and can be made shorter, needing less time while maintaining control. Using Buckingham's pi theorem, deducing the dimensional equation for a problem minimises the number of variables in the experiments. It is evident that if we take the product of the π terms, it will also be a dimensionless number and hence a π term. This idea is used to achieve further reduction in the number of independent π term variables, which further forms a few π terms. An attempt is made to apply the above-discussed matter to form a mathematical model for brake thermal efficiency and brake-specific fuel consumption.

10.9 Case Study on the Engine Performance by Using Alternative Fuels

The chemical stability of oil is influenced by three factors: heat, the availability of oxygen, and the existence of a catalyst. Oil deterioration might be induced by the breakdown of hydrocarbons in oil at high temperatures [5, 53, 54]. The oxygen concentration of insulating oil may cause an increase in acidity and the production of sludge. Catalysts such as iron and copper dissolve in oil while ageing and may hasten the procedure.

Because of the probable exhaustion and rising prices of petroleum, as well as environmental problems generated by the burning of fossil fuels, the hunt for alternative fuels has gained prominence [5, 55–57]. The alternative fuel not only evades the petroleum crises but also reduces pollutant gases emitted by an engine. Hence, the blends of various proportions, such as a 10% blend that is 10% treated transformer oil and 90% diesel fuel, likewise 20%, 25%, 30%, and 40%, are made and then compared to the properties of blends with pure diesel fuel.

The performance of the engine that is fuelled with blends is evaluated in terms of the following values:

1) Brake thermal efficiency
2) Brake-specific fuel consumption

The diesel engine run was conducted with a single cylinder diesel engine. The result obtained was fuelled with blends of treated transformer oil and diesel fuel in varying proportions, such as 10:90, 20:80, 25:75, 30:70 and 40:60. The runs were covered under varying loads of 10, 15 and 20 kg. The performance of the engine was evaluated on the basis of brake thermal efficiency (BTE) and brake-specific fuel consumption (BSFC).

The variables affecting the effectiveness of the phenomenon under consideration are blends of treated transformer oil with diesel and performance characteristics. The dependent or the response variables are as follows:

- Brake thermal efficiency (BTH)
- Brake-specific fuel consumption (BSFC)

For dependent and independent variables involved, the performance of diesel engine is presented in Table 10.3.

Independent and dependent π term

$\pi_1 = [\ (Cv)^a\ (Mf)^b\ (Bd)^c\ ]\ L$

$[M^0L^0T^0] = [M^0L^2T^{-2}]^a\ [M^1L^0T^{-1}]^b [M^0L^1T^0]^c [M^1L^0T^0]^1$

M=0=0+b+0+1, b= -1

L=0=2a+c,

T=0=-2a-b,

∴ a= ½ ; c= -1

$\pi_1 = [Cv^{1/2}\ L/Mf\ Bd]$

Table 10.3 Independent and dependent variables.

Sr. No.	Description	Variables	Symbol	Dimension
1	Load on engine	Independent	L	$[M^1L^0T^0]$
2	Blend	Independent	B	$[M^0L^0T^0]$
3	Flashpoint	Independent	Fp	$[M^0L^0T^0]$
4	Aniline point	Independent	Ap	$[M^0L^0T^0]$
5	Kinematic viscosity	Independent	Vi	$[M^0L^2T^{-1}]$
6	Density	Independent	D	$[M^1L^{-3}T^0]$
7	API gravity	Independent	Apig	$[M^0L^0T^0]$
8	Diesel index	Independent	Di	$[M^0L^0T^0]$
9	Cetane number	Independent	CN	$[M^0L^0T^0]$
10	Calorific value	Independent	Cv	$[M^0L^2T^{-2}]$
11	Time	Independent	T	$[M^0L^0T^1]$
12	Mass of fuel	Independent	Mf	$[M^1L^0T^{-1}]$
13	Bore diameter	Independent	Bd	$[M^0L^1T^0]$
14	Stroke length	Independent	Sl	$[M^0L^1T^0]$
15	Cubic capacity	Independent	Cc	$[M^0L^3T^0]$
16	Flue tank capacity	Independent	Fc	$[M^0L^3T^0]$
17	Engine speed	Independent	N	$[M^0L^0T^{-1}]$
18	Brake thermal efficiency	Dependent	Bte	$[M^0L^0T^0]$
19	Brake-specific flue consumption	Dependent	Bsfc	$[M^2L^2T^{-4}]$

$\pi_2 = [\ (Cv)^a\ (Mf)^b\ (Bd)^c\ ]\ B$
$[M^0L^0T^0] = [M^0L^2T^{-2}]^a\ [M^1L^0T^{-1}]^b [M^0L^1T^0]^c [M^0L^0T^0]^1$
$M = 0 = 0+b+0+0,\ b = 0$
$L = 0 = 2a+c,$
$T = 0 = -2a-b, a = 0$
$\therefore\ c = 0$
$\boldsymbol{\pi_2 = [B]}$

$\pi_3 = [\ (Cv)^a\ (Mf)^b\ (Bd)^c\ ]\ Fp$
$[M^0L^0T^0] = [M^0L^2T^{-2}]^a\ [M^1L^0T^{-1}]^b [M^0L^1T^0]^c [M^0L^0T^0]^1$
$M = 0 = 0+b+0+0,\ b = 0$
$L = 0 = 2a+c,$
$T = 0 = -2a-b, a = 0$
$\therefore\ c = 0$
$\boldsymbol{\pi_3 = [Fp]}$

$\pi_4 = [\ (Cv)^a\ (Mf)^b\ (Bd)^c\ ]\ Ap$
$[M^0L^0T^0] = [M^0L^2T^{-2}]^a\ [M^1L^0T^{-1}]^b [M^0L^1T^0]^c [M^0L^0T^0]^1$
$M = 0 = 0+b+0+0,\ b = 0$
$L = 0 = 2a+c,$
$T = 0 = -2a-b, a = 0$
$\therefore\ c = 0$
$\boldsymbol{\pi_4 = [Ap]}$

$\pi_5 = [\ (Cv)^a\ (Mf)^b\ (Bd)^c\ ]\ Vi$
$[M^0L^0T^0] = [M^0L^2T^{-2}]^a\ [M^1L^0T^{-1}]^b [M^0L^1T^0]^c [M^0L^2T^{-1}]^1$
$M = 0 = 0+b+0+0,\ b = 0$
$L = 0 = 2a+c+2,$
$T = 0 = -2a-b-1,$
$\therefore\ a = ½\ ;\ c = -2$

$\boldsymbol{\pi_5 = [Cv^{1/2} Vi/Bd]}$
$\pi_6 = [\ (Cv)^a\ (Mf)^b\ (Bd)^c\ ]\ D$
$[M^0L^0T^0] = [M^0L^2T^{-2}]^a\ [M^1L^0T^{-1}]^b [M^0L^1T^0]^c [M^1L^{-3}T^0]^1$
$M = 0 = 0+b+0+1,\ b = -1$
$L = 0 = 2a+c-3,$
$T = 0 = -2a-b,$
$\therefore\ a = ½;\ c = 2$

$\boldsymbol{\pi_6 = [Cv^{1/2}\ Bd^2\ D/Mf]}$
$\pi_7 = [\ (Cv)^a\ (Mf)^b\ (Bd)^c\ ]\ Apig$
$[M^0L^0T^0] = [M^0L^2T^{-2}]^a\ [M^1L^0T^{-1}]^b [M^0L^1T^0]^c [M^0L^0T^0]^1$
$M = 0 = 0+b+0+0,\ b = 0$
$L = 0 = 2a+c,$
$T = 0 = -2a-b, a = 0$
$\therefore\ c = 0$
$\boldsymbol{\pi_7 = [Apig]}$

$\pi_8 = [\ (Cv)^a\ (Mf)^b\ (Bd)^c\ ]\ Di$

$[M^0L^0T^0] = [M^0L^2T^{-2}]^a\ [M^1L^0T^{-1}]^b [M^0L^1T^0]^c [M^0L^0T^0]^1$

$M=0=0+b+0+0,\ b=0$

$L=0=2a+c,$

$T=0=-2a-b, a=0$

$\therefore\ c = 0$

$\boldsymbol{\pi_8 = [Di]}$

$\pi_9 = [\ (Cv)^a\ (Mf)^b\ (Bd)^c\ ]\ CN$

$[M^0L^0T^0] = [M^0L^2T^{-2}]^a\ [M^1L^0T^{-1}]^b [M^0L^1T^0]^c [M^0L^0T^0]^1$

$M=0=0+b+0+0,\ b=0$

$L=0=2a+c,$

$T=0=-2a-b, a=0$

$\therefore\ c = 0$

$\boldsymbol{\pi_9 = [CN]}$

$\pi_{10} = [\ (Cv)^a\ (Mf)^b\ (Bd)^c\ ]\ T$

$[M^0L^0T^0] = [M^0L^2T^{-2}]^a\ [M^1L^0T^{-1}]^b [M^0L^1T^0]^c [M^0L^0T^1]^1$

$M=0=0+b+0+0,\ b=0$

$L=0=2a+c,$

$T=0=-2a-b+1, a=½$

$\therefore\ c = -1$

$\boldsymbol{\pi_{10} = [Cv^{1/2}\ T/Bd]}$

$\pi_{11} = [\ (Cv)^a\ (Mf)^b\ (Bd)^c\ ]\ Sl$

$[M^0L^0T^0] = [M^0L^2T^{-2}]^a\ [M^1L^0T^{-1}]^b [M^0L^1T^0]^c [M^0L^1T^0]^1$

$M=0=0+b+0+0,\ b=0$

$L=0=2a+c+1,$

$T=0=-2a-b,\ a=0$

$\therefore c=-1$

$\boldsymbol{\pi_{11} = [Sl/Bd]}$

$\pi_{12} = [\ (Cv)^a\ (Mf)^b\ (Bd)^c\ ]\ Cc$

$[M^0L^0T^0] = [M^0L^2T^{-2}]^a\ [M^1L^0T^{-1}]^b [M^0L^1T^0]^c [M^0L^3T^0]^1$

$M=0=0+b+0+0,\ b=0$

$L=0=2a+c+3,$

$T=0=-2a-b,\ a=0$

$\therefore\ a=0\ ;\ c = -3$

$\boldsymbol{\pi_{12} = [Cc/Bd^3]}$

$\pi_{13} = [\ (Cv)^a\ (Mf)^b\ (Bd)^c\ ]\ Fc$

$[M^0L^0T^0] = [M^0L^2T^{-2}]^a\ [M^1L^0T^{-1}]^b [M^0L^1T^0]^c [M^0L^3T^0]^1$

$M=0=0+b+0+0,\ b=0$

$L=0=2a+c+3,$

$T=0=-2a-b,\ a=0$

∴ a= 0; c = -3

π_{13}= **[Fc/Bd^3]**

π_{14}= [(Cv)^a (Mf)^b (Bd)^c] N

$[M^0L^0T^0]$ = $[M^0L^2T^{-2}]^a$ $[M^1L^0T^{-1}]^b$ $[M^0L^1T^0]^c$ $[M^0L^0T^{-1}]^1$

M=0=0+b+0+0, b= 0

L=0=2a+c,

T=0=-2a-b-1, a= 1/2

∴ a= ½ ; c = -1

π_{14}= **[N Bd/Cv$^{1/2}$]**

π_{D1}= [(Cv)^a (Mf)^b (Bd)^c] Bte

$[M^0L^0T^0]$ = $[M^0L^2T^{-2}]^a$ $[M^1L^0T^{-1}]^b$ $[M^0L^1T^0]^c$ $[M^0L^0T^0]^1$

M=0=0+b+0+0, b= 0

L=0=2a+c,

T=0=-2a-b, a= 0

∴c= 0

π_{D1}= **[Bte]**

π_{D2} = [(Cv)^a (Mf)^b (Bd)^c] Bsfc

$[M^0L^0T^0]$ = $[M^0L^2T^{-2}]^a$ $[M^1L^0T^{-1}]^b$ $[M^0L^1T^0]^c$ $[M^0L^0T^0]^1$

M=0=0+b+0+0, b= 0

L=0=2a+c,

T=0=-2a-b, a= 0

∴ c= 0

π_{D2}= **[Bsfc]**

10.10 Establishment of Dimensionless Group of Π Terms

These independent variables have been reduced into a group of π terms. The list of the independent and dependent π terms of the face drilling activity is shown in Tables 10.4 and 10.5.

10.10.1 Creation of Field-data-based Model

Four independent π terms (π_1, π_2, π_3 and π_4) and two dependent π terms (Z_1 and Z_2) were identified for the model formulation of the field study.

Each pi term is a function of the output terms [15, 22],

$$Z_1 = \text{function of}\left(\Pi_{1,} \Pi_{2,} \Pi_{3,} \Pi_4\right)$$

$$Z_2 = \text{function of}\left(\Pi_{1,} \Pi_{2,} \Pi_{3,} \Pi_4\right)$$

Table 10.4 Independent dimensionless π terms.

Sr. No.	Independent dimensionless ratios	Nature of basic physical quantities
01	$\pi_1 = [([Cv^{1/2} Vi/ Bd])([Cv^{1/2} T / Bd])([B])/([Fp])]$	Specification related to blend formation and time
02	$\pi_2 = [([Cv^{1/2} Bd^2 D / Mf])/([Cv^{1/2} L/ Mf\ Bd])([Ap])]$	Specifications of fuel consumption and engine load
03	$\pi_3 = [([Apig])([Di])/([CN])]$	Specifications of fuel characteristics
04	$\pi_4 = [([Sl/ Bd])([Cc / Bd^3])([Fc / Bd^3])([N\ Bd / Cv^{1/2}])]$	Engine specification

Table 10.5 Dependent dimensionless π terms.

Sr. No.	Dependent dimensionless ratios or π terms	Nature of basic physical quantities
01	$Z_1 = [Bte]$	Brake thermal efficiency
02	$Z_2 = [Bsfc]$	Brake-specific flue consumption

where

$$Z_1 = \Pi_{D1}, \text{first dependent } \pi \text{ term} = Bte$$

$$Z_2 = \Pi_{D2}, \text{second dependent } \pi \text{ term} = Bsfc$$

The most likely accurate mathematical form for the dimensions equations of the phenomenon might be connections considered to be of an exponentially nature.

$$(Z) = K * \left[\left(\left[Cv^{1/2} Vi / Bd\right]\right)\left(\left[Cv^{1/2} T/Bd\right]\right)\left(\left[B\right]\right)/\left(\left[Fp\right]\right)\right]^a, \left[\left(\left[Cv^{1/2} Bd^2\ D/Mf\right]\right)/\left(\left[Cv^{1/2} L / Mf\ Bd\right]\right)\left(\left[Ap\right]\right)\right]^b, \left[\left(\left[Apig\right]\right)\left(\left[Di\right]\right)/\left(\left[CN\right]\right)\right]^c, \left[\left(\left[Sl/Bd\right]\right)\left(\left[Cc/Bd^3\right]\right)\left(\left[Fc/Bd^3\right]\right)\left(\left[N\ Bd/Cv^{1/2}\right]\right)\right]^d$$

10.10.2 Model Formulation by Identifying the Curve-fitting Constant and Various Indices of π Terms

By taking into account four independent variables and one dependent π term, multiple regression analysis assists in identifying the indices of various π terms in the model targeted [58]. Let the model aimed to be of the following form:

$$(Z_1) = K_1 * \left[(\pi_1)^{a1} * (\pi_2)^{b1} * (\pi_3)^{c1} * (\pi_4)^{d1}\right]$$

$$(Z_2) = K_2 * \left[(\pi_1)^{a2} * (\pi_2)^{b2} * (\pi_3)^{c2} * (\pi_4)^{d2}\right]$$

To find the values of a_1, b_1, c_1 and d_1, the equations are presented as follows:

$$\Sigma Z_1 = nK_1 + a_1 * \Sigma A + b_1 * \Sigma B + c_1 * \Sigma C + d_1 * \Sigma D$$

$$\Sigma Z_1 * A = K_1 * \Sigma A + a_1 * \Sigma A * A + b_1 * \Sigma B * A + c_1 * \Sigma C * A + d_1 * \Sigma D * A$$

$$\Sigma Z_1 * B = K_1 * \Sigma B + a_1 * \Sigma A * B + b_1 * \Sigma B * B + c_1 * \Sigma C * B + d_1 * \Sigma D * B$$

$$\Sigma Z_1 * C = K_1 * \Sigma C + a_1 * \Sigma A * C + b_1 * \Sigma B * C + c_1 * \Sigma C * C + d_1 * \Sigma D * C$$

$$\Sigma Z_1 * D = K_1 * \Sigma D + a_1 * \Sigma A * D + b_1 * \Sigma B * D + c_1 * \Sigma C * D + d_1 * \Sigma D * D$$

In the above set of equations, the values of K_1, a_1, b_1, c_1 and d_1 are substituted to compute the values of the unknowns. After substituting these values into the equations, one will obtain a set of five equations, which are to be solved simultaneously to obtain the values of K_1, a_1, b_1, c_1 and d_1. The above equations can be transferred used in the matrix form, and subsequently values of K_1, a_1, b_1, c_1 and d_1 can be obtained by adopting matrix analysis.

$$X_1 = \text{inv}(W) \times P_1$$

$W = 5 \times 5$ matrix of the multipliers of K_1, a_1, b_1, c_1 and d_1
$P_1 = 5 \times 1$ matrix on the LHS and
$X_1 = 5 \times 1$ matrix of solutions
Then, the matrix evaluated is given as follows:
Matrix

$$Z_1 x \begin{bmatrix} 1 \\ A \\ B \\ C \\ D \end{bmatrix} = \begin{bmatrix} n & A & B & C & D \\ A & A^2 & BA & CA & DA \\ B & AB & B^2 & CB & DB \\ C & AC & BC & C^2 & DC \\ D & AD & BD & CD & D^2 \end{bmatrix} x \begin{bmatrix} K_1 \\ a_1 \\ b_1 \\ c_1 \\ d_1 \end{bmatrix}$$

In the above equations, n is the number of sets of readings, and A, B, C and D represent the independent π terms π_1, π_2, π_3 and π_4, respectively, while Z represents the dependent π term.

10.10.3 Basis for Arriving at the Number of Observations

The number of observations taken is 15 based on the probability concept of degree of uncertainty. The formula for calculating the number of readings is as follows:

$$\sqrt{\mathbf{N}} = \left[\{x/Zc\} - \mu \right] \sigma$$

where

x = mean,
μ = median
σ = standard deviation
N = number of readings

For $N \geq 15$, Zc = 2.58 for certainty (confidence level 99%)

$$Zc = 1.96 \text{ for certainty} \left(\text{confidence level} 95\%\right)$$

$$Zc = 1.645 \text{ for certainty} \left(\text{confidence level} 90\%\right)$$

Selecting Zc = 2.58 for certainty with the confidence level 99% satisfies the number of readings:

$$N = [x/\{Ţc\} - \mu]^2 \sigma$$

For N < 30, Ţc = 2.48 for certainty (confidence level 99%)

Ţc = 1.71 for certainty (confidence level 95%)

Ţc = 1.32 for certainty (confidence level 90%)

Selecting Ţc = 2.48 for certainty with the confidence level 99% satisfies the number of readings. Various pie terms formulated based on the readings are presented in Tables 10.6–10.25.

Table 10.6 Data related to the dimension term π_1.

Load (Lo) (kg)	CV (cm^2/s^2)	Mf (kg/s)	Bore diameter (cm)	P1
5	4.61481E+11	0.000139276	8.75	2 787 167 569
5	4.60912E+11	0.000123	8.75	3 154 029 973
5	4.61481E+11	0.000109162	8.75	3 556 041 381
5	4.61481E+11	0.000103564	8.75	3 748 259 834
5	4.60912E+11	0.000101475	8.75	3 823 066 634
10	4.61481E+11	0.000175609	8.75	4 421 024 419
10	4.60912E+11	0.00016236	8.75	4 778 833 293
10	4.61481E+11	0.00014425	8.75	5 382 116 684
10	4.61481E+11	0.000134633	8.75	5 766 553 590
10	4.60912E+11	0.000123	8.75	6 308 059 946
15	4.61481E+11	0.000186129	8.75	6 256 710 645
15	4.60912E+11	0.000176478	8.75	6 594 789 944
15	4.61481E+11	0.000168292	8.75	6 919 864 308
15	4.61481E+11	0.000155346	8.75	7 496 519 667
15	4.60912E+11	0.000144964	8.75	8 028 439 932

Table 10.7 Data related to the dimension terms π_2, π_3 and π_4.

P2	P3	P4
10	38	70
20	38.5	71.11
25	39	72.22
30	44.5	75
40	45	76.11
10	38	70
20	38.5	71.11
25	39	72.22
30	44.5	75
40	45	76.11
10	38	70
20	38.5	71.11
25	39	72.22
30	44.5	75
40	45	76.11

Table 10.8 Data related to the dimension term π_5.

Vi (cm^2/s)	CV (cm^2/s^2)	Bore diameter (cm)	P5
3.372	4.61481E+11	8.75	5.67287E-07
3.772	4.60912E+11	8.75	6.34972E-07
3.85	4.61481E+11	8.75	6.47703E-07
4.33	4.61481E+11	8.75	7.28455E-07
5	4.60912E+11	8.75	8.41692E-07
3.372	4.61481E+11	8.75	5.67287E-07
3.772	4.60912E+11	8.75	6.34972E-07
3.85	4.61481E+11	8.75	6.47703E-07
4.33	4.61481E+11	8.75	7.28455E-07
5	4.60912E+11	8.75	8.41692E-07
3.372	4.61481E+11	8.75	5.67287E-07
3.772	4.60912E+11	8.75	6.34972E-07
3.85	4.61481E+11	8.75	6.47703E-07
4.33	4.61481E+11	8.75	7.28455E-07
5	4.60912E+11	8.75	8.41692E-07

Table 10.9 Data related to the dimensions term π_6.

CV (cm^2/s^2)	Bore diameter (cm)	Mf (kg/s)	Vi (cm^2/s)	P6
4.61481E+11	8.75	0.000139276	3.372	3.02E+11
4.60912E+11	8.75	0.000123	3.772	3.43E+11
4.61481E+11	8.75	0.000109162	3.85	3.85E+11
4.61481E+11	8.75	0.000103564	4.33	4.06E+11
4.60912E+11	8.75	0.000101475	5	4.16E+11
4.61481E+11	8.75	0.000175609	3.372	2.39E+11
4.60912E+11	8.75	0.00016236	3.772	2.6E+11
4.61481E+11	8.75	0.00014425	3.85	2.91E+11
4.61481E+11	8.75	0.000134633	4.33	3.12E+11
4.60912E+11	8.75	0.000123	5	3.43E+11
4.61481E+11	8.75	0.000186129	3.372	2.26E+11
4.60912E+11	8.75	0.000176478	3.772	2.39E+11
4.61481E+11	8.75	0.000168292	3.85	2.5E+11
4.61481E+11	8.75	0.000155346	4.33	2.7E+11
4.60912E+11	8.75	0.000144964	5	2.91E+11

Table 10.10 Data related to the dimension terms π_7, π_8 and π_9.

P7	P8	P9
70	61.91704	54.5803
71.11	61.3888	54.1999
72.22	63.48456	55.7089
75	65.44396	57.11965
76.11	70.642	60.86224
70	61.91704	54.5803
71.11	61.3888	54.1999
72.22	63.48456	55.7089
75	65.44396	57.11965
76.11	70.642	60.86224
70	61.91704	54.5803
71.11	61.3888	54.1999
72.22	63.48456	55.7089
75	65.44396	57.11965
76.11	70.642	60.86224

Table 10.11 Data related to the dimension term π_{10}.

CV (cm^2/s^2)	Time (s)	Bore diameter (cm)	P10
4.61481E+11	290	8.75	22514740
4.60912E+11	330	8.75	25604415
4.61481E+11	370	8.75	28725702
4.61481E+11	390	8.75	30278443
4.60912E+11	400	8.75	31035655
4.61481E+11	230	8.75	17856518
4.60912E+11	250	8.75	19397284
4.61481E+11	280	8.75	21738369
4.61481E+11	300	8.75	23291110
4.60912E+11	330	8.75	25604415
4.61481E+11	217	8.75	16847236
4.60912E+11	230	8.75	17845502
4.61481E+11	240	8.75	18632888
4.61481E+11	260	8.75	20185629
4.60912E+11	280	8.75	21724958

Table 10.12 Data related to the dimension term π_{11}.

Bore diameter (cm)	Stroke (cm)	P11
8.75	11	1.257143
8.75	11	1.257143
8.75	11	1.257143
8.75	11	1.257143
8.75	11	1.257143
8.75	11	1.257143
8.75	11	1.257143
8.75	11	1.257143
8.75	11	1.257143
8.75	11	1.257143
8.75	11	1.257143
8.75	11	1.257143
8.75	11	1.257143
8.75	11	1.257143
8.75	11	1.257143

Table 10.13 Data related to the dimension term π_{12}.

Bore diameter (cm)	Cubic capacity (cm^3)	P12
8.75	1323	1.974857
8.75	1323	1.974857
8.75	1323	1.974857
8.75	1323	1.974857
8.75	1323	1.974857
8.75	1323	1.974857
8.75	1323	1.974857
8.75	1323	1.974857
8.75	1323	1.974857
8.75	1323	1.974857
8.75	1323	1.974857
8.75	1323	1.974857
8.75	1323	1.974857
8.75	1323	1.974857
8.75	1323	1.974857

Table 10.14 Data related to the dimension term π_{13}.

Bore diameter (cm)	Fuel tank capacity (cm^3)	P13
8.75	11 000	16.41983
8.75	11 000	16.41983
8.75	11 000	16.41983
8.75	11 000	16.41983
8.75	11 000	16.41983
8.75	11 000	16.41983
8.75	11 000	16.41983
8.75	11 000	16.41983
8.75	11 000	16.41983
8.75	11 000	16.41983
8.75	11 000	16.41983
8.75	11 000	16.41983
8.75	11 000	16.41983
8.75	11 000	16.41983
8.75	11 000	16.41983

Table 10.15 Data related to the dimension term π_{14}.

Speed (Rps)	Bore diameter (cm)	CV (cm^2/s^2)	P14
25	8.75	4.61481E+11	0.000322
25	8.75	4.60912E+11	0.000322
25	8.75	4.61481E+11	0.000322
25	8.75	4.61481E+11	0.000322
25	8.75	4.60912E+11	0.000322
25	8.75	4.61481E+11	0.000322
25	8.75	4.60912E+11	0.000322
25	8.75	4.61481E+11	0.000322
25	8.75	4.61481E+11	0.000322
25	8.75	4.60912E+11	0.000322
25	8.75	4.61481E+11	0.000322
25	8.75	4.60912E+11	0.000322
25	8.75	4.61481E+11	0.000322
25	8.75	4.61481E+11	0.000322
25	8.75	4.60912E+11	0.000322

Table 10.16 Data related to the dimension terms π_{D1} and π_{D2}.

BTE	BSFC
25.82726	100
29.281	110
32.95202	134
34.73321	156
35.49212	120
40.84398	95
44.23152	120
49.72311	136
53.27476	154
58.38561	171
57.86132	124
61.10097	138
63.99409	160
69.41268	174
74.2919	184

Table 10.17 Data related to the dimension term Π_1 (P_{c1}) combined with terms π_5, π_2, π_3 and π_{10}.

π_2	π_3	π_5	π_{10}	Π_1 (P_{c1})
10	38	5.67287E-07	22 514 740	3.36113426
20	38.5	6.34972E-07	25 604 415	8.44576093
25	39	6.47703E-07	28 725 702	11.9267399
30	44.5	7.28455E-07	30 278 443	14.8695437
40	45	8.41692E-07	31 035 655	23.2199546
10	38	5.67287E-07	17 856 518	2.66572718
20	38.5	6.34972E-07	19 397 284	6.39830374
25	39	6.47703E-07	21 738 369	9.02564103
30	44.5	7.28455E-07	23 291 110	11.4381105
40	45	8.41692E-07	25 604 415	19.1564626
10	38	5.67287E-07	16 847 236	2.51505564
20	38.5	6.34972E-07	17 845 502	5.88643944
25	39	6.47703E-07	18 632 888	7.73626374
30	44.5	7.28455E-07	20 185 629	9.91302912
40	45	8.41692E-07	21 724 958	16.2539683

Table 10.18 Data related to the dimension term Π_2 (P_{c2}) combined with terms π_1, π_4 and π_6.

π_1	π_4	π_6	Π_2 (P_{c2})
2 787 167 569	70	3.02E+11	1.54617969
3 154 029 973	71.11	3.43E+11	1.52958115
3 556 041 381	72.22	3.85E+11	1.49865104
3 748 259 834	75	4.06E+11	1.44310104
3 823 066 634	76.11	4.16E+11	1.42909625
4 421 024 419	70	2.39E+11	0.77308984
4 778 833 293	71.11	2.6E+11	0.76479058
5 382 116 684	72.22	2.91E+11	0.74932552
5 766 553 590	75	3.12E+11	0.72155052
6 308 059 946	76.11	3.43E+11	0.71454813
6 256 710 645	70	2.26E+11	0.51539323
6 594 789 944	71.11	2.39E+11	0.50986038
6 919 864 308	72.22	2.5E+11	0.49955035
7 496 519 667	75	2.7E+11	0.48103368
8 028 439 932	76.11	2.91E+11	0.47636542

Table 10.19 Data related to the dimension term Π_3 (P_{c3}) combined with terms π_7, π_8 and π_9.

π_7	π_8	π_9	Π_3 (P_{c3})
70	61.91704	54.5803	79.409472
71.11	61.3888	54.1999	80.541801
72.22	63.48456	55.7089	82.300224
75	65.44396	57.11965	85.930094
76.11	70.642	60.86224	88.339874
70	61.91704	54.5803	79.409472
71.11	61.3888	54.1999	80.541801
72.22	63.48456	55.7089	82.300224
75	65.44396	57.11965	85.930094
76.11	70.642	60.86224	88.339874
70	61.91704	54.5803	79.409472
71.11	61.3888	54.1999	80.541801
72.22	63.48456	55.7089	82.300224
75	65.44396	57.11965	85.930094
76.11	70.642	60.86224	88.339874

Table 10.20 Data related to dimension term Π_4 (P_{c4}) combined with terms π_{11}, π_{12}, π_{13} and π_{14}.

π_{11}	π_{12}	π_{13}	π_{14}	Π_4 (P_{c4})
1.257143	1.974857	16.41983	0.000322	0.01312683
1.257143	1.974857	16.41983	0.000322	0.01313494
1.257143	1.974857	16.41983	0.000322	0.01312683
1.257143	1.974857	16.41983	0.000322	0.01312683
1.257143	1.974857	16.41983	0.000322	0.01313494
1.257143	1.974857	16.41983	0.000322	0.01312683
1.257143	1.974857	16.41983	0.000322	0.01313494
1.257143	1.974857	16.41983	0.000322	0.01312683
1.257143	1.974857	16.41983	0.000322	0.01312683
1.257143	1.974857	16.41983	0.000322	0.01313494
1.257143	1.974857	16.41983	0.000322	0.01312683
1.257143	1.974857	16.41983	0.000322	0.01313494
1.257143	1.974857	16.41983	0.000322	0.01312683
1.257143	1.974857	16.41983	0.000322	0.01312683
1.257143	1.974857	16.41983	0.000322	0.01313494

Table 10.21 Evaluation of constants *A*, *B*, *C* and *D* representing independent π terms.

Z = logZ	A = Logπ1	B = Logπ2	C = Logπ3	D = Logπ4
1.412078	0.526486	0.18926	1.899872	−1.88184
1.466586	0.926639	0.184573	1.906021	−1.88157
1.517882	1.076522	0.175701	1.915401	−1.88184
1.540745	1.172298	0.159297	1.934145	−1.88184
1.550132	1.365861	0.155061	1.946157	−1.88157
1.611128	0.425816	−0.11177	1.899872	−1.88184
1.645732	0.806065	−0.11646	1.906021	−1.88157
1.696558	0.955478	−0.12533	1.915401	−1.88184
1.726521	1.058354	−0.14173	1.934145	−1.88184
1.766306	1.282315	−0.14597	1.946157	−1.88157
1.762388	0.400548	−0.28786	1.899872	−1.88184
1.786048	0.769853	−0.29255	1.906021	−1.88157
1.80614	0.888531	−0.30142	1.915401	−1.88184
1.841439	0.996206	−0.31782	1.934145	−1.88184
1.870941	1.210959	−0.32206	1.946157	−1.88157
25.001	13.862	−1.299	28.805	−28.23

Table 10.22 Evaluation of constants *A*, *B*, *C* and *D* representing independent π terms.

ZA	A	AA	AB	AC	AD
0.743439	0.526486	0.277187	0.099643	1.000256	−0.99076
1.358995	0.926639	0.858659	0.171032	1.766193	−1.74354
1.634033	1.076522	1.158899	0.189145	2.061971	−2.02584
1.806212	1.172298	1.374282	0.186743	2.267394	−2.20608
2.117265	1.365861	1.865577	0.211792	2.65818	−2.56997
0.686044	0.425816	0.181319	−0.04759	0.808995	−0.80132
1.326567	0.806065	0.649741	−0.09387	1.536377	−1.51667
1.621024	0.955478	0.912938	−0.11975	1.830124	−1.79806
1.827271	1.058354	1.120114	−0.15	2.047011	−1.99165
2.264961	1.282315	1.644333	−0.18718	2.495587	−2.41277
0.70592	0.400548	0.160438	−0.1153	0.760989	−0.75377
1.374994	0.769853	0.592673	−0.22522	1.467356	−1.44853
1.604812	0.888531	0.789488	−0.26782	1.701894	−1.67207
1.834453	0.996206	0.992427	−0.31662	1.926808	−1.8747
2.265634	1.210959	1.466423	−0.39	2.356717	−2.27851
23.172	13.862	14.044	−1.055	26.686	−26.08

Table 10.23 Evaluation of constants *A*, *B*, *C* and *D* representing independent π terms.

ZB	B	AB	BB	BC	BD
0.26725	0.18926	0.099643	0.035819	0.35957	−0.35616
0.270691	0.184573	0.171032	0.034067	0.351799	−0.34729
0.266693	0.175701	0.189145	0.030871	0.336537	−0.33064
0.245436	0.159297	0.186743	0.025375	0.308103	−0.29977
0.240366	0.155061	0.211792	0.024044	0.301774	−0.29176
−0.18008	−0.11177	−0.04759	0.012493	−0.21235	0.210333
−0.19166	−0.11646	−0.09387	0.013562	−0.22197	0.219123
−0.21263	−0.12533	−0.11975	0.015707	−0.24006	0.23585
−0.24471	−0.14173	−0.15	0.020088	−0.27413	0.266719
−0.25783	−0.14597	−0.18718	0.021307	−0.28408	0.27465
−0.50732	−0.28786	−0.1153	0.082864	−0.5469	0.541709
−0.52251	−0.29255	−0.22522	0.085585	−0.5576	0.550452
−0.54441	−0.30142	−0.26782	0.090854	−0.57734	0.567226
−0.58525	−0.31782	−0.31662	0.101012	−0.61472	0.598095
−0.60255	−0.32206	−0.39	0.103722	−0.62678	0.605979
−2.559	**−1.299**	**−1.055**	**0.6974**	**−2.498**	**2.4445**

Table 10.24 Evaluation of constants *A*, *B*, *C* and *D* representing independent π terms.

ZC	C	AC	BC	CC	CD
2.682768	1.899872	1.000256	0.35957	3.609515	−3.57526
2.795344	1.906021	1.766193	0.351799	3.632917	−3.58632
2.907353	1.915401	2.061971	0.336537	3.668761	−3.60448
2.980025	1.934145	2.267394	0.308103	3.740918	−3.63975
3.0168	1.946157	2.65818	0.301774	3.787526	−3.66183
3.060938	1.899872	0.808995	−0.21235	3.609515	−3.57526
3.1368	1.906021	1.536377	−0.22197	3.632917	−3.58632
3.249589	1.915401	1.830124	−0.24006	3.668761	−3.60448
3.339343	1.934145	2.047011	−0.27413	3.740918	−3.63975
3.437508	1.946157	2.495587	−0.28408	3.787526	−3.66183
3.348313	1.899872	0.760989	−0.5469	3.609515	−3.57526
3.404246	1.906021	1.467356	−0.5576	3.632917	−3.58632
3.459482	1.915401	1.701894	−0.57734	3.668761	−3.60448
3.56161	1.934145	1.926808	−0.61472	3.740918	−3.63975
3.641145	1.946157	2.356717	−0.62678	3.787526	−3.66183
48.021	28.805	26.686	−2.498	55.319	−54.2

Table 10.25 Evaluation of constants *A*, *B*, *C* and *D* representing independent π terms.

ZD	D	AD	BD	CD	DD
−2.65731	−1.88184	−0.99076	−0.35616	−3.57526	3.541322
−2.75949	−1.88157	−1.74354	−0.34729	−3.58632	3.540313
−2.85641	−1.88184	−2.02584	−0.33064	−3.60448	3.541322
−2.89944	−1.88184	−2.20608	−0.29977	−3.63975	3.541322
−2.91668	−1.88157	−2.56997	−0.29176	−3.66183	3.540313
−3.03189	−1.88184	−0.80132	0.210333	−3.57526	3.541322
−3.09656	−1.88157	−1.51667	0.219123	−3.58632	3.540313
−3.19265	−1.88184	−1.79806	0.23585	−3.60448	3.541322
−3.24904	−1.88184	−1.99165	0.266719	−3.63975	3.541322
−3.32343	−1.88157	−2.41277	0.27465	−3.66183	3.540313
−3.31653	−1.88184	−0.75377	0.541709	−3.57526	3.541322
−3.36058	−1.88157	−1.44853	0.550452	−3.58632	3.540313
−3.39887	−1.88184	−1.67207	0.567226	−3.60448	3.541322
−3.46529	−1.88184	−1.8747	0.598095	−3.63975	3.541322
−3.52031	−1.88157	−2.27851	0.605979	−3.66183	3.540313
−47.04	**−28.23**	**−26.08**	**2.4445**	**−54.2**	**53.114**

10.10.4 Model Formulation

By taking into account four independent variables and one dependent term, multiple regression analysis assists in identifying the indices of various terms in the model targeted. Let the model aimed to be of the following form:

$$(Z_1) = k_1 * \left[(\Pi_1)^{a1} * (\Pi_2)^{b1} * (\Pi_3)^{c1} * (\Pi_4)^{d1} \right]$$

The above equations are shown below to find the values of a_1, b_1, c_1 and d_1 and to arrive at the regression hyperplane [59, 60]:

$$\Sigma Z_1 \ldots = nK_1 + a_1 * \Sigma A + b_1 * \Sigma B + c_1 * \Sigma C + d_1 * \Sigma D$$

$$\Sigma Z_1 * A = K_1 * \Sigma A + a_1 * \Sigma A * A + b_1 * \Sigma B * A + c_1 * \Sigma C * A + d_1 * \Sigma D * A$$

$$\Sigma Z_1 * B = K_1 * \Sigma B + a_1 * \Sigma A * B + b_1 * \Sigma B * B + c_1 * \Sigma C * B + d_1 * \Sigma D * B$$

$$\Sigma Z_1 * C = K_1 * \Sigma C + a_1 * \Sigma A * C + b_1 * \Sigma B * C + c_1 * \Sigma C * C + d_1 * \Sigma D * C$$

$$\Sigma Z_1 * D = K_1 * \Sigma D + a_1 * \Sigma A * D + b_1 * \Sigma B * D + c_1 * \Sigma C * D + d_1 * \Sigma D * D$$

By inserting these values into the equations, one is left with a set of five equations that must be solved concurrently to obtain the values of K_1, a_1, b_1, c_1 and d_1. The above-mentioned equations may be validated in matrix notation, and additional values of K_1, a_1, b_1, c_1 and d_1 can be derived using matrix analysis:

$$Z_1 = W_1 \times X_1$$

Here,

W_1 = 5×5 matrix of the multipliers of K_1, a_1, b_1, c_1 and d_1
P_1 = 5×1 matrix of the terms on LHS and
X_1 = 5×1 matrix of solutions of values of K_1, a_1, b_1, c_1 and d_1
Then, the matrix obtained is given as follows:

Matrix

$$Z_1 x \begin{bmatrix} 1 \\ A \\ B \\ C \\ D \end{bmatrix} = \begin{bmatrix} n & A & B & C & D \\ A & A^2 & BA & CA & DA \\ B & AB & B^2 & CB & DB \\ C & AC & BC & C^2 & DC \\ D & AD & BD & CD & D^2 \end{bmatrix} x \begin{bmatrix} K_1 \\ a_1 \\ b_1 \\ c_1 \\ d_1 \end{bmatrix}$$

In the above equations, n is the number of sets of readings, and A, B, C and D represent the independent π terms π_1, π_2, π_3 and π_4, respectively, while Z_1 matrix represents the dependent π term.

Substituting the values of A, A^2, BA, CA ,. . ., up to D^2 into the above matrix, the value of constant K_1 and indices a_1, b_1, c_1 and d_1 are evaluated by taking the inverse of matrix W_1 and multiplying it with the Z_1 matrix as shown below:

$$\begin{bmatrix} 25.00063 \\ 23.17162 \\ -2.5585 \\ 48.02126 \\ -47.0445 \end{bmatrix} = \begin{bmatrix} 15 & 13.86193 & -1.29908 & 28.80479 & -28.226 \\ 13.86193 & 14.0445 & -1.055 & 26.68585 & -26.0842 \\ -1.29908 & -1.055 & 0.697372 & -2.49815 & 2.444522 \\ 28.80479 & 26.68585 & -2.49815 & 55.31891 & -54.2029 \\ -28.226 & -26.0842 & 2.444522 & -54.2029 & 53.11378 \end{bmatrix} \begin{bmatrix} K_1 \\ a_1 \\ b_1 \\ c_1 \\ d_1 \end{bmatrix}$$

$$K_1 = 1.93, a_1 = 0.1758, b_1 = -0.7380, c_1 = -0.6950 \text{ and } d_1 = -0.4490$$

Taking antilog of K
Antilog (1.93) = 85.1138
The accurate form of the model evaluated is as follows:

$$(Z_1) = 85.1138 * (\Pi_1)^{0.1758} * (\Pi_2)^{-0.7380} * (\Pi_3)^{-0.6950} * (\Pi_4)^{-0.4490}$$

Brake thermal efficiency (Z_1)

$$(\text{Bte}) = 85.1138 * (\Pi_1)^{0.1758} * (\Pi_2)^{-0.7380} * (\Pi_3)^{-0.6950} * (\Pi_4)^{-0.4490}$$

10.10.4.1 Analysis of the Mathematical Model for the Dependent Pi Term Brake Thermal Efficiency

Z_1 is the brake thermal efficiency. The formulated model based on the experimental data is as follows:

$$Z_1 = 85.1138 * (\Pi_1)^{0.1758} * (\Pi_2)^{-0.7380} * (\Pi_3)^{-0.6950} * (\Pi_4)^{-0.4490}$$

where
Z_1 = brake thermal efficiency
π_1 = Pi terms related to the blend formation and time involvement
π_2 = Pi terms related to the fuel consumption and engine load
π_3 = Pi terms related to the fuel characteristics
π_4 = Pi terms related to the engine specifications

The absolute index of π_1 is highest, viz. 0.1758. Thus, the term is related to the blend formation and time involved in the experimental model. The value of the index is positive indicating that the brake thermal efficiency is directly proportional to terms related to the blend formation and time involved. The brake thermal efficiency increases with a decrease in the bore diameter and flashpoint and with an increase in the cetane number, kinematic viscosity, blend formation and time.

The term π_2 has a negative index −0.7380, which is inversely proportional to the brake thermal efficiency. The π_2 term in the brake thermal efficiency increases with an increase in the mass of fuel and blend formation and with a decrease in the stroke length and aniline point.

The term π_3 has a negative index −0.6950, which is inversely proportional to the brake thermal efficiency. The π_3 terms in the brake thermal efficiency increases with a decrease in the aniline point and diesel index and with an increase in the cetane number.

The term π_4 has a negative index −0.4490, which is inversely proportional to the brake thermal efficiency. The π_4 term in the brake thermal efficiency increases with a decrease in the stroke length, cubic capacity, flue tank capacity and engine speed. Table 10.26 presents the comparison of experimentally calculated values and mathematical model equation base values.

In the above equations, n is the number of sets of readings, and A, B, C and D represent the independent π terms π_1, π_2, π_3, and π_4, respectively, while the Z_2 matrix represents the dependent π term.

Substituting the values of A, A^2, BA, CA, . . ., up to D^2 into the above matrix, the value of constant K_1 and indices a_1, b_1, c_1 and d_1 are evaluated by taking the inverse of matrix W_1 and multiplying with the Z_2 matrix as shown below:

$$\begin{bmatrix} 13.34391 \\ 12.26366 \\ -0.76234 \\ 25.61249 \\ -25.1097 \end{bmatrix} = \begin{bmatrix} 15 & 13.86193 & -1.29908 & 28.80479 & -28.226 \\ 13.86193 & 14.0445 & -1.055 & 26.68585 & -26.0842 \\ -1.29908 & -1.055 & 0.697372 & -2.49815 & 2.444522 \\ 28.80479 & 26.68585 & -2.49815 & 55.31891 & -54.2029 \\ -28.226 & -26.0842 & 2.444522 & -54.2029 & 53.11378 \end{bmatrix} \begin{bmatrix} K_1 \\ a_1 \\ b_1 \\ c_1 \\ d_1 \end{bmatrix}$$

Table 10.26 Comparison of experimentally calculated values and mathematical model equation base values.

Sr. No.	Calculated from mathematical model	Experimental observation	Error = experimental observation − calculated value
1	25.5516	25.8273	0.2757
2	29.9799	29.2810	−0.6989
3	31.8660	32.9520	1.0860
4	33.0555	34.7332	1.6777
5	35.3126	35.4921	0.1795
6	40.9147	40.8440	−0.0707
7	47.6204	44.2315	−3.3889
8	50.6068	49.7231	−0.8837
9	52.6469	53.2748	0.6278
10	56.9381	58.3856	1.4475
11	54.6252	57.8613	3.2361
12	63.2968	61.1010	−2.1959
13	66.4347	63.9941	−2.4406
14	69.2473	69.4127	0.1654
15	74.6129	74.2919	−0.3210

$$K_1 = -0.1338, a_1 = -0.1016, b_1 = 0.7070, c_1 = -0.6336 \text{ and } d_1 = -1.2727$$

Taking antilog of *K*
Antilog (−0.1338) = 0.7348
The exact form of the model obtained is as follows:

$$(Z_2) = 0.7348 * (\Pi_1)^{-0.1016} * (\Pi_2)^{0.7070} * (\Pi_3)^{-0.6336} * (\Pi_4)^{-1.2727}$$

Brake thermal efficiency (Z_2)

$$(Bsfc) = 0.7348 * (\Pi_1)^{-0.1016} * (\Pi_2)^{0.7070} * (\Pi_3)^{-0.6336} * (\Pi_4)^{-1.2727}$$

10.10.4.2 Analysis of the Mathematical Model for the Dependent Pi Term Brake-Specific Fuel Consumption

Z2 is the brake-specific fuel consumption. The formulated model based on the experimental data is as follows:

$$Z_2 = 0.7348 * (\Pi_1)^{-0.1016} * (\Pi_2)^{0.7070} * (\Pi_3)^{-0.6336} * (\Pi_4)^{-1.2727}$$

where
Z_2 = brake-specific fuel consumption
π_1 = Pi terms related to the blend formation and time involvement
π_2 = Pi terms related to the fuel consumption and engine load

π_3 = Pi terms related to the fuel characteristics

π_4 = Pi terms related to the engine specifications

The term π_2 has a positive index 0.7070, which is directly proportional to the brake thermal efficiency. The π_2 term in the brake-specific fuel consumption decreases with an increase in the mass of fuel and blend formation and with a decrease in the stroke length and aniline point.

The absolute index of π_1 is −0.1016. Thus, the term is related to the blend formation and time involved in the experimental model. The value of the index is negative indicating that the brake-specific fuel consumption is inversely proportional to terms related to the blend formation and time involved. The brake-specific fuel consumption decreases with a decrease in the bore diameter and flashpoint and with an increase in the cetane number, kinematic viscosity, blend formation and time.

The term π_3 has a negative index −0.6336, which is inversely proportional to brake-specific fuel consumption. The π_3 terms in the brake-specific fuel consumption increases with a decrease in the aniline point and diesel index and with an increase in the cetane number.

The term π_4 has a negative index −1.2727, which is inversely proportional to brake-specific fuel consumption. The π_4 terms in brake-specific fuel consumption increases with a decrease in the stroke length, cubic capacity, flue tank capacity and engine speed. Table 10.27 presents the comparison of experimentally calculated values and mathematical model equation base values.

Table 10.27 Comparison of experimentally calculated values and mathematical model equation base values.

Sr. No.	Calculated from mathematical model	Experimental observation	Error = experimental observation − calculated value
1	13.7323	13.9388	0.2064
2	12.2896	12.2947	0.0051
3	11.5461	10.9250	−0.6211
4	10.6962	10.3647	−0.3315
5	9.9682	10.1431	0.1749
6	8.6128	8.8140	0.2012
7	7.7439	8.1390	0.3951
8	7.2762	7.2401	−0.0361
9	6.7294	6.7574	0.0280
10	6.2270	6.1659	−0.0611
11	6.5045	6.2218	−0.2827
12	5.8633	5.8919	0.0286
13	5.5489	5.6255	0.0766
14	5.1262	5.1864	0.0602
15	4.7537	4.8458	0.0921

10.10.5 Artificial Neural Network Simulation

This chapter discusses the necessity to formulate ANN simulation of the concerned experimental data. Multilayer feedforward topology is decided for the network. MATLAB software is used for training the network for the response variable and to calculate the percentage error plot for prediction of the network for the concerned variable. Comparison graphs of experimental data, mathematical model-based results and ANN responses for the dependent variables are discussed.

10.10.5.1 Introduction

Based on experimental data for the dependent terms for brake thermal efficiency and brake-specific fuel consumption, experimental-data-based modelling was obtained [17, 19, 61]. Because the verification of experimental-data-based models is not in close proximity in such a complicated event, this became important to create ANN simulations of the observational data. Three layers make up the simulation. The first layer is referred to as the input layer. The total number of independent variables equals the total number of neurons in the input layer. The second layer is the hidden layer. The third layer is the output layer.

10.10.5.2 Procedure for the Formation of ANN Simulation

ANN simulation is run on the MATLAB software. The steps are as follows:

1) The observed data from the experimental study is the input data, and the output data are imported into the program, respectively.
2) The prestd function reads and suitably sizes the input and output data. The prestd preprocess function pre-processes the data so that the mean is 0 and the standard deviation is 1.
3) In the pre-processing step, the input and output data are normalised using mean and standard deviation.
4) The input and output data are then classified into three types: testing, verification and learning. The first 75% of the observations for both brake thermal efficiency and brake-specific fuel consumption are chosen for training, the final 75% for validating and the middle overlapping 50% for tests.
5) The data is then stored in structures for training, testing and validation.
6) Looking at the pattern of the data, a feedforward backpropagation-type neural network is chosen.
7) This network is then trained using the training data.
8) The regression analysis and representations are carried out using standard functions. The values of the regression coefficients and the equations of the regression lines are depicted on a single graph drawn for the single dependent pi terms.

10.10.5.3 ANN Program for Brake Thermal Efficiency

MATLAB software is selected for developing ANN. The program executed for brake thermal efficiency operation is as follows:

```
clear all;
close all;
inputs3=[
```

```
 3.361134       1.546179688     79.409      0.01312683
 8.445761       1.529581151     80.542      0.01313494
11.92674        1.49865104      82.3        0.01312683
14.86954        1.443101042     85.93       0.01312683
23.21995        1.42909625      88.34       0.01313494
 2.665727       0.773089844     79.409      0.01312683
 6.398304       0.764790575     80.542      0.01313494
 9.025641       0.74932552      82.3        0.01312683
11.43811        0.721550521     85.93       0.01312683
19.15646        0.714548125     88.34       0.01313494
 2.515056       0.515393229     79.409      0.01312683
 5.886439       0.509860384     80.542      0.01313494
 7.736264       0.499550347     82.3        0.01312683
 9.913029       0.481033681     85.93       0.01312683
16.25397        0.476365417     88.34       0.01313494

]
 a1=inputs3
 a2=a1
input_data=a2;
output3=[

 25.8273
 29.2810
 32.9520
 34.7332
 35.4921
 40.8440
 44.2315
 49.7231
 53.2748
 58.3856
 57.8613
 61.1010
 63.9941
 69.4127
 74.2919

]
 y1=output3
 y2=y1
```

```
size(a2);
size(y2);
p=a2';
sizep=size(p);
t=y2';
sizet=size(t);
[S Q]=size(t)
[pn,meanp,stdp,tn,meant,stdt] = prestd(p,t);
 net = newff(minmax(pn),[15 1],{'logsig' 'purelin'},'trainlm');
net.performFcn='mse';
net.trainParam.goal=.99;
net.trainParam.show=200;
net.trainParam.epochs=50;
net.trainParam.mc=0.05;
net = train(net,pn,tn);
an = sim(net,pn);
[a] = poststd(an,meant,stdt);
error=t-a;
x1=1:15;
plot(x1,t,'rs-',x1,a,'b-')
legend('Experimental','Neural');
title('Output (Red) and Neural Network Prediction
(Blue) Plot');
xlabel('Experiment No.');
ylabel('Output');
grid on;
figure
error_percentage=100*error./t
plot(x1,error_percentage)
legend('percentage error');
axis([0 15 -100 100]);
title('Percentage Error Plot in Neural Network Prediction');
xlabel('Experiment No.');
ylabel('Error in %');
grid on;
for ii=1:15
xx1=input_data(ii,1);
yy2=input_data(ii,2);
zz3=input_data(ii,3);
xx4=input_data(ii,4);
pause
yyy(1,ii)=85.1138*power(xx1,0.1758)*power(
yy2,-0.738)*power(zz3,-0.695)*power(xx4,-0.449);
yy_practical(ii)=(y2(ii,1));
yy_eqn(ii)=(yyy(1,ii))
```

```
yy_neur(ii)=(a(1,ii))
yy_practical_abs(ii)=(y2(ii,1));
yy_eqn_abs(ii)=(yyy(1,ii));
yy_neur_abs(ii)=(a(1,ii));
pause
end
figure;
plot(x1,yy_practical_abs,'r-',x1,yy_eqn_abs,'b-',x1,yy_neur_
abs,'k-');
legend('Practical','Equation','Neural');
title('Comparision between practical data, equation based data
and neural based data');
xlabel('Experimental');
grid on;
figure;
plot(x1,yy_practical_abs,'r-',x1,yy_eqn_abs,'b-');
legend('Practical',' Equation');
title('Comparision between practical data, equation based data
and neural based data');
xlabel('Experimental');
grid on;
figure;
plot(x1,yy_practical_abs,'r-',x1,yy_neur_abs,'k-');
legend('Practical','Neural');
title('Comparision between practical data, equation based data
and neural based data');
xlabel('Experimental');
grid on;
error1=yy_practical_abs-yy_eqn_abs
figure
error_percentage1=100*error1./yy_practical_abs;
plot(x1,error_percentage,'k-',x1,error_percentage1,'b-');
legend('Neural','Equation');
axis([0 15 -100 100]);
title('Percentage Error Plot in Equation (blue), Neural Network
(black) Prediction');
xlabel('Experiment No.');
ylabel('Error in %');
grid on;
meanexp=mean(output3)
meanann=mean(a)
meanmath=mean(yy_eqn_abs)
mean_absolute_error_performance_function = mae(error)
mean_squared_error_performance_function = mse(error)
```

The performance graph for the brake thermal efficiency observed during ANN simulation is presented from Figures 10.2–10.5.

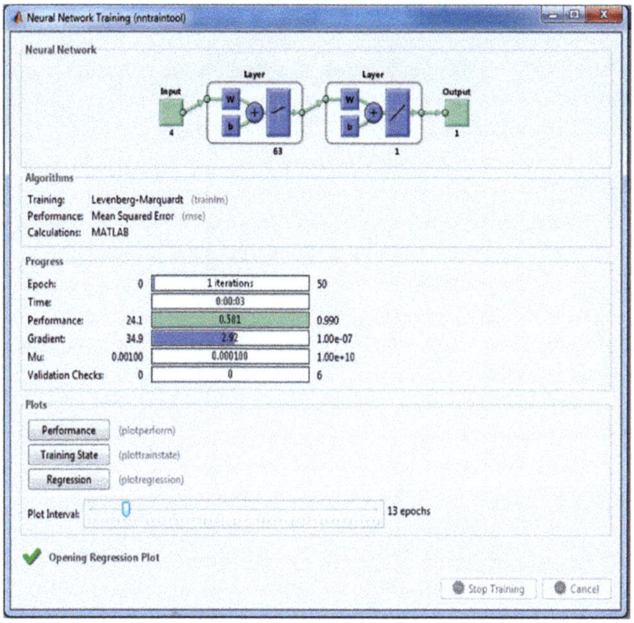

Figure 10.2 Training of the network for the brake thermal efficiency.

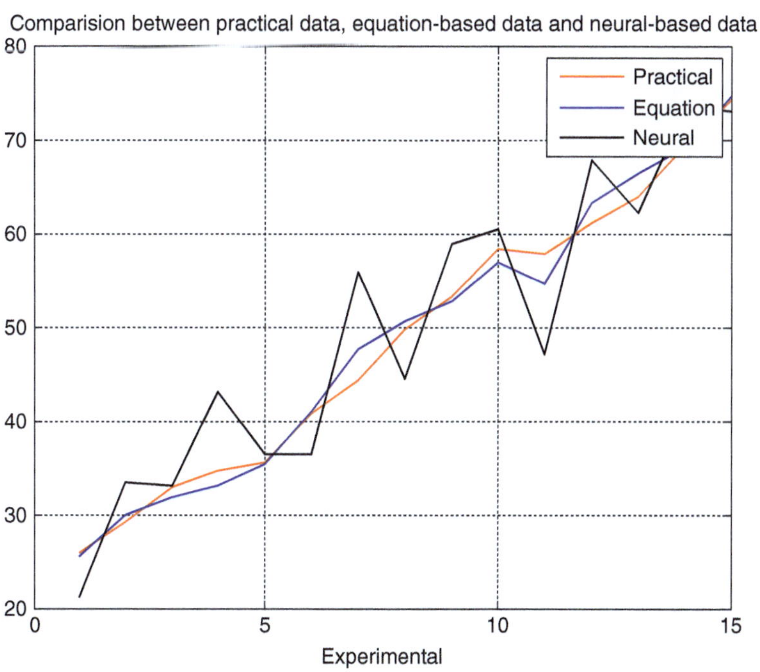

Figure 10.3 Graph of comparison with experimental database, neural network prediction and equation base prediction for the network for the brake thermal efficiency.

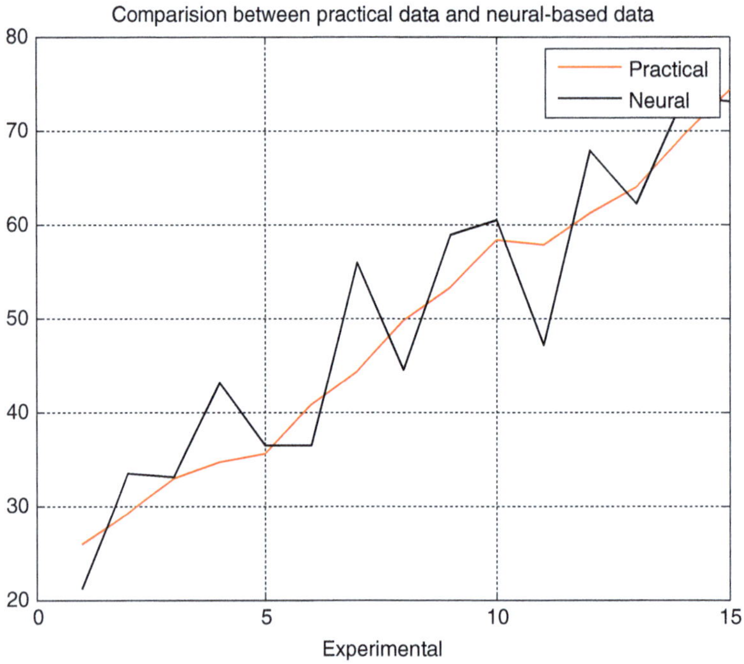

Figure 10.4 Graph of comparison with experimental database and neural network prediction for the network for the brake thermal efficiency.

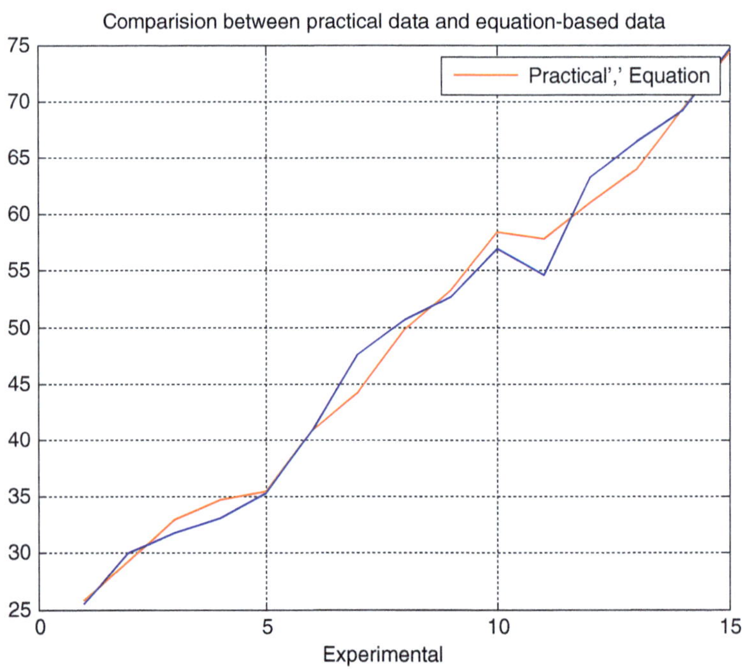

Figure 10.5 Graph of comparison with experimental database and equation base prediction for the network for the brake thermal efficiency.

Table 10.28 Comparison of experimental calculated values, equation base values and ANN based values for the brake thermal efficiency.

Sr. No.	Calculated from mathematical model	Experimental observation	Calculated from ANN
1	25.5516	25.8273	21.2003
2	29.9799	29.2810	33.3832
3	31.8660	32.9520	33.0989
4	33.0555	34.7332	43.0639
5	35.3126	35.4921	36.3728
6	40.9147	40.8440	36.3376
7	47.6204	44.2315	55.9187
8	50.6068	49.7231	44.4205
9	52.6469	53.2748	58.8392
10	56.9381	58.3856	60.3852
11	54.6252	57.8613	47.1562
12	63.2968	61.1010	67.8485
13	66.4347	63.9941	62.1327
14	69.2473	69.4127	73.3544
15	74.6129	74.2919	73.0846

Table 10.29 Comparison between observed and computed values of the dependent pi term for the brake thermal efficiency.

Dependent pi term	π_{D1} brake thermal efficiency
Mean experiment	48.7604
Mean ANN	49.7771
Mean model	48.8473 s

From the above comparison of the phenomenal response by a conventional approach and ANN simulation, it seems to be that the curve obtained by dependent pi terms for brake thermal efficiency π_D overlaps due to the lower percentage of error, which is on the positive side. The comparison of experimental calculated values, equation base values and ANN-based values for the brake thermal efficiency is presented in Tables 10.28 and 10.29.

10.10.5.4 ANN Program for Brake-Specific Fuel Consumption

MATLAB software is selected for developing ANN. The program executed for brake-specific fuel consumption operation is as follows:

```
clear all;
close all;
inputs3=[
```

3.361134	1.546179688	79.409	0.01312683
8.445761	1.529581151	80.542	0.01313494
11.92674	1.49865104	82.3	0.01312683
14.86954	1.443101042	85.93	0.01312683
23.21995	1.42909625	88.34	0.01313494
2.665727	0.773089844	79.409	0.01312683
6.398304	0.764790575	80.542	0.01313494
9.025641	0.74932552	82.3	0.01312683
11.43811	0.721550521	85.93	0.01312683
19.15646	0.714548125	88.34	0.01313494
2.515056	0.515393229	79.409	0.01312683
5.886439	0.509860384	80.542	0.01313494
7.736264	0.499550347	82.3	0.01312683
9.913029	0.481033681	85.93	0.01312683
16.25397	0.476365417	88.34	0.01313494

```
]
 a1=inputs3
 a2=a1
input_data=a2;
output3=[
```

13.9388
12.2947
10.9250
10.3647
10.1431
8.8140
8.1390
7.2401
6.7574
6.1659
6.2218
5.8919
5.6255
5.1864
4.8458

```
]
 y1=output3
 y2=y1
size(a2);
```

```
size(y2);
p=a2';
sizep=size(p);
t=y2';
sizet=size(t);
[S Q]=size(t)
[pn,meanp,stdp,tn,meant,stdt] = prestd(p,t);
net = newff(minmax(pn),[15 1],{'logsig' 'purelin'},'trainlm');
net.performFcn='mse';
net.trainParam.goal=.99;
net.trainParam.show=200;
net.trainParam.epochs=50;
net.trainParam.mc=0.05;
net = train(net,pn,tn);
an = sim(net,pn);
[a] = poststd(an,meant,stdt);
error=t-a;
x1=1:15;
plot(x1,t,'rs-',x1,a,'b-')
legend('Experimental','Neural');
title('Output (Red) and Neural Network Prediction
(Blue) Plot');
xlabel('Experiment No.');
ylabel('Output');
grid on;
figure
error_percentage=100*error./t
plot(x1,error_percentage)
legend('percentage error');
axis([0 15 -100 100]);
title('Percentage Error Plot in Neural Network Prediction');
xlabel('Experiment No.');
ylabel('Error in %');
grid on;
for ii=1:15
xx1=input_data(ii,1);
yy2=input_data(ii,2);
zz3=input_data(ii,3);
xx4=input_data(ii,4);
pause
yyy(1,ii)=0.7348*power(xx1,-0.1016)*power(yy2,0.7070)*power(
zz3,-0.6336)*power(xx4,-1.2727);
yy_practical(ii)=(y2(ii,1));
yy_eqn(ii)=(yyy(1,ii))
yy_neur(ii)=(a(1,ii))
yy_practical_abs(ii)=(y2(ii,1));
```

```
yy_eqn_abs(ii)=(yyy(1,ii));
yy_neur_abs(ii)=(a(1,ii));
pause
end
figure;
plot(x1,yy_practical_abs,'r-',x1,yy_eqn_abs,'b-',x1,yy_neur_
abs,'k-');
legend('Practical','Equation','Neural');
title('Comparision between practical data, equation based data
and neural based data');
xlabel('Experimental');
grid on;
figure;
plot(x1,yy_practical_abs,'r-',x1,yy_eqn_abs,'b-');
legend('Practical',' Equation');
title('Comparision between practical data, equation based data
and neural based data');
xlabel('Experimental');
grid on;
figure;
plot(x1,yy_practical_abs,'r-',x1,yy_neur_abs,'k-');
legend('Practical','Neural');
title('Comparision between practical data, equation based data
and neural based data');
xlabel('Experimental');
grid on;
error1=yy_practical_abs-yy_eqn_abs
figure
error_percentage1=100*error1./yy_practical_abs;
plot(x1,error_percentage,'k-',x1,error_percentage1,'b-');
legend('Neural','Equation');
axis([0 15 -100 100]);
title('Percentage Error Plot in Equation (blue), Neural Network
(black) Prediction');
xlabel('Experiment No.');
ylabel('Error in %');
grid on;
meanexp=mean(output3)
meanann=mean(a)
meanmath=mean(yy_eqn_abs)
mean_absolute_error_performance_function = mae(error)
mean_squared_error_performance_function = mse(error)
```

The performance graph for the brake-specific fuel consumption observed during ANN simulation is presented in Figures 10.6–10.9.

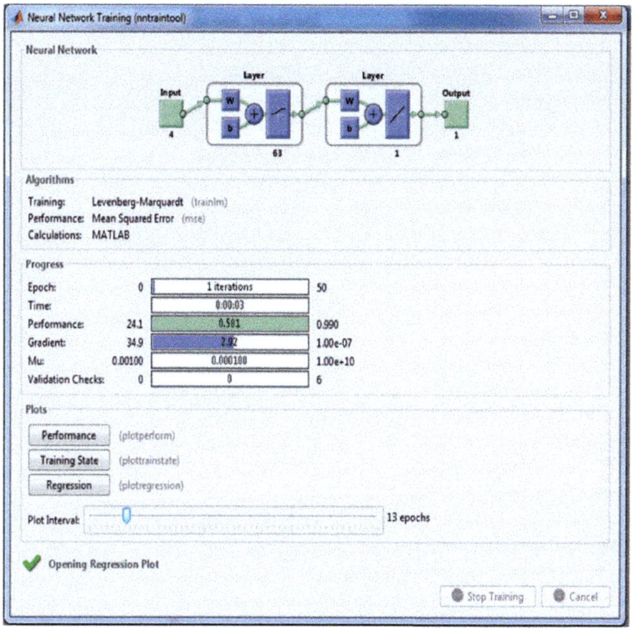

Figure 10.6 Training of the network for brake-specific fuel consumption.

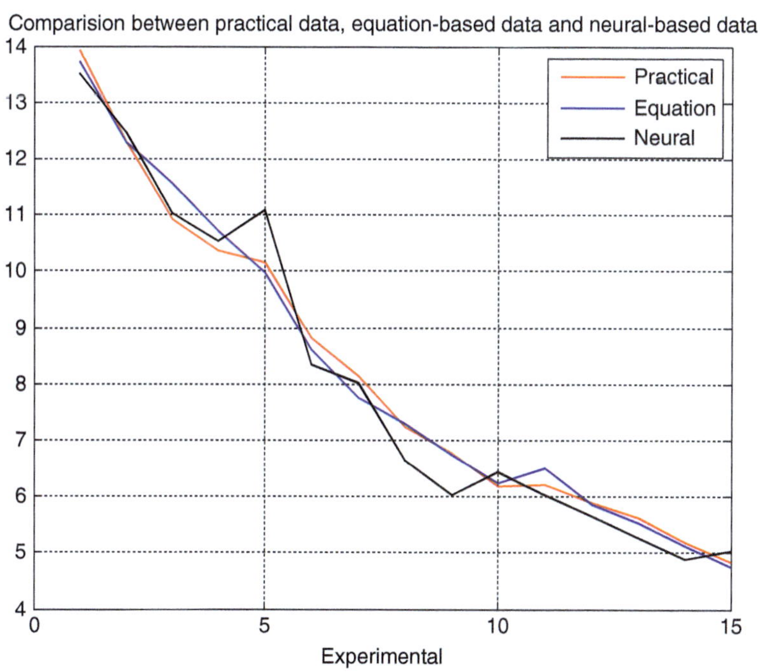

Figure 10.7 Graph of comparison with experimental database, neural network prediction and equation base prediction for the network for brake-specific fuel consumption.

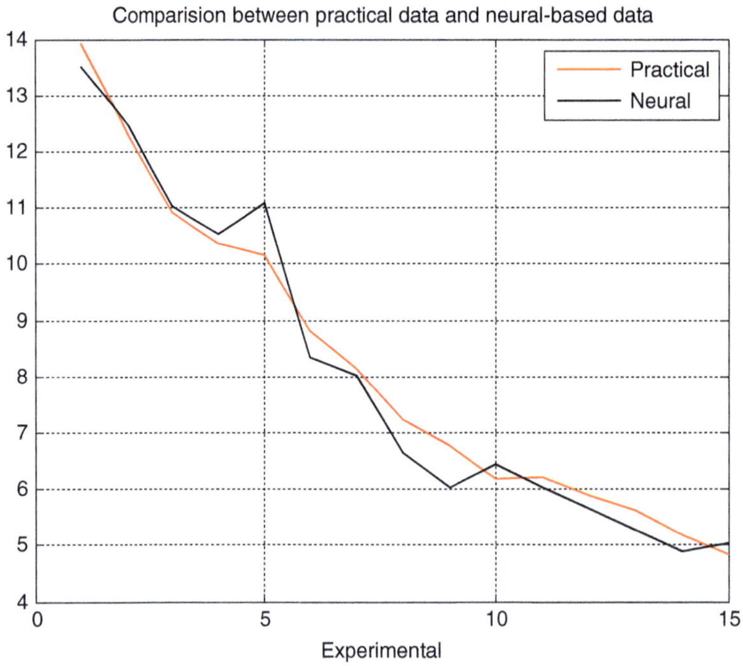

Figure 10.8 Graph of comparison with experimental database and neural network prediction for the network for brake-specific fuel consumption.

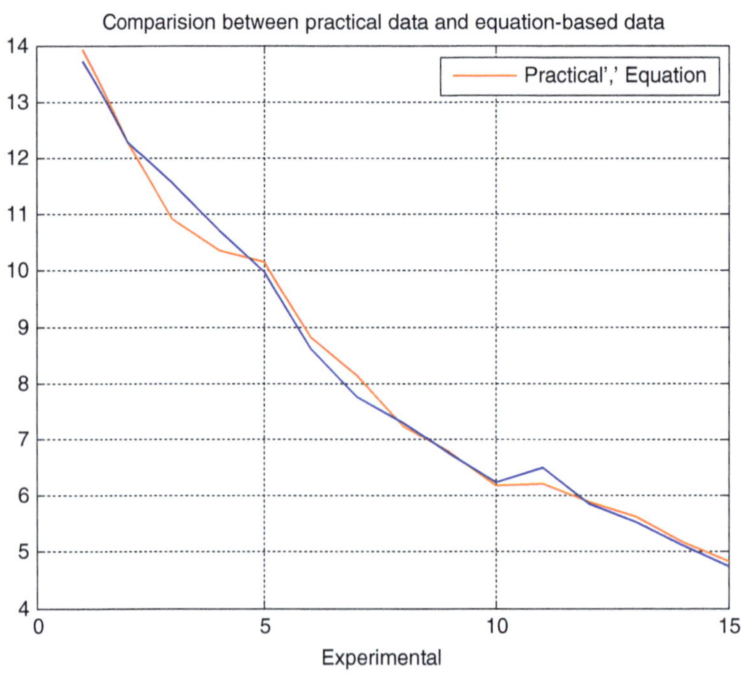

Figure 10.9 Graph of comparison with experimental database and equation base prediction for the network for brake-specific fuel consumption.

From the above comparison of the phenomenal response by a conventional approach and ANN simulation, it seems to be that the curve obtained by dependent pi terms for brake-specific fuel consumption π_{D2} overlaps due to the lower percentage of error, which is on the positive side. The comparison of experimental calculated values, equation base values and ANN-based values for brake-specific fuel consumption is presented in Tables 10.30 and 10.31.

10.10.6 Sensitivity Analysis

Sensitivity analysis is used to evaluate the sensitiveness of each independent pi term involved in the model. The result of the dependent pi term is discussed by introducing the percentage change in each independent pi term. The procedure of sensitivity analysis is a change of ±10% in individual pi term independently introduced (one at time) to obtain the corresponding change in the dependent pi term [62, 63]. The percentage

Table 10.30 Comparison of experimental calculated values, equation base values and ANN based values for brake-specific fuel consumption.

Sr. No.	Calculated from mathematical model	Experimental observation	Calculated from ANN
1	13.7323	13.9388	13.5266
2	12.2896	12.2947	12.4723
3	11.5461	10.9250	11.0329
4	10.6962	10.3647	10.5268
5	9.9682	10.1431	11.0893
6	8.6128	8.8140	8.3383
7	7.7439	8.1390	8.0254
8	7.2762	7.2401	6.6492
9	6.7294	6.7574	6.0256
10	6.2270	6.1659	6.4281
11	6.5045	6.2218	6.0248
12	5.8633	5.8919	5.6491
13	5.5489	5.6255	5.2756
14	5.1262	5.1864	4.8848
15	4.7537	4.8458	5.0318

Table 10.31 Comparison between observed and computed values of the dependent pi term for brake-specific fuel consumption.

Dependent pi term	π_{D2} brake-specific fuel consumption
Mean experiment	8.1703
Mean ANN	8.0654
Mean model	8.1746

change is plotted, and the effects of the introduced change on the response variable for brake thermal efficiency and brake-specific fuel consumption are analysed. Further optimisation of the developed model to determine the best set of input quantities, which will result in maximisation/minimisation of the response variables, (Limiting values) is explained. The response variable terms obtained by field study observation are calculated, i.e. model-based values and ANN-based values are tabulated and compared for acceptability.

10.10.6.1 Sensitivity Analysis of Brake Thermal Efficiency and Brake Specific Fuel Consumption

The influences of various independent π terms have been studied by analysing the indices of various independent pi terms in the models. Through the technique of sensitivity analysis, the change in the value of a dependent π term caused due to an introduced change in the value of an individual independent π term is evaluated. In this case, the change of ±10% is introduced in the individual π terms independently (one at a time). Thus, the total range of the introduced change is 20%. The effect of this introduced change on the percentage change value of the dependent π term is evaluated. The average value of the change in the dependent π terms due to the introduced change of 20% in each π term is shown in Tables 10.32–10.35. These values are plotted in the graph shown in Figures 10.10 and 10.11.

The sensitivity as evaluated is represented and discussed below.

Table 10.32 Sensitivity analysis for the brake thermal efficiency (Z1).

	Pi 1	Pi 2	Pi 3	Pi 4	Z1
Mean	10.18748	0.910141	83.30429	0.01313	45.4482
10% above	11.2062	0.910141	83.30429	0.01313	46.21616146
10% below	9.16873	0.910141	83.30429	0.01313	44.61417545
% Change					0.035248613
Mean	10.18748	0.910141	83.30429	0.01313	45.44823666
10% above	10.18748	1.00116	83.30429	0.01313	42.36129577
10% below	10.18748	0.81913	83.30429	0.01313	49.12313392
% Change					−0.14878109
Mean	10.18748	0.910141	83.30429	0.01313	45.44823666
10% above	10.18748	0.910141	91.6347	0.01313	42.53526291
10% below	10.18748	0.910141	74.9739	0.01313	48.90108483
% Change					−0.14006752
Mean	10.18748	0.910141	83.30429	0.01313	45.44823666
10% above	10.18748	0.910141	83.30429	0.01444	43.54434098
10% below	10.18748	0.910141	83.30429	0.01182	47.64991728
% Change					−0.09033522

Table 10.33 Percentage change in the brake thermal efficiency (Z1) with +/− 10% variation in the mean value of each pi term.

Sensitivity chart	
Pi terms	**Z1**
Pi 1	3.52%
Pi 2	−14.87%
Pi 3	−14.00%
Pi 4	−9.03%

Table 10.34 Sensitivity analysis for brake-specific fuel consumption (Z2).

	Pi 1	Pi 2	Pi 3	Pi 4	Z1
Mean	10.18748	0.910141	83.30429	0.01313	8.180616082
10% above	11.2062	0.910141	83.30429	0.01313	8.101781284
10% below	9.16873	0.910141	83.30429	0.01313	8.268656919
% Change					−0.02039891
Mean	10.18748	0.910141	83.30429	0.01313	8.180616082
10% above	10.18748	1.00116	83.30429	0.01313	8.750858069
10% below	10.18748	0.81913	83.30429	0.01313	7.5933858
% Change					0.141489621
Mean	10.18748	0.910141	83.30429	0.01313	8.180616082
10% above	10.18748	0.910141	91.6347	0.01313	7.701221353
10% below	10.18748	0.910141	74.9739	0.01313	8.745365353
% Change					−0.12763635
Mean	10.18748	0.910141	83.30429	0.01313	8.180616082
10% above	10.18748	0.910141	83.30429	0.01444	7.246120322
10% below	10.18748	0.910141	83.30429	0.01182	9.354521335
% Change					−0.25773132

Table 10.35 Percentage change in brake-specific fuel consumption (Z2) with +/− 10% variation in the mean value of each pi term.

Sensitivity chart	
Pi terms	**Z2**
Pi 1	−2.03%
Pi 2	14.14%
Pi 3	−12.76%
Pi 4	−25.77%

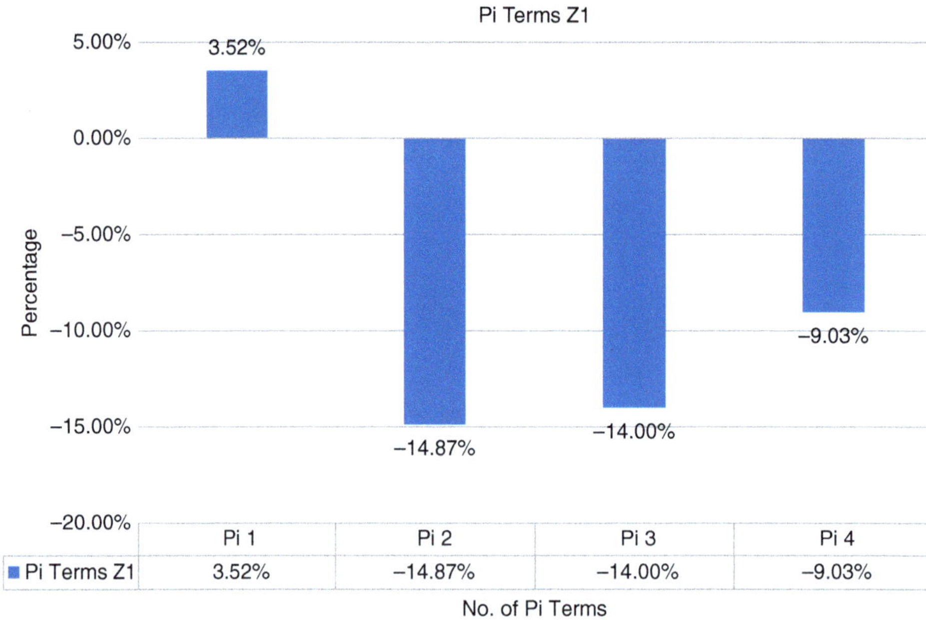

Figure 10.10 Percentage change in pi terms with variations in the dependent term Z_1 (brake thermal efficiency).

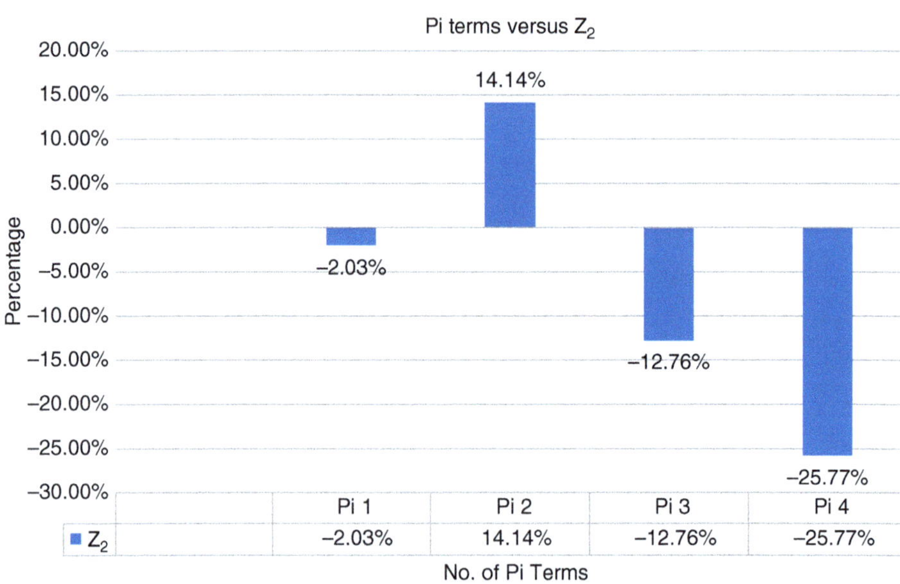

Figure 10.11 Percentage change in pi terms with variations in the dependent term Z_2 (brake-specific fuel consumption).

10.10.6.1.1 Effect of an Introduced Change on the Dependent π Term- Brake Thermal Efficiency

In this model, when a total change of 20% is introduced in the value of the independent π term π_1, a change of 3.52% occurs in the value of Z_1, i.e. brake thermal efficiency (computed from the model). The change brought in the value of Z_1 because of the change in the value of the other independent π terms, namely π_2, is −14.57%. Similarly, the changes of about −14.00%, and −9.03% take place in the value of Z_1 because of the change in the values of π_3 and π_4, respectively.

It can be seen that the highest change takes place in Z_1 because of the π term π_2. Thus, π_2 related to fuel consumption and engine load is the most sensitive π term. The sequence of various π terms in the descending order of sensitivity is π_3, π_4 and π_1.

The sensitivity as evaluated is represented and discussed below.

In this model, when a total change of 20% is introduced in the value of the independent π term π_1, a change of −2.03% occurs in the value of Z_1, i.e. brake thermal efficiency (computed from the model). The change brought in the value of Z_1 because of the change in the value of the other independent π terms, namely π_2, is 14.14%. Similarly, the changes of about −12.76% and −25.77% take place in the value of Z_1 because of the change in the values of π_3 and π_4, respectively.

It can be seen that the highest change takes place in Z_1 because of the π term π_2. Thus, π_4 related to engine specification is the most sensitive π term. The sequence of various π terms in the descending order of sensitivity is π_3, π_4 and π_1.

10.10.7 Optimisation of Models

Models for the phenomena of brake thermal efficiency and brake-specific fuel consumption of the mechanism have been created. The eventual aim of this effort is not only to build models but also to identify the optimum collection of independent variables that will result in the maximisation/minimisation of the optimal solution [10, 27, 30, 64]. In this case, models correspond to the brake thermal efficiency. The objective function corresponds to the modes. The model for the brake thermal efficiency (Z_1) must be maximised. Because the models are non-linear, they must be transformed to a linear form for optimisation. Taking the log on both sides of the model, it can be obtained. We may utilise the linear programming approach, as demonstrated below, to maximise the linear function.

For the dependent π term(Z_1), we have

$$(Z_1) = K1 * (\pi_1)^{a1} * (\pi_2)^{b1} * (\pi_3)^{c1} * (\pi_4)^{d1}$$

Taking the log on both sides of the equation, we have

$$\log(Z_1) = \log K1 + a_1 * \log(\pi_1) + b1 * \log(\pi_2) + c1 * \log(\pi_3) + d1 * \log(\pi_4)$$

Let Log(Z_1) = Z, LogK1 = K1', log(π_1) = X1, log(π_2) = X2, log(π_3) = X3 and log(π_4) = X4, then the linear model in the form of a first-degree polynomial can be written as follows:

$$Z = K1 + a_1 * X1 + b_1 * X2 + c_1 * X3 + d_1 * X4$$

Thus, the equation will be the objective function for the optimisation or to be very specific for maximisation for the purpose of constructing the linear programming problem. If we denote the maximum and minimum values of a dependent π term Z_1 by π_1 max. and π_1 min., then the first two constraints for the problem will be obtained by taking the log of these quantities and by substituting the values of multipliers of all other variables, except the one under consideration, equal to zero. Let the log of the limits be defined as C1 and C2 {i.e. C1 = log(π_1 max.) and C2 = log(π_1 min.)}. Thus, the equations of the constraints will be as follows:

$$1 * X1 + 0 * X2 + 0 * X3 + 0 * X4 \leq C1$$

$$1 * X1 + 0 * X2 + 0 * X3 + 0 * X4 \geq C2$$

The other constraints can be likewise found as follows:

$$0 * X1 + 1 * X2 + 0 * X3 + 0 * X4 \leq C3$$

$$0 * X1 + 1 * X2 + 0 * X3 + 0 * X4 \geq C4$$

$$0 * X1 + 0 * X2 + 1 * X3 + 0 * X4 \leq C5$$

$$0 * X1 + 0 * X2 + 1 * X3 + 0 * X4 \geq C6$$

$$0 * X1 + 0 * X2 + 0 * X3 + 1 * X4 \leq C7$$

$$0 * X1 + 0 * X2 + 0 * X3 + 1 * X4 \geq C8$$

After solving this linear programming problem, one obtains the maximum value of Z1 as well as the set of variable values required to accomplish this maximum value. Finding the antilog of the values of Z1, X1, X2, X3 and X4 yields the values of the independent pi terms. In this scenario, the real values of the multipliers and variables are discovered and replaced in the previous equations, and the actual issue may be presented as follows. This may now be solved as a linear programming problem in MS Excel using the MS Solver. Thus, the actual problem is to maximise Z1, where

$$Z1 = K1 + a_1 * X1 + b_1 * X2 + c_1 * X3 + d_1 * X4$$

$$Z1 = \log(85.1138) + 0.1738 * \log(\pi_1) - 0.7380 * \log(\pi_2) - 0.695 * \log(\pi_3) - 0.449 * \log(\pi_4)$$

$$Z1 = \log(85.1138) + (0.1738 * X1) + (-0.7380 * X2) + (-0.6950 * X3) + (-0.4490 * X4)$$

Subject to the following constraints:

$$1 * X1 + 0 * X2 + 0 * X3 + 0 * X4 \leq 1.3658$$

$$1 * X1 + 0 * X2 + 0 * X3 + 0 * X4 \geq 0.4005$$

The other constraints can be likewise found as follows:

$$0*X1+1*X2+0*X3+0*X4 \leq 0.1892$$

$$0*X1+1*X2+0*X3+0*X4 \geq -0.3221$$

$$0*X1+0*X2+1*X3+0*X4 \leq 1.9461$$

$$0*X1+0*X2+1*X3+0*X4 \geq 1.8998$$

$$0*X1+0*X2+0*X3+1*X4 \leq -1.8820$$

$$0*X1+0*X2+0*X3+1*X4 \geq -1.8820$$

On solving the above problem with the MS solver, we obtain

$$X1 = 1.3688, X2 = 0.1892, X3 = 1.8998, X4 = -1.882 \text{ and } Z = 1.3854$$

Thus, Z1 Max. = Antilog(1.5229) = 33.33, and corresponding to this values of the Z1 Max., the values of independent π terms are obtained by taking the antilog of X1, X2, X3 and X4. These values are 23.2166, 1.5459, 88.3283 and 0.01312, respectively.

A similar procedure has been adopted to optimise the models for Z2.

Z2 Min. = Antilog(0.7892) = 6.1554, and corresponding to this value of Z2 Min., the values of the independent π terms are obtained by taking the antilog of 2.5147, 0.4763, 79.396 and 0.01312, respectively.

10.10.8 Reliability of the Models

Therefore, prior to proceeding to the stage of sensitivity of inputs, the validity of the model must be determined [19, 65]. This is because even if observed data is purified using recognised statistics-based approaches, there is a probability that some impure data will enter the mathematical processing of the data, even if MATLAB is used.

The strategy to determine validity would be to replace known inputs into the model for each observation and determine the difference in response by model and actually observed response. This will provide a pattern of the error distribution and the frequency with which it occurs. The model's dependability is established using this distribution and reliability literature.

10.10.8.1 Reliability of Model for Z1

The error calculated during the brake thermal efficiency is 4.34%.

The reliability of the brake thermal efficiency mathematical model is 100−4.34 = 95.66%.

10.10.8.2 Reliability of Model for Z2

The error calculated during the brake thermal efficiency is 5.25%.

The reliability of the brake-specific fuel consumption mathematical model is 100−5.25 = 94.75%.

10.11 Summary

The following is the conclusion of an experimental result obtained from the single-cylinder water-cooled diesel engine fuelled with treated transformer oil–diesel fuel blends. Blends of treated transformer oil and diesel fuel can be the best alternative for the diesel engine, as they possess the same calorific value as diesel fuel. The brake thermal efficiency of blends up to 20% TTO is higher than that of diesel fuel. It decreases with an increase in the proportion of treated transformer oil and increases with an increase in load. The brake-specific fuel efficiency (BSFC) decreases as the load increases with the proportion of TTO in blends. BSFC is a constant at a load of 15 kg, which indicates the existence of the BSFC value at the 15 kg load. The exhaust gas temperature of 10% and 20% TTO blends is less than that of pure diesel, but as the percentage of treated transformer oil increases, the exhaust gas temperature also increases.

References

1 Waghmare, S.N., Shelare, S.D., Tembhurkar, C.K., and Jawalekar, S.B. (2020). Pyrolysis system for environment-friendly conversion of plastic waste into fuel. *Lect. Notes Mech. Eng.* 131-138: https://doi.org/10.1007/978-981-15-4748-5_13.

2 Aglawe, K.R., Yadav, R.K., and Thool, S.B. (2021). Preparation, applications and challenges of nanofluids in electronic cooling: a systematic review. *Mater. Today Proc.* 43: 366–372. https://doi.org/10.1016/j.matpr.2020.11.679.

3 Belkhode, P.N., Shelare, S.D., Sakhale, C.N. et al. (2021). Performance analysis of roof collector used in the solar updraft tower. *Sustainable Energy Technol. Assess.* 48: 101619. https://doi.org/10.1016/j.seta.2021.101619.

4 Belkhode, P.N., Ganvir, V.N., Shende, A.C., and Shelare, S.D. (2021). Utilization of waste transformer oil as a fuel in diesel engine. *Mater. Today Proc.* 49: 262–268. https://doi.org/10.1016/j.matpr.2021.02.008.

5 Atmanli, A., Ileri, E., Yuksel, B., and Yilmaz, N. (2015). Extensive analyses of diesel-vegetable oil-n-butanol ternary blends in a diesel engine. *Appl. Energy* 145: 155–162. https://doi.org/10.1016/j.apenergy.2015.01.071.

6 Waghmare, S., Shelare, S., Sirsat, P. et al. (2020). Development of an innovative multi-operational furnace. *Int. J. Sci. Technol. Res.* 9: 885–889.

7 Simsek, S. (2020). Effects of biodiesel obtained from Canola, sefflower oils and waste oils on the engine performance and exhaust emissions. *Fuel* 265: 117026. https://doi.org/10.1016/j.fuel.2020.117026.

8 Belkhode, P.N. (2018). Mathematical modelling of liner piston maintenance activity using field data to minimize overhauling time and human energy consumption. *J. Inst. Eng. Ser. C* 99: 701–709.

9 Undirwade, S.K. (2018). Development and optimization of experimental data based models for bamboo sliver cutting by using human powered flywheel motor. *Int. J. Mech. Prod. Eng. Res. Dev.* 8: 1007–1020. https://doi.org/10.24247/ijmperdfeb2018121.

10 Belkhode, P., Sakhale, C., and Bejalwar, A. (2020). Evaluation of the experimental data to determine the performance of a solar chimney power plant. *Mater. Today Proc.* 27: 102–106. https://doi.org/10.1016/j.matpr.2019.09.006.

11 Modak, J.P., Belkhode, P.N., Bodhankar, D. et al. (2008). Modeling and analysis of front suspension for improving vehicle ride and handling. *First Int. Conf. Emerg. Trends Eng. Technol.*, IEEE 731–734.

12 Sahu, P., Shelare, S., and Sakhale, C. (2020). Smart cities waste management and disposal system by smart system: a review. *Int. J. Sci. Technol. Res.* 9: 4467–4470.
13 Urunkar, R.U. and Deshpande, P.P. (2014). Study of drive mechanisms of bicycle, tricycle or like vehicles to optimize operating performance -a review. *J. Eng. Res. Appl.* 4: 214–219.
14 Khope, P.B. and Modak, J.P. (2014). Development of mathematical model for forage cutting process energized by human powered flywheel motor. *Int. J. Sci. Technol. Res.* 2 (3): 4–10.
15 Kadu, R.S., Awari, G.K., Sakhale, C.N., and Modak, J.P. (2014). Formulation of mathematical model for the investigation of tool wears in boring machining operation on cast iron using carbide and CBN tools. *Procedia Mater. Sci.* 6: 1710–1724. https://doi.org/10.1016/j.mspro.2014.07.157.
16 Zakiuddin, K.S. and Modak, J.P. (2012). Post harvest crop processing machine. *Agric. Eng. Int.: CIGR J.* 14: 99–104.
17 Zakiuddin, K.S. and Modak, J.P. (2010). Formulation of data based ann model for the human powered fodder - chopper. *J. Theor. Appl. Inf. Technol.* 15: 79–85.
18 Ghuge, V.D. and Modak, J.P. (2013). Developing an instrumentation system for measurement of rotational speed and slide mesh of spur gears of a novel gearbox for human powered flywheel motor (Hpfm). *Int. J. Sci. Technol. Res.* 2: 40–43.
19 Chandak, P.A., Lende, A., and Modak, J. (2018). Modeling of human power flywheel motor through artificial neural network- a novel approach. *Procedia Comput. Sci.* 125: 77–84. https://doi.org/10.1016/j.procs.2017.12.012.
20 Bhatkulkar, H.S. and Modak, J.P. (2014). Design & development of nursery fertilizer mixer energized by human powered flywheel motor. *Int. J. Res. Emerg. Sci. Technol.* 1: 69–73.
21 Dhale, A.D. and Modak, J.P. (2012). Formulation of experimental data based model for oil press using human powered flywheel motor as energy source. *Agric Eng Int: CIGR J.* 14: 218–229.
22 Dhutekar, P., Mehta, G., Modak, J. et al. (2021). Establishment of mathematical model for minimization of human energy in a plastic moulding operation. *Mater. Today Proc.* 47 (9): 1–6. https://doi.org/10.1016/j.matpr.2021.05.330.
23 Modak, J.P. and Bapat, A.R. (1994). Formulation of a generalized experimental model for a manually driven flywheel motor and its optimization. *Appl. Ergon* 25: 119–122. https://doi.org/10.1016/0003-6870(94)90075-2.
24 Khope, P., Modak, J., and Singh, M. (2014). Development of human powered energy unit for fodder cutting. *J. Process. Energy Agric.* 18 (4): 158–160.
25 Shelare, S.D., Kumar, R., and Khope, P.B. (2021). Formulation of a mathematical model for quantity of deshelled nut in charoli nut deshelling machine. Lecture Notes. *Adv. Metrol. Meas. of Engineering Surfaces, Lect. Notes Mech. Eng.* 89–97. https://doi.org/10.1007/978-981-15-5151-2_9.
26 Shelare, S.D., Kumar, R., and Khope, P.B. (2021). Flywheel energy application in commercial and agricultural field: a typical review. In: *Lecture Notes in Mechanical Engineering*, 177–186. Springer https://doi.org/10.1007/978-981-16-1079-0_19.
27 Undirwade, S.K., Singh, M.P., and Sakhale, C.N. (2015). Experimental investigation of processing time, number of slivers and resistive torque required for human powered bamboo sliver cutting operation. *J. Bamboo Ratt.* 14: 33–51.
28 Mali, P.K., Sakhale, D.C.N., and Shelare, S.D. (2015). A literature review on design and devlopement of maize thresher. *Int. J. Pure Appl. Res. Eng. Technol.* 3: 9–14.

29 Sonde, V.M., Belkhode, P.N., and Sakhale, C.N. (2014). Approach to formulate mathematical model for wood chipper cutter using HPFM concept. *Int. Conf. Multidiscip. Res. Pract.* I: 509–511.

30 Sakhale, C.N., Waghmare, S.N., Undirwale, S. et al. (2014). Formulation and comparison of experimental based mathematical model with artificial neural network simulation and RSM (response surface methodology) model for optimal performance of sliver cutting operation of bamboo. *Procedia Mater. Sci.* 6: 877–891. https://doi.org/10.1016/j.mspro.2014.07.105.

31 Bejalwar, A. and Belkhode, P. (2018). Analysis of experimental setup of a small solar chimney power plant. *Procedia Manuf.* 20: 481–486. https://doi.org/10.1016/j.promfg.2018.02.071.

32 Mowade, S., Waghmare, S., Shelare, S., and Tembhurkar, C. (2020). Mathematical model for convective heat transfer coefficient during solar drying process of green herbs. *Adv. Intell. Syst. Comput.* 1025: 867–877. https://doi.org/10.1007/978-981-32-9515-5_81.

33 Nwakuba NR, Ejesu PK, Okafor VC. A mathematical model for predicting the drying rate of cocoa bean (*Theobroma cacao* L.) in a hot air dryer *Agric. Eng. J. CIGR J.* 2017, 19 (3), 195–202.

34 Yurdem, H., Demir, V., and Degirmencioglu, A. (2010). Development of a mathematical model to predict clean water head losses in hydrocyclone filters in drip irrigation systems using dimensional analysis. *Biosyst. Eng.* 105: 495–506. https://doi.org/10.1016/j.biosystemseng.2010.02.001.

35 Jekayinfa, S.O. and Bamgboye, A.I. (2006). Estimating energy requirement in cashew (*Anacardium occidentale* L.) nut processing operations. *Energy* 31: 1305–1320. https://doi.org/10.1016/j.energy.2005.07.001.

36 Bidabadi, M., Panahifar, P., and Sadeghi, S. (2018). Analytical development of a model for counter-flow non-premixed flames with volatile biofuel particles considering drying and vaporization zones with finite thicknesses. *Fuel* 231: 172–186. https://doi.org/10.1016/j.fuel.2018.05.090.

37 Shelare, S., Kumar, R., and Khope, P. (2021). Assessment of physical, frictional and aerodynamic properties of charoli (*Buchanania lanzan* Spreng) nut as potentials for development of processing machines. *Carpathian J. Food Sci. Technol.* 13 (2): 174–191. https://doi.org/10.34302/crpjfst/2021.13.2.16.

38 Waghmare SN, Shelare SD, Tembhurkar CK, Jawalekar SB. Development of a model for the number of bends during stirrup making process. Lect. Notes in Mech. Eng., 2021, 69–78. https://doi.org/10.1007/978-981-15-5151-2_7.

39 Khope PB, Shelare SD. Prediction of torque and cutting speed of pedal operated chopper for silage making. Lect. Notes in Mech. Eng., Springer, 2021, 243–249. https://doi.org/10.1007/978-981-16-1769-0_22.

40 Belkhode, P.N. (2019). Analysis and interpretation of steering geometry of automobile using artificial neural network simulation. *Engineering* 11 (4): 231–239. https://doi.org/10.4236/eng.2019.114016.

41 Ko, E. and Park, J. (2019). Diesel mean value engine modeling based on thermodynamic cycle simulation using artificial neural network. *Energies* https://doi.org/10.3390/en12142823.

42 Belkhode, P.N. (2019). Development of mathematical model and artificial neural network simulation to predict the performance of manual loading operation of underground mines. *J. Mater. Res. Technol.* 8 (2): 2309–2315. https://doi.org/10.1016/j.jmrt.2019.04.015.

43 Undirwade, S.K., Singh, M.P., and Sakhale, C.N. (2017). Formulation of mathematical model for processing time required for bamboo sliver cutting using HPFM. *Mater. Today Proc.* 4: 10174–10178. https://doi.org/10.1016/j.matpr.2017.06.343.

44 Dehghani, H. (2011). International journal of rock mechanics & mining sciences development of a model to predict peak particle velocity in a blasting operation. *Int. J. Rock Mech. Min. Sci.* 48: 51–58. https://doi.org/10.1016/j.ijrmms.2010.08.005.

45 Birkhoff, G. and Equations, P.D. (1948). Mathematics for engineers — III dimensional analysis of partial differential equations. *Electrical Engineering* 67 (12): 1185–1188. https://doi.org/10.1109/EE.1948.6444503.

46 Kumar, J. and Khamba, J.S. (2010). Modeling the material removal rate in ultrasonic machining of titanium using dimensional analysis. *Int. J. Adv. Manuf. Technol.* 48: 103–119. https://doi.org/10.1007/s00170-009-2287-1.

47 Strasik, M., Hull, J.R., Mittleider, J.A. et al. (2010). An overview of Boeing flywheel energy storage systems with high-temperature superconducting bearings. *Supercond. Sci. Technol.* 23: https://doi.org/10.1088/0953-2048/23/3/034021.

48 Miragliotta, G. (2011). The power of dimensional analysis in production systems design. *Int. J. Prod. Econ.* 131: 175–182. https://doi.org/10.1016/j.ijpe.2010.08.009.

49 Singh, R. (2012). Mathematical modeling for surface hardness in investment casting applications. 26: 3625–3629. https://doi.org/10.1007/s12206-012-0854-0.

50 Singh, M.C.R. (2014). Mathematical modeling of surface roughness of castings produced using ZCast direct metal casting. 96 (2): 145–155. https://doi.org/10.1007/s40032-014-0139-8.

51 Tran, T.A. (2021). Comparative analysis on the fuel consumption prediction model for bulk carriers from ship launching to current states based on sea trial data and machine learning technique. *J. Ocean Eng. Sci.* 6 (4): 317–339, https://doi.org/. doi: 10.1016/j.joes.2021.02.005.

52 Everett, E. (1945). Some engineering applications of the Buckingham Pi theorem. *Ohio State Engineer* 28 (3): 22–24.

53 Khan, J. and Arsalan, M.H. (2016). Solar power technologies for sustainable electricity generation – A review. *Renewable Sustainable Energy Rev.* 55: 414–425. https://doi.org/10.1016/j.rser.2015.10.135.

54 Demirbas, A. (2007). Progress and recent trends in biofuels. *Prog. Energy Combust. Sci.* 33: 1–18. https://doi.org/10.1016/j.pecs.2006.06.001.

55 Yilmaz, N. and Atmanli, A. (2017). Experimental assessment of a diesel engine fueled with diesel-biodiesel-1-pentanol blends. *Fuel* 191: 190–197. https://doi.org/10.1016/j.fuel.2016.11.065.

56 Yilmaz, N., Atmanli, A., and Vigil, F.M. (2018). Quaternary blends of diesel, biodiesel, higher alcohols and vegetable oil in a compression ignition engine. *Fuel* 212: 462–469. https://doi.org/10.1016/j.fuel.2017.10.050.

57 Atmanli, A. (2016). Comparative analyses of diesel-waste oil biodiesel and propanol, n-butanol or 1-pentanol blends in a diesel engine. *Fuel* 176: 209–215. https://doi.org/10.1016/j.fuel.2016.02.076.

58 Oppong, P.K., Addo, A., Mao, H. et al. (2018). Comparative evaluation of Japanese rice thresher and locally built thresher. *African J. Food Integr. Agric.* 2: 8–14. https://doi.org/10.25218/ajfia.2018.01.001.02.

59 Zain, A.M., Haron, H., Qasem, S.N., and Sharif, S. (2012). Regression and ANN models for estimating minimum value of machining performance. *Appl. Math. Model.* 36: 1477–1492. https://doi.org/10.1016/j.apm.2011.09.035.

60 Asiltürk, I. and Çunkaş, M. (2011). Modeling and prediction of surface roughness in turning operations using artificial neural network and multiple regression method. *Expert Syst. Appl.* 38: 5826–5832. https://doi.org/10.1016/j.eswa.2010.11.041.

61 Math, R.G., Ramesh, G., Nagender, A., and Satyanarayana, A. (2016). Design and development of annatto (*Bixa orellana* L.) seed separator machine. *J. Food Sci. Technol.* 53: 703–711. https://doi.org/10.1007/s13197-015-2019-5.

62 Prashanth, N. and Sulaiman, S. (2013). To study the effect of seismic loads and wind load on hyperbolic cooling tower of varying dimensions and RCC shell thickness. *Int. J. Emerg. Trends Eng. Dev.* 4: 2249–6149.

63 Milani, K., Mirzakhanlari, S., Mamourian, M., and Kalogirou, S.A. (2017). Optimization of effective parameters on solar updraft tower to achieve potential maximum power output : a sensitivity analysis and numerical simulation. *Appl. Energy* 195: 725–737. https://doi.org/10.1016/j.apenergy.2017.03.057.

64 Mofidian, R., Barati, A., Jahanshahi, M., and Hassan, M. (2019). Optimization on thermal treatment synthesis of lactoferrin nanoparticles via Taguchi design method. *SN Appl. Sci.* 2019: 1339. https://doi.org/10.1007/s42452-019-1353-z.

65 Waghmare, S.N., Sakhale, C.N., Tembhurkar, C.K., and Shelare, S.D. (2020). Assessment of average resistive torque for human-powered stirrup making process. *Adv. Intell. Syst. Comput.* 1025. https://doi.org/10.1007/978-981-32-9515-5_79.

11

Challenges to Biofuel Development

11.1 Introduction

Rapid industrialisation is destroying the environment. Energy is the most important sector in developing as well as developed countries. Demand and availability of fossil fuels vary inversely, and thus the reserves of fossil fuel will come to an end in the near future. Environmental pollution and global warming are the two major alarming issues on which renewable energy can be the immediate solution. Maximum cost effective and limited emission are the two deciding keys in the success of any technology. The chief data to be considered, related to energy, are social, economic, geographical, environmental and political policies [1]. While planning and building any industry, its aspect in all those fields should be critically considered. The new technologies must be cost effective to make it more sustainable. The technology developments and global economy should go hand in hand for the steady sustainable progress of the industry.

While developing the technologies at the commercial level, it has to cross certain barriers:

Policy barriers: The existing policies of any country should always be renewed and upgraded as per the market demand and the fuel available. The taxes should be curtailed as per the scenario [2].

Market barrier: The market should also promote the newly reliable biofuels according to their cost.

Cost barrier: Fossil fuel production is at a lower cost than that of renewable fuels. Production of fuels at a larger scale and the integration of technologies will help in reducing the cost of biofuels [3, 4].

Other barriers, such as social, cultural and geographical, must be considered with care. Technologies must be upgraded with the advancement of cutting the production cost. Biofuels in this way can be helpful to create more local job opportunities and replace the traditional fossil fuels. If the proper production process is used, it will control the carbon emissions in the environment and thus can be a greener solution to future demand. There are still problems with these biofuel production as it results in deforestation for land, higher cost and emissions of more NOx than the traditional fuels. Nowadays, attention has been paid to electro-fuels that do not require biomass for the production. In this, fuel storing electrical energy or solar heat is used to proceed the chemical reaction as in the case

Sustainability in Biofuel Production Technology, First Edition. Pratibha S. Agrawal, Pramod N. Belkhode, and Samuel Lalthazuala Rokhum.

of solar fuels [5, 6]. Thus, this alternative fuel needs extensive research and thinking process.

If biofuels are produced properly, the carbon emissions can be controlled and global biofuel production can be increased manifold. Brazil is one of the leading countries in using bioethanol and replacing the traditional fuel [7]. The newly sold cars in Brazil have an option to use desired fuels.

Apart from this, if fuel is spilled accidently in the ecosystem, then its contamination would cause another dangerous issue. If biofuels are spilled accidently, they biodegrade avoiding long-term damage to the environment. This is a large achievement solving problems associated with the transportation of fuels. This transportation risk is again minimised by producing the biofuel at the site of requirement. Traditional fuels are exported from its origin, while biofuels can be locally produced leading to a lesser import–export process [8]. This will also be helpful to create job opportunities for the local area people improving their living standards. It provides more economic stability to the nation.

Biofuel needs minimum mechanical adjustment in the engines when used. This makes the process user friendly. Biofuels also need overall less maintenance. Consumers will always prefer to use such engines rather than switching directly to electric cars or hydrogen running cars, which need completely different engines while configuring the vehicle in the factory itself. Biofuel engines have the potential to go much down in prices once integrated properly and accepted globally [9]. If any extra penny is sent in while using biofuel, it will otherwise also be utilised for cleaner earth. With increased attention, research interest, government funding and integrated technology, biofuel can focus major attention in the long run.

11.2 Key Issues and Challenges in Biofuel Production Pathways

There are certain issues to be addressed properly in the biofuel production, which are as follows.

11.2.1 Production from Biomass

As discussed earlier, biofuel produced from crops and vegetable oil will be in direct conflict with the global food demand and seems to be unethical. Energy is converted from wood to fuel under optimum conditions to make the process energetically and environmentally favourable.

The problem associated with this is the biochemical structure of wood, which contains tissues of lignin, cellulose and hemicellulose [10]. During fermentation, it must be broken into simpler monomer or dimer. Often, this production causes difficulty in times of stress. Many researchers have proposed solutions for overcoming issues, such as genetically modifying the lignin composition or pre-treatment of biomass to make the tissues more penetrable, departing it from the cell wall easily and promoting the biofuel production pathway.

11.2.2 Transportation of Goods

The main challenges during biofuel transportation occur before and after biodiesel production, when feedstock is supplied to the plant processing unit and when the produced biofuel is supplied to the consumer. It has been observed by Hamelinck and Faaij [11] that the transportation cost can be reduced when feedstock is imported in the pellet form, which is always desirable. The optimal transportation method always depends on the quantity and the distance.

Biodiesel so formed is blended with gasoline and supplied the same to the consumer. Again, the challenges to the transportation occur. Such blended gasoline can tend to separate during transportation. The corrosive nature of such blends can also impose the problems in pipelines. Again, such transportation is favourable for shorter distances. Thus, it requires much more efficient ways in transportation for its use in the long term.

11.2.3 Economic Effects

One of the major concerns is the cost analysis of oil, gasoline and biofuel, which affects the economy of the country. The use of agro-waste as feedstock is one of the suggested ways, which also benefits farmers.

Currently, low oil prices make it difficult for biodiesel to maintain its place in the market. Government also provides subsidies to the biodiesel plants. High biofuel availability and too much supply in spite of less demand further drop the price of oil and biodiesel. This situation is currently very small, but with increasing industrial growth in the biodiesel sector, this problem can become massive [12–14].

The use of vegetable crops as feedstock is beneficial for the commodities living in the nearby area. As demand for these products increases, it becomes economically beneficial for the farmers in that area, but for the others, it becomes difficult to fight in the food versus fuel debate.

11.2.4 Switching to Biofuels

When the consumer switches to the biofuel option, the engine needs little or no modification at all, making it user friendly. However, the problems associated with its use are that biodiesel contains methanol that can ruin the rubber parts of the engines, which are to be replaced constantly. The compatibility of polymer parts of the engines and biodiesel generates technical issues. Biodiesel tends to swell rubber parts, resulting in the faster degradation of these parts. It is found that the mechanical properties of the rubber decrease as the biodiesel content in the blend increases [15]. Many automobile companies have already started designing the engines for biodiesel use. However, many countries are banning the use of biodiesel; at the same time, 'non-electric' vehicles' future is seen to be in danger in the coming few decades. This will result in poor cost returns and planning for massive investments.

11.2.5 Environmental Effects

Biodiesel is degradable and responsible for decreased CNG emissions. However, apart from these advantages, biodiesel is responsible for increased NOx emissions. Production of biodiesel on a large scale results in increased deforestation and increased use of water and other fertilisers. Consumption of water increases from 5 to 2000 l of ethanol production. Chemical fertilisers such as nitrogen can pollute the surrounding; nitrogen deposition can also cause dead zones in waterways. After combustion, biofuel produces less carbon monoxide and hydrocarbons as compared to traditional fossil fuel [16]. However, the cause of increased nitrous oxide emission still cannot be determined. Production of biodiesel on a global level requires a large land area that results in biodiversity loss, which again disturbs the environmental balance.

A debate for food versus fuel continues for the country. Farmers have to take steps for the production of crops either for food or for fuel [17]. Two independent markets exist for supply. The battle for supply between these two markets eventually increases the crop selling price. Rising food prices in the market will again risk the livelihood of the society.

Another long-term effect is that farmers are shifting their regular crops towards biodiesel-producing crops. It disturbs the harmony in the agricultural sector. The issues are getting raised between farmers, suppliers and consumers. The food chain is also getting disturbed.

11.3 Other Glycerol Derivatives in Diesel Production

Countries all over the world decided to use their own energy policies to reduce greenhouse gas (GHG) emissions until 2030. Reducing the use of fossil fuel is one of the major steps towards this. This leads to the search for alternative fuel that are still in the inventory step. Electric motors cannot compete with spark ignition or compression engines. Thus, the use of biodiesel blends is essential to achieve minimum GHG emissions [18]. Vehicle engines were improved and adopted for replacing fossil fuel with biodiesel blends. To date, B20 (20% biodiesel and 80% diesel) is accepted worldwide, which is thought to be improved to B30 till 2030 [19].

Despite the above goal, it is not so easy to replace fossil fuels with biofuels. One of the major issues is that though bioethanol seems to be competent, but its production harms the agricultural sector. The use of seed oil is also not feasible due to its high viscosity. Amongst various routes, pyrolysis is not very useful due to its technical complexity, and the microemulsion-based technique is not advisable due to its solubility problem. Thus, transesterification for non-toxic, biodegradable biodiesel production proved to be the most commonly used method. Biodiesel seems to be a sustainable solution, but its generation is associated with the production of undesired glycerol [20]. Removal of this glycerol is necessary before the use of biodiesel in engines; otherwise, it will be converted into toxic polyglycols or acrolein at high temperatures.

To remove glycerol, washing is performed several times, which makes the process expensive. After washing, biodiesel is purified by various methods of neutralisation, evaporation and drying. This excess glycerol produced lowered its market value [21–23]. This excess glycerol produced must be turned to value-added products; otherwise, cost in the biodiesel production will be one of the major hurdles in making its use at the global level. To avoid this, different chemical routes have been investigated, which can generate eco-fuel with the properties similar to those of fossil fuel.

To overcome the above difficulties shown in Figure 11.1, researchers have also tried to produce some of the derivatives of glycerol, such as glycerol triacetate, glycerol carbonate

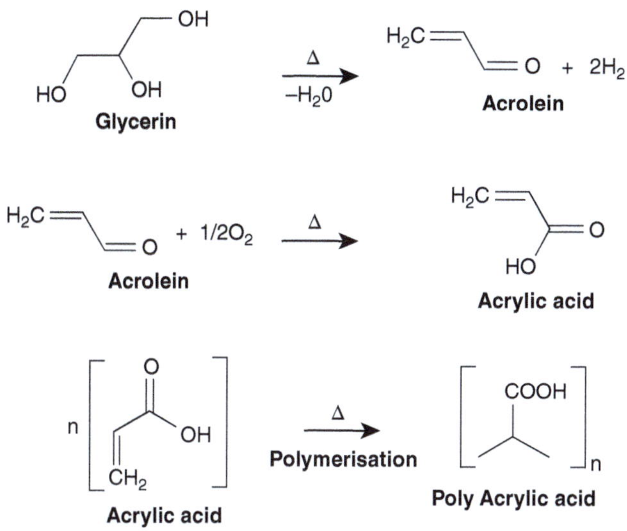

Figure 11.1 Production of toxic chemicals in engines from glycerol present.

or monoglycerides [24]. This can be performed by using ester-substituted alcohol instead of methanol/ethanol in the transesterification process.

The interesterification reaction of TG with methyl acetate in the presence of a strong acid results in the production of FAME and glycerol triacetate (triacetin) and traces of diacetin and monoacetin. All these side products formed are soluble in fossil fuel and thus need not be separated. The only process required will be then purification and neutralisation of biodiesel in an acid-catalysed process.

$$\text{TG} + 3\text{Methyl Acetate} \rightarrow \text{Glycerol Triacetate} + 3\text{FAME (Gliperol)}$$

Such a mixture of 3 moles of FAME and 1 mole of triacetin, termed as glycerol, exhibits good fuel characteristics compared to the routine biodiesel fuel. The molar ratio of oil/methyl acetate studied was in the range 1:3 to 1:9 and in the temperature range of 40–200 °C. The use of ethyl acetate has also been studied by several researchers, in which the yield is FAEE and triacetin instead of FAME [25]. Although results indicate similar behaviour, it is still less studied.

The presence of triacetin in the fuel has also been studied as it stands against the traditional methods. It has been observed that triacetin is an anti-knocking agent when used in diesel engine, improving the performance and emission properties. Thus, this method can compete well in the discussion forum for excess glycerol production and its alternative approach.

Another widely used transesterification reagent is dimethyl carbonate (DMC), again producing biodiesel soluble coproducts. This is a more attractive reaction as raw material DMC is produced from methane and CO, which are in turn produced in the thermochemical conversion of biomass, making the process fully renewable.

$$\text{TG} + \text{DMC} \rightarrow \text{FAGE (fatty acid glycerol carbonate)} + 3\text{FAME} + \text{Glycerol carbonate}$$

The reaction scheme is as presented in Figure 11.2.

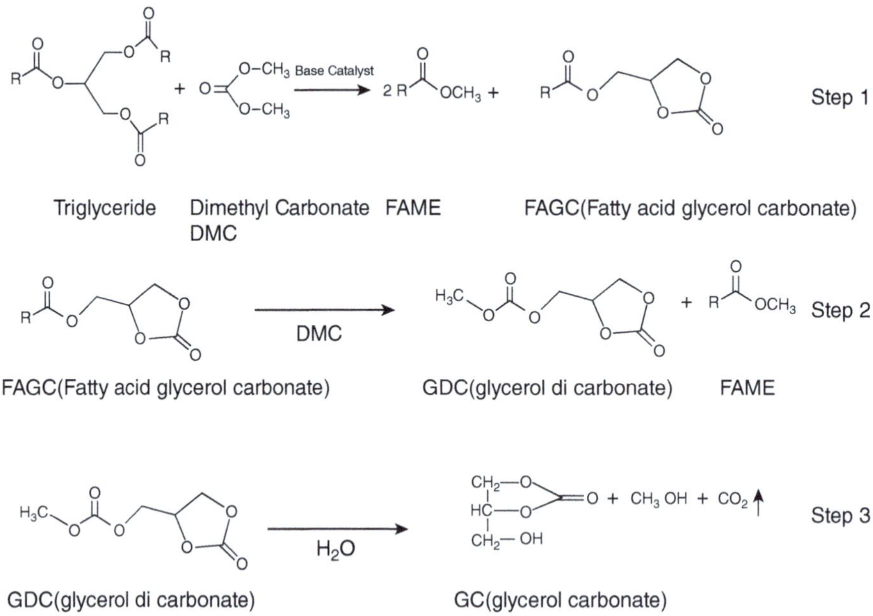

Figure 11.2 Base-catalysed FAME production using DMC.

The base catalyst used in the above irreversible process may be KOH, sodium methoxide, sodium hydride or amines. The formation of FAGCs positively influences the fuel properties. An observable decrease up to 50% in total hydrocarbons (THCs), carbon monoxide (CO), carcinogenic emissions such as benzene and 1,3-butadiene and particulate matter (PM) using 96% diesel and 4% DMC blend is observed.

Another pathway considered is the use of 1,3-selective lipases, which stops the transesterification process midway, giving 2 moles of FAEE and 1 mole of MG in the presence of ethanol. This incomplete process prevents the production of glycerol.

$$\text{TG} + \text{ethanol} \rightarrow \text{MG} + \text{FAEE}$$

The MG obtained is not removed, and the mixture shows compatible fuel properties. The presence of MG in the fuel enhances the lubricity of biodiesel. Unused ethanol is also fed directly to engine as a fuel. Though there are no significant changes observed in the efficiency of engines and the COx and NOx emissions, a reduction in the particulate matter is surely seen when using such biofuels [26]. This type of ethanol-based biodiesel has its own limitation when used at low temperatures. Such problems should be dealt with careful blending of biodiesel–petrodiesel–ethanol under optimised conditions, which can be formulated by mathematical modelling and verified in the laboratory for better performance. High-cost enzymes are still a main hurdle for globalising this process.

11.4 Biofuel Blends and Future Trends

Nowadays, vegetable oils are directly mixed with certain low viscous solvents to match the biodiesel properties instead of giving chemical treatments to TG. Low viscosity eucalyptus oil, orange oil or camphor oil with low cetane numbers can be mixed in oils to improve the performance of biodiesel-fuelled engine. Such blending controls the cost involved in chemical treatments and the related separation and purification processes.

The use of fuel additives in blending is also recommended to improve the cetane index and ignition capacity and solve the cold start problems [27]. Blends of ternary mixture of straight vegetable oils, biodiesel and higher alcohols have been combined and studied by many researchers for decreasing the smoke emissions. Usually, bioethanol is avoided in such a ternary mixture due to its low solubility in vegetable oil as well as fossil diesel. Addition of surfactant into it can tackle such a phase separation problem. Blends with diethyl ether increase engine performance properties and with dimethyl carbonate reduce the viscosity [28]. Diethyl ether and dimethyl carbonate are the cetane index improver. B25–B40 is possible by this method depending on the type of feedstock used.

11.5 Environmental Effects of Biofuels

The AEO 2020 [29] reference case states that in the jet fuel market, gasoline and diesel and biofuels share a small but relatively increasing rate as shown in Figure 11.3. In gasoline, diesel and jet fuel, the blended percentage of biofuels (ethanol, biodiesel, renewable diesel and biobutanol) is expected to increase from 7.3% in 2019 to 9% in 2040 before slightly declining through 2050.

11.5.1 Biofuel Impact on Food Security

Biofuels are the liquid fuels that are derived from the biological sources such as trees, agricultural waste, crops or grass. These are vital parts of renewable energy family. As biofuel are produced from agricultural crops, such as sugar cane, cassava, corn, rapeseed, soya bean, palm oil, wheat and others, this may adversely affect food availability if the production of food is switched to biofuels. Renzaho et al. [30] reported that 40% of the maize crops in the United States which produce biofuels are deflected from global food markets. This will lead to the problem of a shortage of food supplies and high levels of chronic hunger in many cases. Thus, if the soil resources are converted into energy plant fields, then the production of food is weakened, and this is the cause for rising food market [31], which

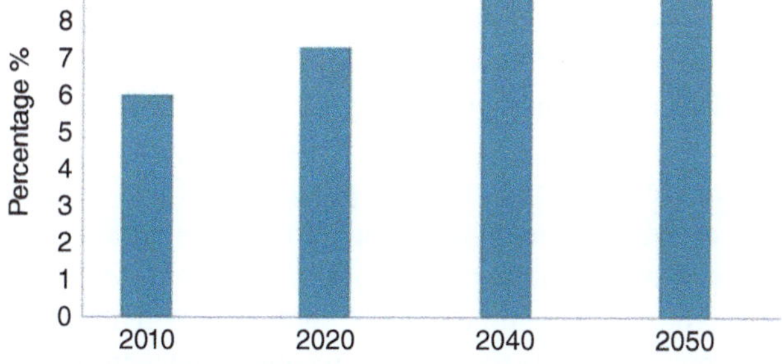

Figure 11.3 Biofuel as a percentage of gasoline, diesel and jet fuel consumption increases in the AEO 2020 reference case.

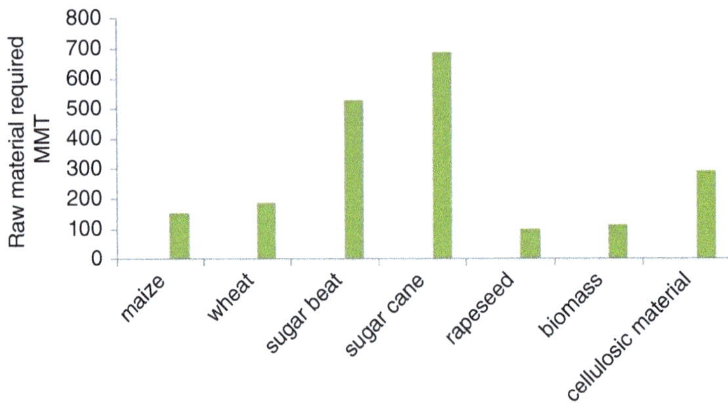

Figure 11.4 Feedstock requirement for 1000 l ethanol alcohol or biodiesel.

means that increasing the production of biofuel leads to an increase in food prices, and hence people will face difficulties in buying the food.

Subramaniam et al. [32] suggested that industrialisation and urbanisation are already affecting land and labour for the agricultural sector, and now biofuel is creating a new problem to this field. Nowadays, the benefit that the farmers are receiving is motivating them to grow more biofuel crops rather than food crops. At this point of time, the income earned is more of concern than growing edible crops. This results in a shortage of food supplies, which leads to an increase in the prices of food. Spiess [33] shows in Figure 11.4 the requirement of feedstock for the production of 1000 l ethanol alcohol or biodiesel and explains that only 86% of the cereals will be produced by developing countries, which are needed to meet the daily requirements by 2030. Hence, the import will rise from the current levels of 103 MMT (million metric tons) to 265 MMT by 2030, and in this way, the production and consumption will be affected, which will have a negative effect on the general price level and so the impoverished population is greatly affected in developing countries.

11.5.2 Bioenergy Effect on Water (Quantity and Quality)

The bioenergy production also has an impact on the quality and quantity of water for washing plants and seeds and for their growth. Lin et al. [34] reported that in developing countries the water supply is affected by the consumption of water in the cultivation of biofuel crops. These growing biofuel crops would change the availability of clean and potable water, which results in a decrease in the level of water resources. In many countries, for agricultural use the resources of water are becoming insufficient for the demand because of the large scale of domestic and industrial uses. Water resources are already deficient due to climate change, which results in reduced rainfall, and the production of biofuels puts more pressure on it. Crops such as corn and palm oil for biofuel production require more water to grow. Dimitriou et al. [35] suggested that there are crops that have a lower impact on the quality of water, and these crops are perennial grasses, such as switchgrass, miscanthus, elephant grass and trees growing in short rotations than conventional crops.

Table 11.1 Blue water consumption for biofuels by different countries.

Country	Blue water consumption (m^3 GJ^{-1})
Portugal (bioethanol)	86
UK, Poland, Germany (Biodiesel), Spain (crop-based biodiesel)	90
Europe (bioethanol)	3.3
Europe (biodiesel)	1.9
Netherlands (algae)	8

Wu et al. [36] demonstrate that corn compared to other crops, such as wheat and soya bean, requires more water in every growing stage, and the cultivation of corn has a large potential for water consumption as it is estimated that its production capacity of 100 million gal/per year uses water as a community of 5000 people. Conversion of woodlands into crop fields influences the problems such as soil erosion, water yield, surface runoff and minerals entering the ground water due to increased use of fertilisers. Production of biodiesel results in contaminated wastewater, which when left untreated increases the content of minerals and nutrients in surface water. The quality of water is of concern due to increased bioenergy crop cultivation, which is a nutrient pollution resulting from surface water runoff. Hammond and Li [37] reported that for biofuels the increased agricultural biomass production causes an increase in the total global water consumption by 2050 in those areas that are already water stressed, and the additional demand of water further increases the overall impact of biofuels on the environment. Berger et al. [38] in Table 11.1 showed blue water consumption of biofuels, which depends upon the growing crops and country.

11.5.3 Emissions of Greenhouse Gases

In the production of bioenergy, the main factor that is of concern is the reduction in GHG emissions. Zhang et al. [39] suggested that when fossil fuels are burnt, they produce a large amount of greenhouse gases in the atmosphere, and these gases increase the temperature of the climate resulting in global warming. GHGs including CO_2, nitrous oxide (N_2O), methane (CH_4), carbon monoxide (CO) and air pollutants – including ammonia (NH_3), nitrogen oxides (NO_x) and sulphur oxides (SO_x) – are emitted during the use of biofuels. Nitrous oxide and carbon dioxide are major components due to their large quantity. The emission of carbon dioxide when fossil fuels are burnt is greater than that emitted from the direct use of biofuels.

Joshi et al. [40] reported that the CO_2 produced after the burning of biofuels is utilised by the feedstock crops for the photosynthesis process and lowers the emission of GHGs. Sharma et al. [41] proposed that in 2020 the CO_2 level is 394.5 ppm (parts per million volume), whereas in 2050 it will rise to 500 ppm if the emissions occur and not controlled, resulting in global warming. To solve this problem, bioenergy is a proper substitute that is renewable and eco-friendly. Pereira et al. [42] pointed out that corn ethanol reduced GHG emission by 40%, which is more than Brazilian sugar cane ethanol.

Collotta et al. [43] suggested that the biomass production effect on soil chemistry, which includes the removal and storage of carbon from atmosphere, and N_2O emissions are the key factors in environmental issues. Nitrous oxide is the dominant source that occurs due to the use of nitrogen fertilisers. For the growth of feedstock, climate is responsible for temperature and precipitation, which increases the biomass yield and N_2O emission [44]. Crutzen et al. [45] proposed that in order to boost yield, nitrogen present in fertiliser can be emitted as NO_2 when applied to crops. Its potential is 296 times greater than CO_2 in global warming. Biofuel is produced from crops, and their growth becomes better when fertilisers are used. Most of the fertilisers contain nitrogen and phosphorous. Therefore, when biofuels produced from these crops are burnt, they release more nitrous oxide gas. For example, when rapeseed oil is burnt, it releases more nitrous oxide (N_2O) gas as higher fertilisers are required for its production. During biofuel production, the amount of energy required has a great impact on GHGs, and precautions must be taken to avoid an increase in nitrous oxide emission. The corn expansion required for the increased production of ethanol will stimulate the emission of N_2O. For the cultivation of corn, more fertilisers are required, mainly nitrogen containing fertiliser that is responsible for the direct emission of N_2O. Kesieme et al. [46] suggested that the cultivation location has a strong impact on environment, and the choice of path has the lowest impact due to difference in the yield and LUC (land use changes) (Figure 11.5).

Mandade and Shastri [48] in Figures 11.6 and 11.7 showed the biofuel production and greenhouse emission with different lignocellulosic feedstock and focused on the optimal selection of feedstock for the Indian biofuel sector.

The European Environment Agency in Figure 11.8 shows that in 2018 the reduction is 23%, which is more than the target of 20% in 2020, and it is estimated that by 2030 and by 2050 the target of 40 and 80% reductions in GHG emission will be achieved [49].

11.5.4 Effect of Biofuel on Biodiversity

Biodiversity is the term used to describe the enormous variety of life on earth; it is the measure of variation at the genetic species and the ecosystem level. Elshout et al. [22, 50] pointed out that ecological characteristics, such as landscape configuration, habitat loss,

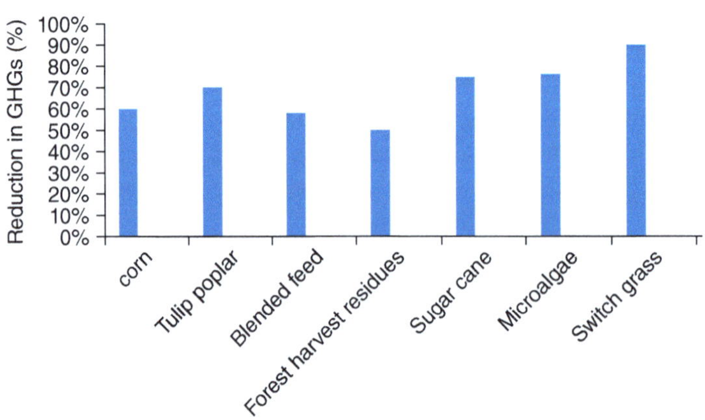

Figure 11.5 Reduction in greenhouse gas emission by different feedstock. *Source:* Adapted from [47].

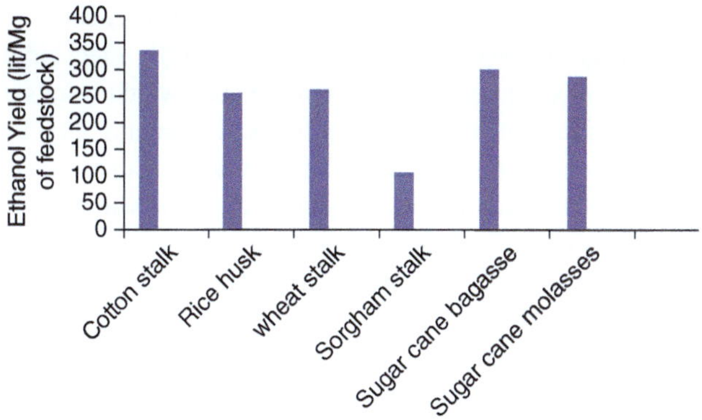

Figure 11.6 Estimated production of ethanol from lignocellulosic feedstock.

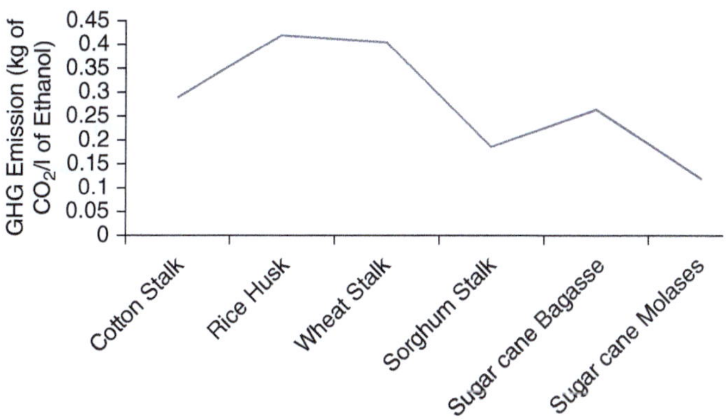

Figure 11.7 GHG emission with different lignocellulosic feedstock.

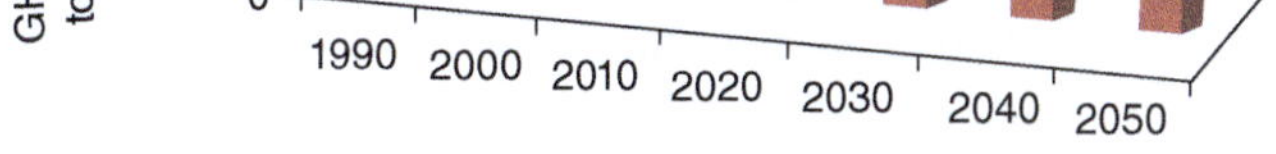

Figure 11.8 GHG emission trend and target per year.

excessive use of resources for harvesting, and different forms of pollution for the formation of biofuel, are the key factors that affect biodiversity. Native species, which is originated and developed in its surrounding habitat, and invasive species, which is a species of plants and animals and crops, support them in a different way; for example, migrating birds get food from cereal residues, whereas maize has a negative effect on bird diversity.

Feedstock used for the production of biofuel also have an impact on biodiversity since the synthetic fertilisers and other chemical growth agents used in feedstock directly affect biodiversity. For biofuel, the production of crops requires expansion of the area, which results in habitat loss since forest and grasslands are converted into crop fields. Correa et al. [24, 51] suggested that the demand for food and energy further creates a pressure on biodiversity. By 2050, with a population projection of 9000 million people, the demand of food will increase by more than 60%, and for this, the expansion of around 70 million ha of crop lands is required, which will have an adverse effect on biodiversity. Jeswani et al. [52] pointed out that agricultural biodiversity is also affected by biofuel production as it involves monocropping in which a single crop is growing year after year and the rotation of other crops is absent on the same land. In this way, the natural balance of soils is disturbed, resulting in decreasing varieties of bacteria and microorganisms that are essential to maintain the fertility of soil. Liu et al. [53] suggested that the impact of algal biofuels on biodiversity is uncertain. The cultivation of algae in the coastal areas will have adverse effects on the coastal biodiversity due to its rapid production and multiplication. Its effect on the coastal biodiversity include mud flats, salt marshes, sea grass beds, coral reefs and mangroves. The negative effect of biofuel feedstock can be lowered by using degraded land or residues of agriculture and forest. Currently, the impact of biofuel production on biodiversity is a major concern for environmentalists.

As the crude oil prices are increasing rapidly, an alternate substitute for this is required. Biofuels fulfil this purpose and are the proper substitute for these oils, and by blending them with crude oil the problem will be overcome. Biofuels are obtained from agricultural crops, which have different impact on environment, and the challenges that arise are food security, water issues such as its quantity and quality, GHG emissions and biodiversity. By selecting a proper cultivating area, feedstock and optimal management, the challenges related to biofuels can be minimised.

11.6 Economic Impact of Biofuels

India is one of the fastest growing economies and the third largest consumer of primary energy in the world after the United States and China. India will overtake Europe as the third largest consumer of energy by 2030 [54]. The government of India makes efforts to improve energy security by reducing import dependence by decreasing the use of fossil fuel 10% from current levels by 2022.

The national policy of biofuel targets 20% blending of ethanol and 5% blending of biodiesel by 2030 [55]. The benefits to imply this policy in the future include reduced import dependency, cleaner environment, health benefits, MSW management, infrastructural investment in rural areas, employment generation and additional income to farmers. In India, there are many operational biofuel plants, and some are planned. India's new biofuel policy encourages the use of wasteland and non-edible oil seed to make biofuel by Pongamiapinnata (Karanja), Meliaazadirachta (Neem), castor, *Jatropha curcas*, *Calophyllum inophyllum*, *Simarouba glauca* and *Hibiscus cannabinus* [54]. The national biodiesel mission identified that jatropha is the most suitable inedible oil to make a blend with 20% biodiesel. Farmers are now encouraged to grow a variety of different biomass crops as intercrops [56].

Pryshliak et al. [57] provided the data on biofuel production. It was estimated that approximately 187 thousand barrels of oil equivalent per day was produced in 2000, and this production was increased to 1841 thousand barrels of oil equivalent per day in 2020 as shown in Table 11.2. At the global level, it is predicted that the biofuel market will increase and reach 153.8 billion US dollars by 2024.

Pryshliak et al. [57] provided the data on the use of biofuel in the US as shown in Table 11.3. The use of biofuel from 2017 onwards decreases, which may be due to rising oil prices, regulation such as renewable fuel standard, use of biofuel releases air pollutant other than GHGs, decrease in water resources, monoculture, use of fertilisers, shortage of food, industrial pollution, future rise in price and changes in land use.

Pryshliak et al. [57] explained the biofuel production in petajoules of different countries in Table 11.4. He explained that the United States is the largest manufacturer of biofuel in the world and accounts for about 38% of global biofuel production in the United States. The national army in the United States is the largest consumer of biofuel.

Table 11.2 Global biofuel production from 2000 to 2019 [56].

Year	Production in thousand barrels of oil equivalent per day
2000	187
2001	204
2002	241
2003	298
2004	337
2005	396
2006	504
2007	684
2008	923
2009	1025
2010	1183
2011	1250
2012	1268
2013	1381
2014	15051
2015	1511
2016	1560
2018	1631
2019	1787
2020	1841.5

Source: Adapted from [56].

Table 11.3 Biofuel consumption in the United States from 2009 to 2020.

Year	Use of biofuel in trillion British thermal units
2009	1553
2010	1821
2011	1934
2012	1890
2013	2014
2014	2077
2015	2153
2016	2287
2017	2304
2018	2283
2019	2255
2020	2000

Table 11.4 Global biofuel production.

Name of the countries	Global biofuel production in petajoules
United States	1557.1
Brazil	992.2
Indonesia	275.5
Germany	143.5
France	113
China	111.3
Argentina	102.8
Thailand	95.6
Netherland	79.2
Spain	66.7

11.6.1 Jobs in the Field of Biofuel

The data of jobs in different countries is given in Table 11.5, and Brazil is the leading country of biofuel-related jobs, which may be due to Brazil being the second largest producer of biofuel in the world. It is the largest producer in the world due to its advanced agro-industrial technology and its enormous amount of arable land available.

Biofuel feedstock comes from different types of sources, such as crop wastes, forest residues, urban wood waste and energy crops. To obtain the approximate data of job creation and economic feedstock production, we must consider advanced biofuel production from dedicated energy crops. By the Biomass Research and Development Initiative assumptions, jobs related to feedstock production increased 88 000 by 2022, while the feedstock produced approximately 11 billion that year as shown in Figures 11.9 and 11.10.

Table 11.5 Biofuel-related jobs worldwide by region.

Name of different nations	Number of jobs (in thousands)
Brazil	839
United States	297
European Union	239
China	51
India	35
Worldwide	2475

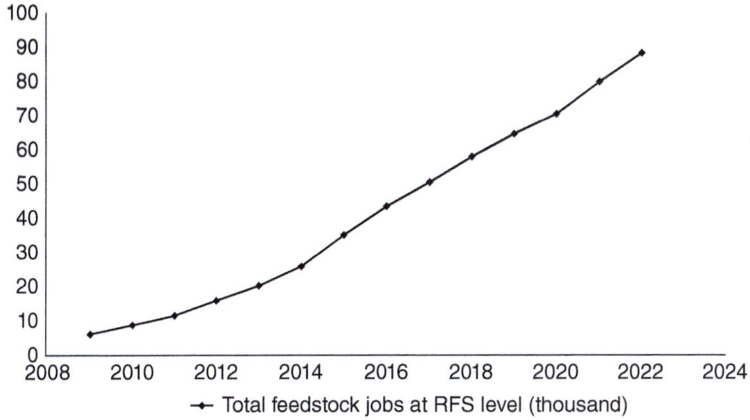

Figure 11.9 Total feedstock jobs at the RFS level.

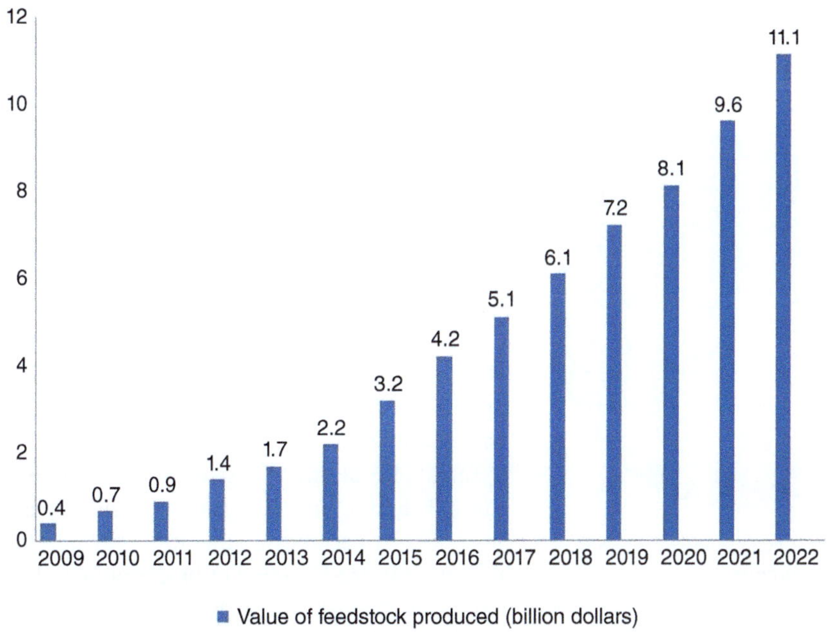

Figure 11.10 Economic value of cellulosic feedstock production.

11.6.2 Bioethanol

Ethanol is a renewable biofuel as it can be produced from agricultural crops, such as sugar cane, corn, potatoes and also from heavy molasses [58]. The US government is making efforts to produce ethanol from the non-conventional method that uses less energy than the conventional fermentation method. Rosales-Calderon and Arantes [59] explained the synthesis of cellulosic ethanol from lignocellulosic material, such as native prairie grasses, fast-growing trees, sawdust and even waste paper. However, it is a more expensive process due to enzymes. Biofuel ethanol obtained from both crops and waste material is environmentally friendly as it reduces carbon monoxide and many pollutants from the tailpipe of vehicles and makes air cleaner.

According to the Ministry of Petroleum and Natural Gas, the availability of ethanol for different purposes is given [60] in Figure 11.11.

India's Ethanol Blending Programme (EBP) Performance in different years is explained in Figure 11.12. This data shows that the blending of ethanol is increased and decreased due to the up and down production of agricultural waste and crops.

Bajpai et al. [61] explained that there are some research challenges in the field of bioethanol:

- Low production cost of enzyme.
- Developing robust fermenting organisms that ferment all sugar in the raw material.
- Reduce the number of steps.

Very high gravity (VHG) [62] processes try to significantly improve the economy by performing research efforts, such as optimisation of improved mixing alternatives, new strategies for oxygen supply, more robust organisms towards microbial contamination and more thermotolerant organisms.

Esteves et al. [63] explained in her report that in Brazil it is mandatory for the blending 27% ethanol in gasoline, and in other countries it has a small blend rate, that is, 12% in Argentina, 10% in Colombia and 7.8% in Peru.

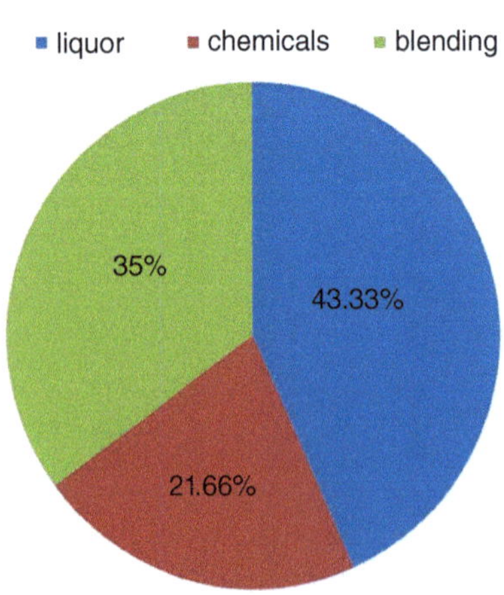

Figure 11.11 Availability of ethanol for different purposes.

11.6.3 Biodiesel

The production method of biodiesel is explained above. The biodiesel industries play an important role in the economy of every country. By the use of biodiesel, we became less dependent on fossil fuel, and simultaneously, it increased employment and tax revenue, increasing every nation's GDP. Figure 11.13 shows the data of India's Biodiesel Blending Performance in kilolitre (KL) and the biodiesel percentage in diesel on the Pan India basis [64].

Table 11.6 shows the sale price of biodiesel through bimonthly auctions. The value of sale price almost becomes double in a year due to a reduction in the blending of biodiesel from 13 to

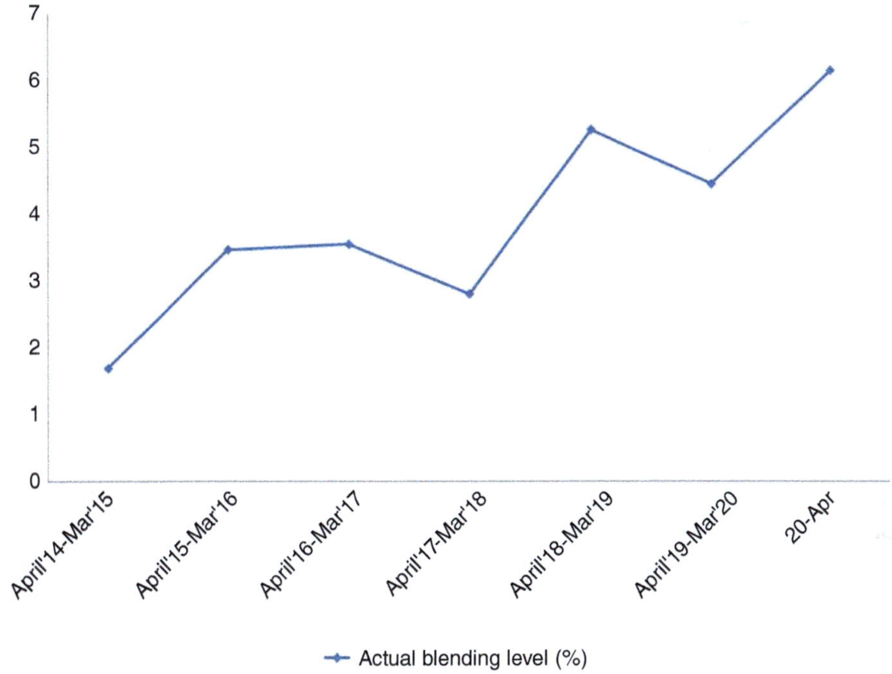

Figure 11.12 India's bioethanol blending performance in kilolitre (KL).

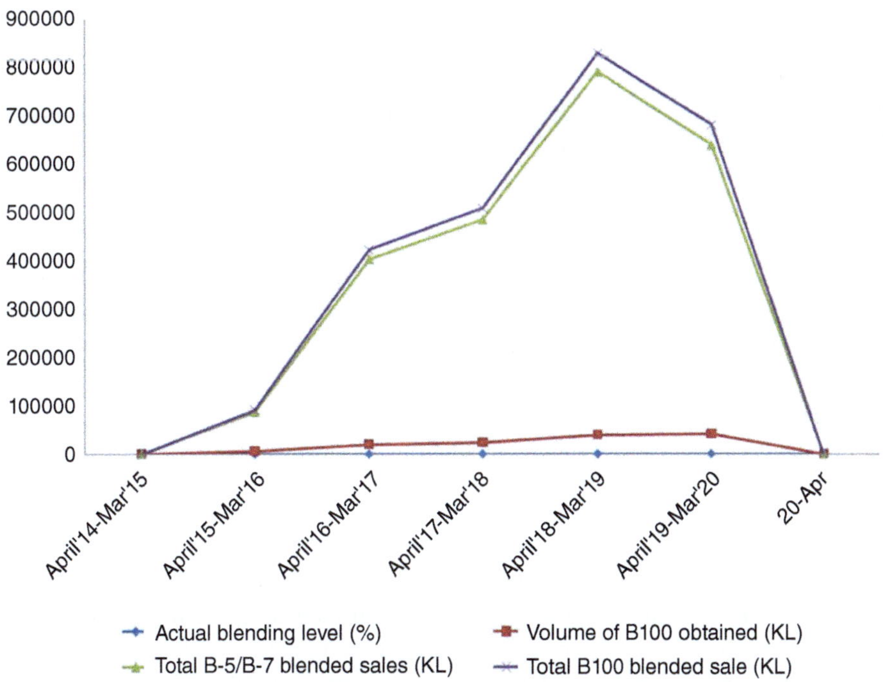

Figure 11.13 India's biodiesel blending performance in kilolitre (KL).

Table 11.6 The sale price of biodiesel in Brazil.

Data per month	Price of biodiesel in Brazilian reals per cubic metre
Jan–Feb'19	2635
Mar–Apr'19	2333
May–Jun'19	2287
Jul–Aug'19	2329
Sep–Oct'19	2857
Nov–Dec'19	3076
Jan–Feb'20	3012
Mar–Apr'20	3002
May–Jun'20	2713
Jul–Aug'20	3512
Sep–Oct'20	5043
Nov–Dec'20	5552
Jan–Feb'21	4425
Mar–Apr'21	4708
May–Jun'21	5536

10% and also due to the shortage of raw material in that year. According to the national biodiesel board, when 1 billion gallons of biodiesel is produced, it will give assistance to 39 027 jobs ('Production Statistics', National Biodiesel Board, 2012, http://www.biodiesel.org/production/production-statistics).

11.6.4 Biohydrogen

Biohydrogen is produced from many methods, such as methane-steam reforming (MSR), fermentation, microbial fuel cells and production of biohydrogen by algae. However, 95% of hydrogen is produced by the MSR method [65].

Meloni et al. [66] reported the synthesis of biohydrogen by a steam-reforming process using a nickel catalyst. The nickel-based catalysts are used due to their high activity, low cost and excellent thermal exchange. By the addition of metallic or bimetallic species, the selectivity and durability of nickel catalysts are enhanced. This method reduces the problems of the steam-reforming process.

Biohydrogen has a wide range of applications in various sectors, which include steel and fertiliser production, long-range shipping, transportation fuel and generating electricity. Sadik-Zada [67] promised to increase the production of biohydrogen 50 times more than now in the next six years. In addition, approximately 110 billion US dollars will be spent on the production of green hydrogen, and it will generate more than 120 000 jobs that will help the economy in the crucial period of COVID-19.

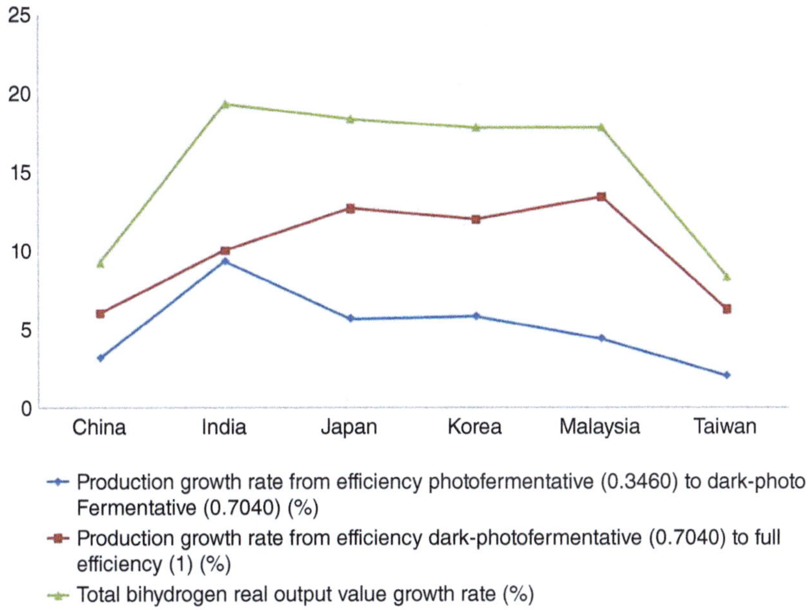

Figure 11.14 Real output growth rate of biohydrogen by production efficiency improvement.

Lee et al. [68] explained the 26 new data envelopment analysis model with 55 biohydrogen production experiments. These experiments contained different processes. The efficiency of some processes is explained in Figure 11.14, and this data impacts the economy of different countries. India and Japan are the two countries that have the maximum advantage by these experiments in their economy. By these experiments, there will be more growth in the biohydrogen industry.

11.6.5 Biogas

Biogas is synthesised by two types of methods: natural and industrial methods, and more biogas is produced through industrial methods (mostly by the anaerobic digestion method) [69].

Scarlat et al. [70] explained the goal of the European Union to build a low carbon economy and to reduce greenhouse gas emission by 80–95% in 2050. Like other biofuels, biogas is also used for many purposes, including generating electricity and transportation vehicles. He also explained the method for storing biogas in grid injection, and this grid injection allows the use of biogas wherever it is needed. The development of biogas plants in the market is supported in several countries by administrative procedures and financial support and mainly for generating electricity. The revised renewable energy directive gives responsibility to fuel suppliers to increase the share of renewable and low carbon fuel. This will provide good opportunities for more use of biogas in transportation, also acting as a spur for technology improvement and cost reduction. Tabatabaei et al. [71] represented in his paper the data of the European Biomass Association, the world biogas market forecast for the year 2025 by geography as shown in Figure 11.15.

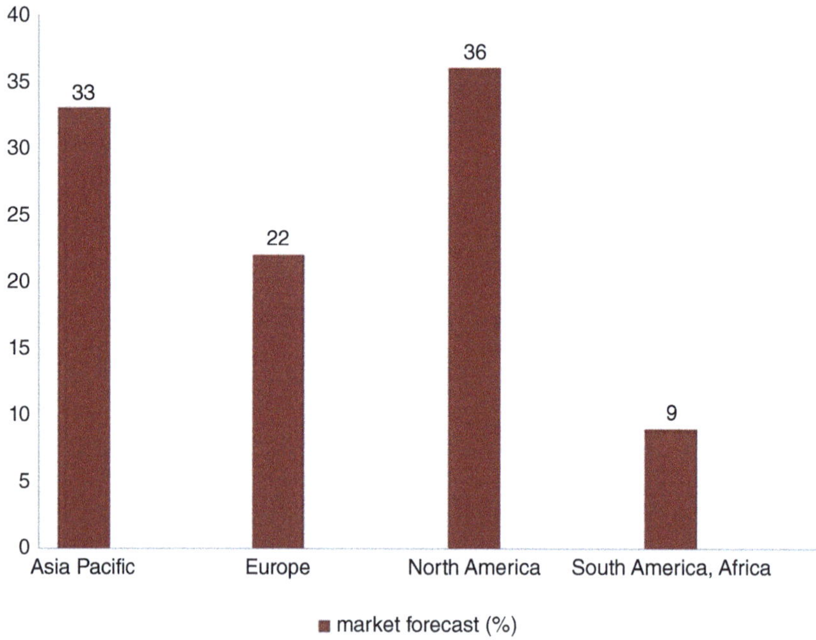

Figure 11.15 The world biogas market forecast for 2025 by geography.

The demand for biofuel has increased day by day because of its environmentally friendly nature. Initially, the cost of biofuel was higher, but nowadays it is lower because biofuel is now made from waste material. Progress in the field of biofuel affects the economy of a particular region. The United States is the largest producer of biofuel, and due to the advanced biofuel industry approximately 400 000 jobs are created, and a total of 1.9 billion jobs are created in the United States. The growth of biofuel industries makes nations less dependent on imported oil. If the biofuel production will increase as estimated, then it will bring great economic and employment benefits.

11.7 Biorefineries

The use of an integrated processing unit combining various fields and technology together will be able to make the biodiesel production more sustainable. Biorefinery is a technology that emphasises on the complete utilisation of biomass in value-added products. Different physicochemical and biochemical/enzymatic processes are clubbed together in a biorefinery as shown in Figure 11.16. The energy or fuel produced can be useful in various sectors. The sustainability of the biorefinery is measured in terms of its economy, energy efficiency, environmental emissions and large-scale production in the long run. Although the integrated biorefinery concept is adopted by almost all countries, a very few industries have been reported thus far. The concept is evolving and being redefined frequently. Therefore, a detailed study of the biorefinery industry, including challenges, opportunities and future prospects, should be considered.

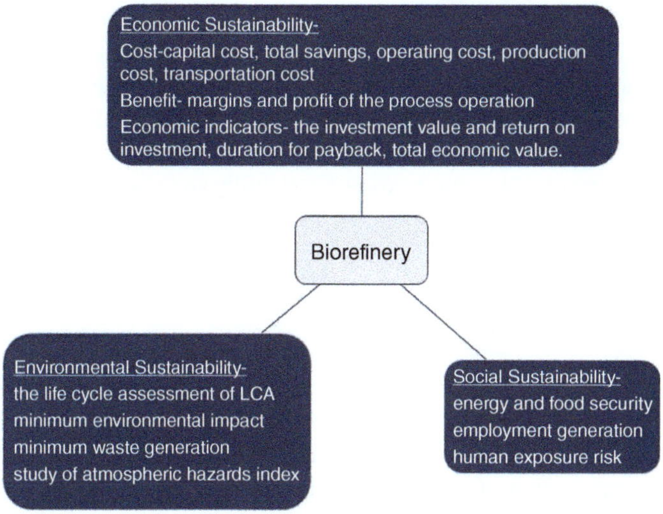

Figure 11.16 Sustainability of biorefineries.

Based on the generation of the feedstock used, a biorefinery can also be termed as first-, second-, and third-generation biorefinery with the same advantages and drawbacks. Various processes involved are as follows:

Thermochemical Processes: Such processes include gasification, pyrolysis and liquefaction techniques. In the gasification process, value-added intermediates, liquid biofuels, electricity and heat are generated. In pyrolysis, the high-temperature treatment of biomass under anaerobic conditions is involved, whereas the liquefaction process is carried out in the presence of H_2O and CO/H_2 producing bio-crudes and important side products. The high processing cost of this technology can be compensated for in the production of value-added by-products in a biorefinery.

Biochemical Processes: Such processes consist of the depolymerisation of cellulose and hemicellulose into monomeric sugars followed by enzymatic fermentations, producing bioethanol, biobutanol and other side products.

Chemical Processes: Such processes include the conversion of lignocellulosic biomass into xylose by acid hydrolysis or to syngas ($CO + H_2$) by a chemical pathway, such as FT process and others.

A biorefinery can be biomethanol-, bioethanol- or biogas-based depending on the technology and pathway followed as per the demand of the regional society. Apart from the crop-based feedstock, other agro-waste materials, such as stems, leaves and roots, are also fed to the reactor for getting converted into biofuel. The concept should be able to reduce the logistics expenses, maximise the use of agro-waste and fully utilise energy. The wastewater is also utilised in the cultivation.

Currently, biorefineries are being developed worldwide in various sectors of energy, transport, food, chemical, health, pharmaceutical, etc. The major challenges for a biorefinery are maintaining a balance between socioeconomic and environmental sustainability. Currently, biorefinery industries face several challenges related to the use of feedstock biomass, process or technology adopted for production and commercial viability. More research needs to be developed to make the process completely profitable in all senses and feasible from the laboratory to pilot scale to industrial and should reach the global level. Proper mathematical modelling and artificial neural networking (ANN) for the assessment of the data is required to cut short the failure attempts. In summary, by integrating various biodiesel technologies, the concept of a biorefinery arises for viability and sustainability of the biofuel production in the long term.

11.8 Summary

It can be concluded that the tolerable technologies are already in existence and have improved continuously in the last five decades as substitutes for fossil fuels. However, in the past decade, studies have mainly focused on finding the best catalysts and optimal operating conditions for the production of hydrogenated green fuels. In this sense, the green diesel can be obtained without generating glycerol as a by-product. Such biofuels do not require any separation, cleaning or refining process for use. Elimination of all these steps reduces the cost for production at the industrial scale.

Another interesting aspect is that the use of renewable reagents, such as ethanol, acetate or ethyl carbonate, makes the process 100% renewable and can contribute to the minimisation of emissions. The use of lipase-selective enzymes is always recommended, but its high cost pulls the technology behind. Undoubtedly, a vast scope in the research in this area is still open for all types of blends, catalysts and pathways that reduce the economic price of the biofuels.

In summary, it would be of great interest to obtain heterogeneous catalysts for the interesterification processes that produce glycerol derivatives, together with the methyl or ethyl esters of fatty acids that constitute oils and fats. Finally, we must consider the possibilities that the most recent strategy opens the door to achieve the application of vegetable oils as biofuels in today's diesel CI engines to reduce the viscosity and obtain a suitable fuel mixture.

As a general conclusion, any alternative methods that are able to compete with conventional biodiesel apart from the initially developed solutions should be considered through the LCA method for long-term sustainability and at the industrial level.

References

1 Chilakamarry, C.R., Sakinah, A.M.M., Zularisam, A.W., and Pandey, A. (2021). Glycerol waste to value added products and its potential applications. *Syst. Microbiol. Biomanufacturing* 1 (4): 378–396. https://doi.org/10.1007/s43393-021-00036-w.

2 Howlett, M., Mukherjee, I., and Koppenjan, J. (2017). Policy learning and policy networks in theory and practice: the role of policy brokers in the Indonesian biodiesel policy network. *Policy Soc.* 36 (2): 233–250. https://doi.org/10.1080/14494035.2017.1321230.

3 Liu, Y., Yang, X., Adamu, A., and Zhu, Z. (2021). Economic evaluation and production process simulation of biodiesel production from waste cooking oil. *Curr. Res. Green Sustain. Chem.* 4: 100091. https://doi.org/10.1016/j.crgsc.2021.100091.

4 Haas, M.J., McAloon, A.J., Yee, W.C., and Foglia, T.A. (2006). A process model to estimate biodiesel production costs. *Bioresour. Technol.* 97 (4): 671–678. https://doi.org/10.1016/j.biortech.2005.03.039.

5 Zailani, S., Iranmanesh, M., Hyun, S.S., and Ali, M.H. (2019). Barriers of biodiesel adoption by transportation companies: a case of Malaysian transportation industry. *Sustainabillity* 11 (3): 931. https://doi.org/10.3390/su11030931.

6 Gracia, A., Barreiro-Hurlé, J., and Pérez y Pérez L. (2018). Overcoming the barriers for biodiesel use in Spain: an analysis of the role of convenience and price. *J. Clean. Prod.* 172: 391–401. https://doi.org/10.1016/j.jclepro.2017.10.013.

7 Cardoso, B.F., Shikida, P.F.A., and Finco, A. (2017). Development of Brazilian biodiesel sector from the perspective of stakeholders. *Energies* 10 (3): 399. https://doi.org/10.3390/en10030399.

8 Guo, M. and Song, W. (2019). The growing U.S. bioeconomy: drivers, development and constraints. *New Biotechnol.* 49: 48–57. https://doi.org/10.1016/j.nbt.2018.08.005.

9 Johnston, M. and Holloway, T. (2007). A global comparison of national biodiesel production potentials. *Environ. Sci. Technol.* 41 (23): 7967–7973. https://doi.org/10.1021/es062459k.

10 Hamza, M., Ayoub, M., Bin, S.R. et al. (2021). A review on the waste biomass derived catalysts for biodiesel production. *Environ. Technol. Innov.* 21: 101200. https://doi.org/10.1016/j.eti.2020.101200.

11 Hamelinck, C.N. and Faaij, A.P.C. (2006). Outlook for advanced biofuels. *Energy Policy* 34 (17): 3268–3283. https://doi.org/10.1016/j.enpol.2005.06.012.

12 Naylor, R.L. and Higgins, M.M. (2017). The political economy of biodiesel in an era of low oil prices. *Renew. Sust. Energ. Rev.* 77: 695–705. https://doi.org/10.1016/j.rser.2017.04.026.

13 Dey, S., Reang, N.M., Das, P.K., and Deb, M. (2021). A comprehensive study on prospects of economy, environment, and efficiency of palm oil biodiesel as a renewable fuel. *J. Clean. Prod.* 286: 124981. https://doi.org/10.1016/j.jclepro.2020.124981.

14 Al-Muhtaseb, A.H., Osman, A.I., Murphin Kumar, P.S. et al. (2021). Circular economy approach of enhanced bifunctional catalytic system of CaO/CeO2 for biodiesel production from waste loquat seed oil with life cycle assessment study. *Energy Convers. Manag.* 236: 114040. https://doi.org/10.1016/j.enconman.2021.114040.

15 Casas-Godoy, L., Barrera-Martínez, I., Ayala-Mendivil, N. et al. (2020). Biofuels. In: *Biobased Products and Industries* (ed. C.M. Galanakis), 125–170. Elsevier https://doi.org/10.1016/B978-0-12-818493-6.00004-X.

16 Chen, H., Ding, M., Li, Y. et al. (2020). Feedstocks, environmental effects and development suggestions for biodiesel in China. *J. Traffic Transp. Eng. (English Ed)* 7 (6): 791–807. https://doi.org/10.1016/j.jtte.2020.10.001.

17 Nurkhoiry, R., Azahari, D.H., Amelia, R., and Roosganda, E. (2021). Environmental effect of biodiesel mandatory policy. *IOP Conf. Ser. Earth Environ. Sci.* 892 (1): 012093. https://doi.org/10.1088/1755-1315/892/1/012093.

18 Aldhaidhawi, M., Chiriac, R., Bădescu, V. et al. (2017). Investigation on the mixture formation, combustion characteristics and performance of a diesel engine fueled with diesel, biodiesel B20 and hydrogen addition. *Int. J. Hydrog. Energy* 42 (26): 16793–16807. https://doi.org/10.1016/j.ijhydene.2017.01.222.

19 Sentanuhady, J., Majid, A.I., Prashida, W. et al. (2020). Analysis of the effect of biodiesel B20 and B100 on the degradation of viscosity and total base number of lubricating oil in

diesel engines with long-term operation using ASTM D2896 and ASTM D445-06 methods. *TEKNIK* 41 (3): 269–274. https://doi.org/10.14710/teknik.v41i3.32515.

20 Singh, R., Singh, S., and Kumar, M. (2020). Impact of n-butanol as an additive with eucalyptus biodiesel-diesel blends on the performance and emission parameters of the diesel engine. *Fuel* 277: 118178. https://doi.org/10.1016/j.fuel.2020.118178.

21 Gonzalez-Garay, A., Gonzalez-Miquel, M., and Guillen-Gosalbez, G. (2017). High-value propylene glycol from low-value biodiesel glycerol: a techno-economic and environmental assessment under uncertainty. *ACS Sustain. Chem. Eng.* 5 (7): 5723–5732. https://doi.org/10.1021/acssuschemeng.7b00286.

22 Ji, L., Lei, F., Zhang, W. et al. (2019). Enhancement of bioethanol production from Moso bamboo pretreated with biodiesel crude glycerol: substrate digestibility, cellulase absorption and fermentability. *Bioresour. Technol.* 276: 300–309. https://doi.org/10.1016/j.biortech.2019.01.017.

23 Li, C., Lesnik, K.L., and Liu, H. (2013). Microbial conversion of waste glycerol from biodiesel production into value-added products. *Energies* 6 (9): 4739–4768. https://doi.org/10.3390/en6094739.

24 Brock, D., Koder, A., Rabl, H.P. et al. (2020). Optimising the biodiesel production process: implementation of glycerol derivatives into biofuel formulations and their potential to form hydrofuels. *Fuel* 264: 116695. https://doi.org/10.1016/j.fuel.2019.116695.

25 Mufrodi, Z., Rochmadi, S., and Budiman, A. (2014). Synthesis acetylation of glycerol using batch reactor and continuous reactive distillation column. *Eng. J.* 18 (2): 29–40. https://doi.org/10.4186/ej.2014.18.2.29.

26 Zong, L., Ramanathan, S., and Chen, C.C. (2010). Predicting thermophysical properties of mono- and diglycerides with the chemical constituent fragment approach. *Ind. Eng. Chem. Res.* 49 (11): 5479–5484. https://doi.org/10.1021/ie901948v.

27 Zheng, Y., Shadloo, M.S., Nasiri, H. et al. (2020). Prediction of viscosity of biodiesel blends using various artificial model and comparison with empirical correlations. *Renew. Energy* 153: 1296–1306. https://doi.org/10.1016/j.renene.2020.02.087.

28 Razzaq, L., Farooq, M., Mujtaba, M.A. et al. (2020). Modeling viscosity and density of ethanol-diesel-biodiesel ternary blends for sustainable environment. *Sustainability* 12 (12): 5186. https://doi.org/10.3390/su12125186.

29 Gulen, G. (2015). Commodity Price cycles: explained by an expert. *Way Ahead* 11 (2): 13–15. https://doi.org/10.2118/0215-013-twa.

30 Renzaho, A.M.N., Kamara, J.K., and Toole, M. (2017). Biofuel production and its impact on food security in low and middle income countries: implications for the post-2015 sustainable development goals. *Renew. Sust. Energ. Rev.* 78: 503–516. https://doi.org/10.1016/j.rser.2017.04.072.

31 Kurowska, K., Marks-Bielska, R., Bielski, S. et al. (2020). Food security in the context of liquid biofuels production. *Energies* 13 (23): 6247. https://doi.org/10.3390/en13236247.

32 Subramaniam, Y., Masron, T.A., and Azman, N.H.N. (2020). Biofuels, environmental sustainability, and food security: a review of 51 countries. *Energy Res. Soc. Sci.* 68: 101549. https://doi.org/10.1016/j.erss.2020.101549.

33 Spiess, W.E.L. (2013). Does biofuel production threaten food security? *Food Eng. Ser.* 629–644. https://doi.org/10.1007/978-1-4614-7906-2_31.

34 Lin, L., Cunshan, Z., Vittayapadung, S. et al. (2011). Opportunities and challenges for biodiesel fuel. *Appl. Energy* 88 (4): 1020–1031. https://doi.org/10.1016/j.apenergy.2010.09.029.

35 Cuéllar-Franca, R., Dimitriou, I., García-Gutierrez, P. et al. (2015). Carbon capture and utilisation: application of life cycle thinking to process design. *Comput. Aided Chem. Eng.* 37. https://doi.org/10.1016/B978-0-444-63577-8.50088-7.

36 Wu, Y., Zhao, F., Liu, S. et al. (2018). Bioenergy production and environmental impacts. *Geosci. Lett.* 5 (1): 1–9. https://doi.org/10.1186/s40562-018-0114-y.

37 Hammond, G.P. and Li, B. (2016). Environmental and resource burdens associated with world biofuel production out to 2050: footprint components from carbon emissions and land use to waste arisings and water consumption. *GCB Bioenergy* 8 (5): 894–908. https://doi.org/10.1111/gcbb.12300.

38 Berger, M., Pfister, S., Bach, V., and Finkbeiner, M. (2015). Saving the planet's climate or water resources? The trade-off between carbon and water footprints of European biofuels. *Sustainability* 7 (6): 6665–6683. https://doi.org/10.3390/su7066665.

39 Zhang, Y., Heath, G., Carpenter, A., and Fisher, N. (2016). Air pollutant emissions inventory of large-scale production of selected biofuels feedstocks in 2022. *Biofuels Bioprod. Biorefin.* 10 (1): 56–69. https://doi.org/10.1002/bbb.1620.

40 Joshi, G., Pandey, J.K., Rana, S., and Rawat, D.S. (2017). Challenges and opportunities for the application of biofuel. *Renew. Sust. Energ. Rev.* 79: 850–866. https://doi.org/10.1016/j.rser.2017.05.185.

41 Sharma, S., Kundu, A., Basu, S. et al. (2020). Sustainable environmental management and related biofuel technologies. *J. Environ. Manag.* 273: 111096. https://doi.org/10.1016/j.jenvman.2020.111096.

42 Pereira, L.G., Cavalett, O., Bonomi, A. et al. (2019). Comparison of biofuel life-cycle GHG emissions assessment tools: the case studies of ethanol produced from sugarcane, corn, and wheat. *Renew. Sust. Energ. Rev.* 110: 1–12. https://doi.org/10.1016/j.rser.2019.04.043.

43 Collotta, M., Champagne, P., Tomasoni, G. et al. (2019). Critical indicators of sustainability for biofuels: an analysis through a life cycle sustainabilty assessment perspective. *Renew. Sust. Energ. Rev.* 115: 109358. https://doi.org/10.1016/j.rser.2019.109358.

44 Dunn, J.B. (2019). Biofuel and bioproduct environmental sustainability analysis. *Curr. Opin. Biotechnol.* 57: 88–93. https://doi.org/10.1016/j.copbio.2019.02.008.

45 Crutzen, P.J., Mosier, A.R., Smith, K.A., and Winiwarter, W. (2016). N2O Release from Agro-biofuel Production Negates Global Warming Reduction by Replacing Fossil Fuels. In: *Paul J. Crutzen: A Pioneer on Atmospheric Chemistry and Climate Change in the Anthropocene*, SpringerBriefs on Pioneers in Science and Practice, vol. 50 (ed. P. Crutzen and H. Brauch), 227–238. Cham: Springer https://doi.org/10.1007/978-3-319-27460-7_12.

46 Kesieme, U., Pazouki, K., Murphy, A., and Chrysanthou, A. (2019). Attributional life cycle assessment of biofuels for shipping: addressing alternative geographical locations and cultivation systems. *J. Environ. Manag.* 235: 96–104. https://doi.org/10.1016/j.jenvman.2019.01.036.

47 Kargbo, H., Harris, J.S., and Phan, A.N. (2021). "Drop-in" fuel production from biomass: critical review on techno-economic feasibility and sustainability. *Renew. Sust. Energ. Rev.* 135: 110168. https://doi.org/10.1016/j.rser.2020.110168.

48 Mandade, P. and Shastri, Y. (2019). Multi-objective optimization of lignocellulosic feedstock selection for ethanol production in India. *J. Clean. Prod.* 231: 1226–1234. https://doi.org/10.1016/j.jclepro.2019.05.311.

49 Löffler, K., Burandt, T., Hainsch, K., and Oei, P.Y. (2019). Modeling the low-carbon transition of the European energy system - a quantitative assessment of the stranded assets problem. *Energy Strategy Rev.* 26: https://doi.org/10.1016/j.esr.2019.100422.

50 Elshout, P.M.F., van Zelm, R., van der Velde, M. et al. (2019). Global relative species loss due to first-generation biofuel production for the transport sector. *GCB Bioenergy* 11 (6): 763–772. https://doi.org/10.1111/gcbb.12597.

51 Correa, D.F., Beyer, H.L., Fargione, J.E. et al. (2019). Towards the implementation of sustainable biofuel production systems. *Renew. Sust. Energ. Rev.* 107: 250–263. https://doi.org/10.1016/j.rser.2019.03.005.

52 Jeswani, H.K., Chilvers, A., and Azapagic, A. (2020). Environmental sustainability of biofuels: a review: environmental sustainability of biofuels. *Proc. R. Soc. A Math Phys. Eng. Sci.* 476 (2243): 20200351. https://doi.org/10.1098/rspa.2020.0351.

53 Liu, Y., Xu, Y., Zhang, F. et al. (2014). The impact of biofuel plantation on biodiversity: a review. *Chin. Sci. Bull.* 59 (34): 4639–4651. https://doi.org/10.1007/s11434-014-0639-1.

54 Demirbas, A., Bafail, A., Ahmad, W., and Sheikh, M. (2016). Biodiesel production from non-edible plant oils. *Energy Explor. Exploit.* 16 (6): 3621–3647. https://doi.org/10.1177/0144598716630166.

55 Das, S. (2020). The National Policy of biofuels of India – a perspective. *Energy Policy* 143: 111595. https://doi.org/10.1016/j.enpol.2020.111595.

56 Kumar, S., Chaube, A., and Jain, S.K. (2012). Sustainability issues for promotion of Jatropha biodiesel in Indian scenario: a review. *Renew. Sust. Energ. Rev.* 16 (2): 1089–1098. https://doi.org/10.1016/j.rser.2011.11.014.

57 Pryshliak, N., Tokarchuk, D., and Shevchuk, H. (2021). The socio-economic and environmental importance of developing biofuels: the Ukrainian case on the international arena. *Polityka Energ. – Energy Policy J.* 24 (1): 133–152. https://doi.org/10.33223/epj/131829.

58 Zhang, Y., San, X., Tsubaki, N. et al. (2010). Novel ethanol synthesis method via C1 chemicals without any agriculture feedstocks. *Ind. Eng. Chem. Res.* 49 (11): 5485–5488. https://doi.org/10.1021/ie901882s.

59 Rosales-Calderon, O. and Arantes, V. (2019). A review on commercial-scale high-value products that can be produced alongside cellulosic ethanol. *Biotechnol. Biofuels* 12 (1): 1–58. https://doi.org/10.1186/s13068-019-1529-1.

60 Ebadian, M., McMillan, J.D., and Saddler, J. (2019) Commercializing Conventional and Advanced Transport Biofuels from Biomass and other Renewable Feedstocks, Implementation Agendas: 2018–2019 Update Compare and Contrast Transport Biofuels Policies. IEA Bioenergy Task 39, 256, 128.

61 Bajpai, P. (2021). *Development in Bioethanol*, 1–222. Singapore: Springer https://doi.org/10.1007/978-981-15-8779-5.

62 Cruz, M.L., de Resende, M.M., and Ribeiro, E.J. (2021). Improvement of ethanol production in fed-batch fermentation using a mixture of sugarcane juice and molasse under very high-gravity conditions. *Bioprocess Biosyst. Eng.* 44 (3): 617–625. https://doi.org/10.1007/s00449-020-02462-x.

63 Esteves, R.A. and Pereira, R.G. (2016). Análise sobre a evolução do biodiesel no Brasil. *Espacios* 37 (N° 02): 2016.

64 Lobo, A., Ramesh, D.K., Chowdhury, D.R., and Mitra, A. (2020). Experimental investigation on influence of additives on emissions, combustion and performance of diesel engine along with EGR fueled with waste cooking oil derived biodiesel. *Int. J. Emerg. Technol.* 10 (1): 01–03.

65 Wu, X., Wu, C., and Wu, S. (2015). Dual-enhanced steam methane reforming by membrane separation of H2 and reactive sorption of CO2. *Chem. Eng. Res. Des.* 96: 150–157. https://doi.org/10.1016/j.cherd.2015.02.010.

66 Meloni, E., Martino, M., and Palma, V. (2020). A short review on Ni based catalysts and related engineering issues for methane steam reforming. *Catalysts* 10 (3): 352. https://doi.org/10.3390/catal10030352.

67 Sadik-Zada, E.R. (2021). Political economy of green hydrogen rollout: a global perspective. *Sustainability* 13: 13464. https://doi.org/10.3390/su132313464.

68 Lee, D.H. (2021). Biohydrogen yield efficiency and the benefits of dark, photo and dark-photo fermentative production technology in circular Asian economies. *Int. J. Hydrog. Energy* 46 (27): 13908–13922. https://doi.org/10.1016/j.ijhydene.2020.08.275.

69 Igliński, B., Piechota, G., Iwański, P. et al. (2020). 15 years of the polish agricultural biogas plants: their history, current status, biogas potential and perspectives. *Clean Techn. Environ. Policy* 22 (2): 281–307. https://doi.org/10.1007/s10098-020-01812-3.

70 Scarlat, N., Dallemand, J.F., and Fahl, F. (2018). Biogas: developments and perspectives in Europe. Renew. *Energy* 129 (A): 457–472. https://doi.org/10.1016/j.renene.2018.03.006.

71 Tabatabaei, M., Aghbashlo, M., Valijanian, E. et al. (2020). A comprehensive review on recent biological innovations to improve biogas production, part 2: mainstream and downstream strategies. Renew. *Energy* 146: 1392–1407. https://doi.org/10.1016/j.renene.2019.07.047.

12

Greener Catalytic Processes in Biofuel Production

12.1 Introduction

The production of biodiesel has been under pressure due to increasing issues related to sustainability because of feedstock and high processing cost [1, 2]. Homogeneous catalysts have been used instead of heterogeneous catalysts in the production of biodiesel using the conventional process [3]. Biofuels are non-toxic, biodegradable and nature friendly as compared to the conventional fuels, such as fossil fuels, creating pollution and environmental issues [4]. Biofuel production from the biomass is the best substitute for the fossil fuel. Biodiesel production is mainly produced from the transesterification process with the homogenous catalysts or the basic catalysts in which free fatty acids (FFAs) are low, as in the case of acid catalysts in which free fatty acids are more than 2 wt%, resulting in a variation in the saponification reaction and thus changing the process [5]. Fats and oil obtained from the bio-based materials in the form of residues are used for the analysis of biofuels, which are composed of high-energy triglycerides with a lower carbon source [6]. Depending upon the feedstock used for the production, Jatropha curcas, Sterculia foetida, and many other species are selected as they are nature friendly, in which fungal lipases is used as the catalyst [7]. Methanol or ethanol is used as alcohol; ethanol is safe to use with better solubility in the oil, whereas methanol is lower in cost and more inclined to the esterification process [8]. Biodiesel production using the fungal fermentation process is a good approach, but the duration of the pre-treatments takes a long period [9]. Microbial enzymes are used for quick catalysis and maximum yields to obtain the renewable fuels [10]. The pre-treatment process is used for treating the biomass, such as thermochemical, physicochemical and biological methods use microbial enzymes as the catalysts for treatment [11]. Lipases possess the capacity to catalyse triglycerides and glycerol fatty acid as they release the free acid; therefore, they have the capacity for the transesterification of fatty acids at a large scale for the production of biofuel [12, 13]. Homogeneous catalysts, such as KOH and H_2SO_4, are mostly used for processing both the acid and base due to a sustainable approach [14]. Technologies, such as ultrasonic, microwaves and hydrodynamic cavitation, are used for the transesterification process [15]. Processing time and the utilisation of energy are minimised using the sustainable catalysts, resulting in a greener catalytic process.

Sustainability in Biofuel Production Technology, First Edition. Pratibha S. Agrawal, Pramod N. Belkhode, and Samuel Lalthazuala Rokhum.

12.2 Sustainable Catalysts

The greener and cheaper route for the production of biodiesel includes various catalysts:

A) Chemical catalyst
B) Industrial wastes
C) Biological catalysts

Reaction timing and energy consumption can be improved by the selection of suitable catalysts that are more precisely sustainable catalysts, which can be reused for the number of cycles.

12.2.1 Chemical Catalyst

Various catalysts used in the processes are presented in Table 12.1.

Table 12.1 Catalysts for biodiesel preparation.

System	Substrate	Catalyst	Catalyst loading	Alcohol:Oil[a]	Conversion	
Conventional	Soya bean oil	K_2CO_3/Al-Ca hydrotalcite	2.0 wt%	13:01	95	[16]
Conventional	*Mesuaferrea* oil	Co@ZnO	2.5 wt%	09:01	98	[17]
Conventional	Rapeseed oil	Na_2CO_3	5.0 wt%	11:01	96	[18, 19]
Conventional	Sunflower oil	K^+@MgO	3.0 wt%	10:03	98	[20]
MW	WCO	CaO	1.3 wt%	9.6:1	90.4	[21]
MW	Waste lard	CaO/zeolite	3.0 wt%/vol%	30:01:00	97.1	[22]
Conventional	Sunflower oil	CaO/zeolite	6.0 wt%	06:01	97.5	[23]
Conventional	Sunflower oil	$K_2CO_3/\gamma\text{-}Al_2O_3$	5.0 wt%	12:1[b]	99.3	[24]
Conventional	Sunflower oil	S-ZrO/MCM-41	5.0 wt%	09:01	96.9	[25, 26]
MW	Yellow horn seed oil	Novozym 435	8.0 wt%	06:01	95	[27]
MW	WCO	Novozym 435	10.0 wt%	6:1[c]	94	[28]
US	Soya bean oil, WCO	Combi-lipases	25.0 wt%	5:1[b]	90.0, 70.0	[29]
US	Lauric acid	CALB 10 000	2.0 wt%	2:1[b]	96.8	[30]
MW	*Nannochloropsis* sp.	[EMIM]MeSO4	–	–	36.7	[31]
MW	*Calophyllum* in.	[BMIM]HSO4	0.5 wt%	09:01	92.8	[32]
Conventional	*Jatropha* seed oil	[β-CD-6-Im-(CH2)3-HSO_3][HSO4]-Fe_3O_4	3.0 wt%	10:01	94.7	[33]
MW	WCO	SO_4/Fe-Al-TiO_2	3.0 wt%	10:01	96	[34]

[a] Methanol.
[b] Ethanol.
[c] Dimethyl carbonate.
Source: From [35]/MDPI/CC BY 4.0.

Heterogeneous catalysts, such as K_2CO_3-loaded Al–Ca hydrotalcite, were successfully applied for the transesterification of soya bean oil with methanol, which resulted in 95% conversion in two hours at 65 °C with 2 wt% catalyst loading that can be reused four times with conversions above 87%, as reported by Sun et al. The catalysts were characterised using spectroscopy, X-ray diffraction, scanning electron microscopy and Hammett indicator methods [16].

Feedstock non-edible oil and heterogeneous nanocatalysts are used for biodiesel production in which synthesised cobalt-doped ZnO nanocatalysts are used, which results in the optimum value achieving 98.03% conversion at 60 °C with a catalyst loading of 2.5 wt%, as reported by Borah et al. The catalysts could be reused twice. The result indicates good catalytic activity for the large-scale production [17]. Similarly, Laskar et al. developed a method to produce biodiesel in 97% conversion using silver-doped ZnO [36].

The performance of K_3PO_4, K_2CO_3, Na_3PO_4 and Na_2CO_3 in the production of biodiesel using the transesterification reaction of rapeseed oil with the alcohols is investigated by Malins. The use of potassium salts was active with less stability, whereas sodium salts was active with the recovery rates above 80%. Na_2CO_3 indicates more stability and can be reused five times in the biodiesel sample [19].

Solid acid catalysts based on metal oxides and magnetic nanoparticles are used in the transesterification reaction of different oils and waste materials for the production of biodiesel [5, 37]. Metal oxides, such as zirconia, zinc, titanium, iron, tungsten and magnetic materials, are used in the reaction for the optimisation of the reaction parameters, such as the amount of catalyst, molar ratio of oil to alcohol, reaction time and temperature with the comparison of yield obtained. Metal-oxide-based catalysts provide more sustainable, green catalytic activity [18].

Basic catalysts are prepared by the alkali activation of mesoporous $MgCO_3$ using KOH. The transesterification of sunflower oil was carried out at 65 °C, with a methanol:oil weight ratio of 3:10 and a catalyst concentration of 3 wt%, referred to as the oil weight. No catalytic activity is shown with original mesoporous carbonate, whereas MgO obtained by $MgCO_3$ heating reached 45% conversion after 15 hours of reaction time, and the K-doped sample, which attained 98% oil conversion in 30 minutes, indicates that the K catalyst was recovered and reused, giving 50% oil conversion after 120 minutes of the reaction time, as studied by Estevez et al. [38].

The development of heterogeneous catalysts from calcium is of much importance due to the large availability, high activity and high catalytic activity when calcined at high temperatures in the biodiesel production. Calcium oxide is a cheap, basic catalyst for the transesterification of vegetal oils with low free fatty acid contents. Sharma et al. compared its activity to that of KOH for biodiesel production from waste cotton seed cooking oil. Process parameters, such as methanol-to-oil ratio, catalyst loading and reaction time, were optimised to enhance the yield of KOH- and CaO-catalysed conditions, that is, 96.55% and 90.41%, respectively, which shows good agreement. The biodiesel yield achieved 90% up to four cycles. The efficiency factor is larger than that of the conventional method in the case of microwave-assisted process and results in an increased yield with a decrease in the reaction time, making the technique energy efficient. Figure 12.1 shows the microwave-assisted biodiesel production process, as studied by Sharma, for the production of biodiesel from waste cotton seed cooking oil using the microwave-assisted transesterification process [20].

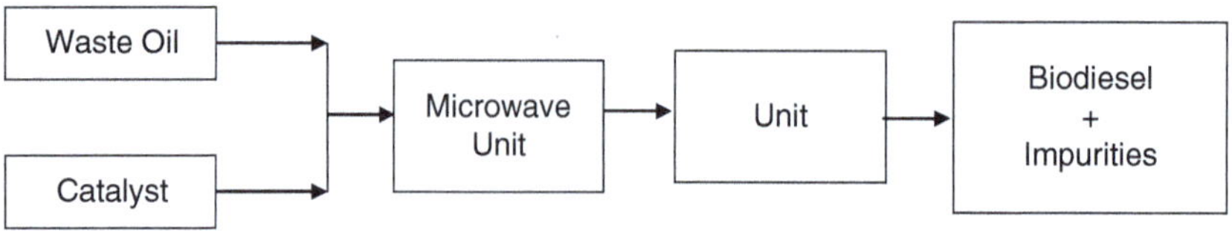

Figure 12.1 MW-assisted biodiesel production process with either CaO or KOH as the catalyst. *Source:* Adapted from [20].

The estimation of activation energy of various catalysts running the reaction at varying temperatures of 45, 50 and 55 °C indicates that KOH is more active with an activation energy of 13.05 KJ/mol as compared to CaO with a value of 28.93 KJ/mol. It was efficiently activated by microwave, and their corresponding activation energies were 34.5 and 50.4 KJ/mol, which were more than those of the conventional experiment. Similarly, catalyst loading and reaction time are significant for the optimum condition. The heterogeneous catalyst was recovered and reused for four cycles, maintaining yields of more than 90%, as confirmed in scale-up tests [35].

The optimisation and modelling of biodiesel production were studied by Lawan et al. using CaO on silica zeolite under microwave with the preparation of different samples of CaO-loaded zeolite, which resulted in the best yield at 35 wt% CaO/zeolite catalyst and reduced the reaction residence time and energy consumption, and thus improving efficiency, costs and sustainability. When heterogenous catalysts are recovered and reused, the biodiesel yield is reduced with the utilisation of reused catalysts many times due to the presence of reacting species that remain on the surface of the catalyst after regeneration [22].

Solid acidic catalysts have also proven themselves to be good candidates for the esterification reaction for biodiesel [39] Metal oxides modified with sulphate ions, in particular, are characterised by high strength, toughness and good corrosion resistance. Dehghani et al. investigated the influence of US irradiation on the catalytic properties and performance of sulphate mesoporous zirconium-doped MCM-41 (Zr/Si molar ratio = 0.2) [26]. The catalyst is prepared via the US-assisted impregnation/hydrothermal hybrid method, as shown in Figure 12.2.

Three samples were prepared at different irradiation power levels, 30, 60 and 90 W, to obtain the highest yield. The sample is characterised by a smaller particle size and highly dispersed particles with the reuse of the catalyst for several cycles. The conversion loss with the sonicated catalyst is less than the conversion loss.

Sustainable, renewable, biodegradable and non-toxic diesel fuel substitutes can easily be produced through base- or acid-catalysed esterification and transesterification reactions [40]. Biodiesel is superior to diesel oil for health and the environment (it is low in sulphur, emits a low level of noxious particles such as HC and CO, and is better in the CO_2 cycle in reducing global warming) as well as for engine performance (better lubrication, high cetane number and more complete combustion) [41, 42]. It has a high flashpoint, better viscosity and similar caloric power to fossil fuels [18]. Biodiesel is usually produced by the transesterification of vegetable oils or animal fats with chemical catalysts as shown in Figure 12.3, especially in the presence of strong acidic or basic solutions, such as hydrochloric acid, sulphuric acid, sodium hydroxide, sodium methoxide and potassium hydroxide. The catalyst is required to initiate the reaction and act as a solubiliser, which is needed because alcohol is sparingly soluble in the oil phase, and non-catalysed reactions are extremely slow. The catalyst promotes an increase in alcohol solubility to allow the reaction to proceed at a reasonable rate [44]. Figure 12.3 show the solid acid-catalysed

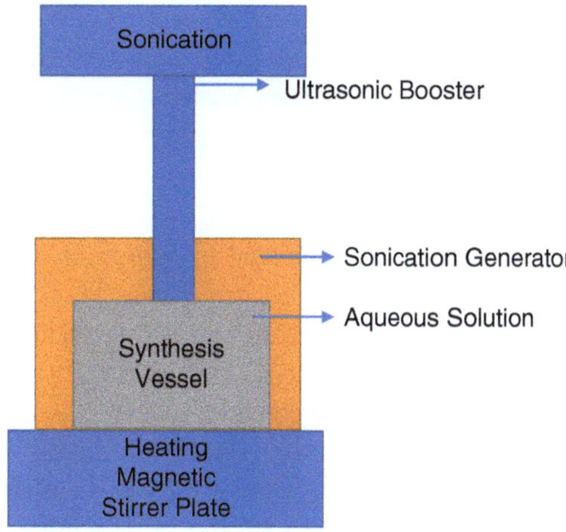

Figure 12.2 Scheme of US horn for biodiesel production catalysed by sulphated zirconia nanocatalyst supported on MCM-41. *Source:* Adapted from [26].

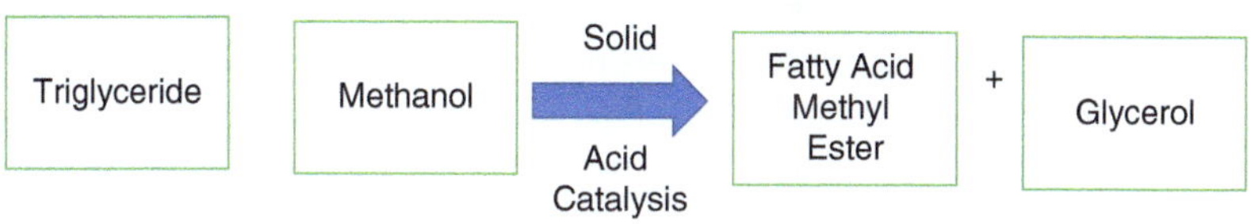

Figure 12.3 Solid acid-catalysed transesterification reaction. *Source:* Adapted from [43].

transesterification reaction. Homogeneous alkali catalysts can convert triglycerides (TGs) to their corresponding fatty acid methyl esters (FAMEs) with high yield, less time and low cost [43]. Separating the catalyst from the product mixture for recycling is technically difficult. After the reaction, the catalyst should be neutralised or removed with a large amount of hot water, which will produce a large amount of industrial wastewater.

Two types of catalytic mechanisms, i.e. heterogeneous and homogeneous processes, can be used for biodiesel production. Heterogeneous catalysis can be identified as solid and enzymatic, where particularly lipases are used to improve reaction rates and yields [45, 46]. Biodiesel production processes based on the use of acid catalysts are good alternatives to conventional processes because of their simplicity and the simultaneous promotion of esterification and transesterification reactions without soap formation [47]. Thus, the acid catalysts can directly produce biodiesel from low-grade, highly acidic, and water-containing oils.

More recently, a 'green' approach to biodiesel synthesis has stimulated the application of sustainable solid acid catalysts as replacements for such liquid acid catalysts such that the use of harmful substances and the generation of toxic wastes are avoided; meanwhile, the ease of catalyst separation after the reactions can be attained.

12.2.2 Industrial Wastes

Homogeneous base catalysts lack in their catalytic activity due to the high FFA content in cooking and any edible oil. Acid catalysts are also not used much due to their corrosive nature. Exploration of eco-friendly material by utilising large-scale industrial waste materials can provide a potential catalyst in biodiesel production. Industrial wastes can be useful in low-cost biodiesel production. In addition, the solid base heterogeneous catalyst requires lower cost and mild conditions in biodiesel production and are hence preferred more. Industry waste material can be explored with the target of developing sustainable catalyst for biodiesel production. Various factories, industries, mills and thermal power plants generate a huge amount of waste every year.

12.2.2.1 Lime Mud

Lime mud, waste generated in pulp mills, is mainly composed of calcium carbonate with trace amounts of magnesium carbonate and other minerals. The chemical composition of waste produced from pulp mill is lime mud, carbide slug, blast furnace slag, red mud and dolomite, all of which contain higher amounts of CaO followed by SiO_2 [48]. High-temperature calcination at 800 °C converts $CaCO_3$ into CaO, which can be useful in further biodiesel production. It was also observed that higher temperatures resulted in reduced conversion, which may be due to the sintering of the catalyst. Hui et al. [49] compared the catalytic activity of calcined lime mud with that of commercial CaO. The calcined catalyst showed higher basicity than the commercially available CaO. Moreover, the calcined catalyst showed resistance to the effect of water and CO_2; otherwise, the exposure of moisture to calcium oxide catalysts would result in an air slaking process converting CaO to Ca(OH)2, thus affecting the catalytic activity of the commercially available CaO [50]. In addition, the presence of CO_2 in the atmosphere converts CaO back into $CaCO_3$, again reducing its activity [51].

The refining of bauxite in aluminium industries produces a large amount of red mud, which is a heterogeneous mixture of different oxides calcium, oxidised iron (imparting red colour), silica and aluminium. With every tonne of alumina, about two tonnes of red mud are produced. Due to the use of sodium hydroxide solution in the refining process, its pH is very high, around 10–13. Many attempts have been made earlier to use this huge amount of waste in various areas, such as ceramics, iron recovery industries [52] and steel-processing industries [53, 54]. The waste treatment methods utilising red mud for wastewater purification, waste gas purification and soil remediation have also been reported by earlier workers [55].

The number and strength of base sites in red mud make it a highly basic active catalyst. The highly increasing demand of limitedly available petroleum diesel has attracted researchers towards the alternative fuels. The cost of the catalyst used in the biodiesel became an important parameter when deciding the sustainability and compatibility of the biodiesel. However, the composition of red mud varies worldwide. Liu et al. [56] used this waste as a basic catalyst. The uncalcined red mud shows the presence of iron and aluminium oxide. After calcination, the structure is affected. In a comparative study with KOH, it has been observed that red mud catalyst shows increased activity due to the presence of active pores [57]. It has been observed that the B50 blend of biodiesel has shown better engine performance characteristics.

Moreover, it has been observed that the red mud catalysts once used in biodiesel production can be deactivated due to the hindrance of the active catalytic sites by the contaminants. The catalyst can be regenerated after washing with hexane followed by calcination treatments. This regeneration method shows increased catalytic activity, improving the properties again [58]. This method of reusing red mud waste as a catalyst is beneficial as it could recycle the waste and thereby reduce the cost of the catalyst. Such a low-cost catalyst could make the biodiesel production economical.

12.2.2.2 Slag

Slag is produced when metals are extracted or separated from their respective ores. Its chemical composition shows the presence of various metals, metal oxides, metal sulphides and silicon dioxide. It can be used as a catalyst in biodiesel production. Millions of tonnes of calcium carbide residue, commonly known as carbide slag, are generated as waste during the industrial production of ethyne gas. Moreover, it is disposed as landfiller creating severe environmental problems. As its chemical composition shows the presence of $Ca(OH)_2$ as the major constituent, it can be successfully used as feedstock to prepare the basic desired CaO catalyst. Thus, developing slag as a catalyst in biodiesel production has attracted the attention of many researchers [59, 60]. Niu et al. [61, 62] compared the catalytic activity of the calcium-rich carbide slag with $Ca(OH)_2$ and found that the carbide slag is superior in the transesterification process used in the production of biodiesel. The calcination of the carbide slag at 650 °C resulted in high basicity, high surface area and large pore volume, which makes it more active when compared with commercially available CaO.

Biomass gasification yields producer gas with the generation of huge amounts of straw, husk or powder as waste. A novel solid base gasified slag catalyst can be developed from these waste materials using simple pulverisation and sieving processes. It has been found that due to high-temperature melting in the biomass gasifier, this catalyst possesses high stability, low leaching and good catalytic activity [63]. The chemical composition after XRF shows the presence of oxides SiO_2, CaO, K_2O, MgO, FeO, and Al_2O_3. It has also been observed that the yield is not much affected after reusing the catalyst for 33 reactions. Only small amounts of K and Mg (<10 ppm) may be leached after some time. Such a reusable catalyst makes the biodiesel production cheaper.

To answer the urgent need to develop a new catalyst with high efficiency and low cost, the blast furnace slag, a high-yield industrial solid waste, is thought to be used. It is produced in large volumes during iron-making processes and thus can be used as a low-cost abundant catalyst for transesterifications of esters including triglycerides. It contains various silicates, alumino-silicates and metals. It has been observed [64] that calcinated slag at 800 °C can efficiently produce 97% yield of FAME. In addition, this catalyst can be recovered even after air exposure for 24 hours. Thus, this catalyst can provide a sustainable chemical pathway for biodiesel production in an economic and ecological way.

12.2.2.3 Rocks

Heterogeneous catalysts have the advantage of easy separation from the product. Numerous naturally available catalysts were tried in biodiesel synthesis. Amongst various calcium and magnesium carbonate rich rocks, dolomite is considered to be the best-suited catalyst [65] in the biodiesel production. The reason for this is the high content of $CaMg(CO_3)_2$, which after calcination produces a mixed oxide of CaO·MgO. The catalytic activity of a KI/KIO3 and KIO3/HAP impregnated hydroxyapatite (HAP) derived from natural phosphate rocks was also tested for biodiesel production [66]; 5% w/w KI/KIO_3 impregnation concentration provided a maximum yield of 91.78%.

12.2.2.4 Fly Ash

Around 25 million tonnes of ash per annum are produced in the countries such as South Africa [67]. The management of such a huge amount of fly ash is again of great concern.

Coal fly ash contains high amounts of silicates and silicate alumina. Fly ash, which is used as a raw material during zeolite synthesis, can be successfully used as a base catalyst in the biodiesel production. Thus, proper chemical treatment of fly ash can make it a strong candidate in the biodiesel production field. FA/Na-X and ion-exchanged FA/K-X were used in the transesterification process [68]. The biodiesel production using waste cooking oil with high contents of both CaO and SO_3 in fly ash was proposed by earlier researchers [69]. It has been reported that the fly ash acts as an acid as well as a base catalyst. This versatile catalyst thereby increases the yield by catalysing both the transesterification and hydroesterification reactions. The fly ash with different loadings of KNO_3 was also used [67], and it has been observed that 5 wt% KNO_3 shows the maximum efficiency. Coal fly ash-zinc oxide composite (CFA/ZnO) was studied for waste cooking oil. A study [70] shows that a catalyst loading of 0.5 wt% was helpful in obtaining 98.14% and that the catalyst can be effectively used for four successive cycles. Arif et al. [71] found that coal fly ash, when chemically modified using HCl and $Ca(NO_3)_2{\cdot}4H_2O$, can provide a better yield of biodiesel. Thus, for quite a long time period, researchers aimed to develop a waste utilisation approach for the effective bioenergy production.

12.2.3 Biological Catalysts

The central idea behind the biological catalyst is that a large amount of unavoidable waste is generated on a daily basis from living organs. Different sources can be employed in meeting the increasing energy demands worldwide. CaO [72] has already proven to be a promising candidate in the transesterification for biodiesel production due to its low cost. It requires moderate reaction conditions. CaO can be effectively obtained from various biological sources, such as shells, bones, eggs, plants, etc. The benefit of using these catalysts is that they provide biodiesel with high purity, and there is less need of purification of biodiesel obtained by using these catalysts. In addition, there is less wastage of water for washing the biodiesel in the last step.

Worldwide, many waste shells are disposed of in landfills as they have no practical usage, but now there are many practical uses of these shells like as a catalyst in biodiesel production, which decreases the cost of biodiesel formation. The composition of mussel shells basically contains calcium carbonate and a very low amount of organic matter or other compounds [73]. For using it as a catalyst, $CaCO_3$ is converted into calcium oxide by calcination temperature higher than 900 °C. The mussel shell is used as a catalyst for the transesterification of FAME after calcination, and activity and stability of the mussel shell catalyst basically depend upon its specific surface area [74].

Mussel shells are both economical and environmentally friendly due to their activation for up to seven cycles for biodiesel production. Until the last cycles, the biodiesel yield is greater than 90%. From an economical point of view, it reduces the biodiesel production cost [75]. Many studies also proved that under particular conditions of temperature, amount of catalyst, ratio of oil and methanol we can obtain almost 100% yield of the biodiesel. There are other methods to prepare waste mussel shell catalysts to increase their activity. Firstly, calcination at 900 °C activates it by adding carbon and then saturates it with a solution of NaOH. This catalyst was used for the transesterification of palm oil into biodiesel with a yield of 95% [76].

By the use of chicken eggshells, two types of catalyst were synthesised: Eggshell–$CaO_{C\text{–}H\text{–}D}$ (calcination–hydration–dehydration) catalyst 1 and Eggshell–CaDG (calcium diglyceroxide) catalyst 2. To convert chicken eggshells into Eggshell–CaO_{C-H-D} catalysts, the reactant was calcined, the calcined product with distilled water was subjected to sonication, and the sonication product was calcined again. By this whole process, Eggshell–CaO_{C-H-D} is prepared. For preparing Eggshell–CaDG, Eggshell–CaO_{C-H-D} was mixed with glycerol and methanol with a particular amount and heated for three hours at 6 °C. Both the catalysts provide good results for the preparation of biodiesel. Catalyst 1 in transesterification provided 93% yield and catalyst 2 provided 96% yield. Catalyst 2 is more active as compared to catalyst 1 because transesterification using a smaller amount of catalyst provides more yield, a lower methanol-to-oil ratio and less reaction time [77].

Another type of catalyst from chicken and guinea fowl egg shells was derived by using a calcination process. These catalysts also provided a good transesterification process with the yields of 95% and 96%. These catalysts are active for up to 12 cycles with yields greater than 91% [78], and they provide good results with the microwave-assisted reaction. There are many benefits of using this catalyst; for example, if commercial Cao is used as a catalyst with waste cooking oil to convert into biodiesel, the yield is approximately 65–67%, but if CaO derived from eggshell is used as waste cooking oil, then the yield is greater than 90% [79]. The only limitation of using eggshell-based catalysts is that it takes more time for the reaction to proceed.

The waste eggshells of quail are utilised as a heterogeneous catalyst with a microwave irradiation technique. Catalysts derived from waste eggshells quail by a calcination method, and the reaction was carried out in a microwave with a power of 800 W. This reaction was performed within five minutes with a methanol-to-oil ratio of 18:1, and the amount of catalyst used was 4 wt%. This catalyst provided good results up to three cycles. The yield decreases more after the third cycle, and therefore there is a need to again activate the catalyst [80].

Waste quail eggshell-based catalyst properties are enhanced by treating it with 0.005 M HCl. When the result was checked after treatment with acid, then it showed that the quantity of phosphate decreased from 1.1 to 0.1 and the quantity of other beneficial metal oxides increased as the composition of $CaCO_3$ increased from 97.3% to 98.4%. The quail eggshell catalyst shows high catalytic activity and a large number of basic sites that match the activity of homogeneous catalyst potassium methoxide. Using this catalyst in the transesterification of palm oil showed 98% conversion into FAME. This catalyst gave good results up to five cycles [81].

Cockle shell is used as a heterogeneous catalyst by the same calcination process, but it provides good yield with crude jatropha oil when the calcination temperature is 700 °C. This catalyst shows 94% conversion of crude jatropha oil into biodiesel [82]. In some other studies, cockle shells were used as a catalyst and compared with commercial CaO for the preparation of biodiesel from palm oil. This study shows that there are many factors that affect the yield of the product, such as methanol-to-oil ratio, reaction time and temperature. By changing all these factors, the results show that CaO derived from cockle shell and commercial CaO provided almost the same result. Therefore, it is better to cockle shell-based catalysts to utilise waste [83].

When a cockle-shell-based catalyst was used for the transesterification of jatropha oil, the yield was 94% when the calcined temperature was 800 °C and the amount of catalyst was 1.5 wt% oil [84]. In some research studies, the response surface methodology (RSM) was performed to check the yield of FAME. According to RSM, the optimal conditions for the reaction were a methanol-to-oil molar ratio of 8.5:1, 3.5 wt% oil and 44 minutes to complete the reaction. The yield obtained by this method was 94% [85]. These cockle-shell-based catalysts were checked for up to four cycles. In the first and second cycles, it provided almost 94% yield; in the third cycle, the yield was 89% due to the moisture content present in the reactant, which changed CaO into $Ca(OH)_2$. However, there was no large change in the yield of the product. The activity of the catalyst was improved by recalcination [86]. Cockle eggshell is also used as a heterogeneous catalyst to convert animal fat (chicken fat) into biodiesel. According to statistical analysis, if we use 4.9 wt% catalyst amount and 0.55:1 methanol-to-oil mass ratio for three hours, then we obtain 98% yield of the product [87].

CaO is derived from calcined river snail shell and used as a heterogeneous catalyst. Using this approach with waste cooking oil, the conversion rate is 98% [88]. Using the snail shell as a catalyst at a calcination temperature of 1000 °C provided 98% yield of product using soya bean oil. Another very good point of this reaction is that the reaction provides 100% conversion rate at room temperature This catalyst is economically very viable due to its use for up to eight cycles without any retreatment on the catalyst [89].

The transesterification of rapeseed oil using snail shells in powder form was directly used as a catalyst, producing 78% of biodiesel. When calcined snail shells were used, 93% yield was obtained; when calcined snail shells were loaded with the KBr solution, they became a more active catalyst. By using this activated catalyst, 99% yield of the product was obtained, so this activated catalyst can be used four times with good yield [90].

Another method was used to prepare a catalyst from waste snail shells that are nano-CaO particles. It was prepared by treating snail with acid and then hydrothermal treatment, and then the calcination–hydration–dehydration method was used. This catalyst is characterised by different methods with a particle size of 40 nm and a surface area of $9.37\,m^2 g^{-1}$. For the production of biodiesel, this CaO nano-particle derived from waste snail shells is treated with scum oil and hydnocarpus wightiana oil, and the yields obtained are 96.929% and 98.93%, respectively [91].

The catalyst is derived from a waste oyster shell by a calcination method. The oyster shell has a very small surface area of approximately $0.2\,m^2 g^{-1}$, and calcination increases the surface by $2\,m^2 g^{-1}$. To increase the activity of this catalyst, it is saturated with KI, which increases the surface area up to 31 times that of the starting material, that is, $5.9\,m^2 g^{-1}$. This catalyst is used to transesterify soya bean oil into biodiesel with $1\,mmol\,g^{-1}$ of catalyst at 50 °C, and the methanol-to-oil ratio of 10:1 for four hours of the reaction time [92]. Some studies showed that when a waste oyster shell was used as a catalyst with waste cooking oil at a temperature of 65 °C with microwave heating and a methanol-to-oil ratio of 9:1, 87.3% yield of FAME was obtained [93].

There is another method to prepare a catalyst from oyster shells by a combustion method at various temperatures. The combustion was performed in an electric furnace in air. The catalyst prepared by this method was used to make biodiesel of soya bean oil under reaction condition of 25 wt% of catalyst, and the reaction time for this reaction was five hours. The biodiesel obtained by this method was 74%, but the purity level of biodiesel obtained was 98.4% [94].

The fish scale used a heterogeneous scale and was prepared by calcination at high temperature. Generally, by calcination, HAP is generated from CaO and H_3PO_4; at high temperature, HAP is converted into β-$Ca_3(PO_4)_2$, which has high basicity and provides a high yield of biodiesel [95]. The fish scale as a catalyst provides good results for up to five cycles.

Table 12.2 Catalyst from animal waste bones.

Animal bone used	Oil used	Yield of biodiesel (%)	Reference
Sheep bones	Palm oil	97	[96]
Sheep bones	Canola oil	95	[97]
Bovine bones	Soya bean oil	97	[98]
Pork bones	Jatropha oil	94	[99]
Chicken bones	Palm oil	95	[100]
Pig bones	Palm oil	96	[101]

Source: Adapted from [102].

Every year around billions of bone waste is generated in slaughter, which is a major issue for the environment if it is not disposed of in a proper way. There are many methods of preparing catalysts from animal waste bones, but calcination is a simple method to synthesise these catalysts as there are other chemicals required. Table 12.2 shows the preparation of catalysts from the used animal bones. There is another method of the preparation of the catalyst to increase the activity of catalyst: firstly, calcination is performed, then salt of the same metal is loaded to fill all the pore sites and then it is thermally activated by removing moisture content [102].

Utilising wood ash as a heterogeneous catalyst is a better way to reduce the cost of biodiesel production, and there are also many studies in which wood ash is used as a catalyst. In some studies, wood ash (babul tree stem) is converted into two types of catalysts: calcined wood ash catalyst (CWC) and activated wood ash catalyst (AWC). The first one was simply prepared by calcination at different temperatures, but in the second one wood ash and K_2CO_3 were mixed together in mortar for half an hour and dried in an oven, then calcined at various temperatures and 97–99% yield was obtained by using these catalysts [103]. In an attempt to use waste biomass as a sustainable catalyst for biodiesel production, Laskar et al. prepared a highly active catalyst by open air burning of mango peel (*Mangifera indica*) for the conversion of soya bean oil to biodiesel in 98% yield under the optimised reaction conditions. The high catalytic activity of the peel ash catalyst was attributed to the presence of highly basic metal oxides, such as K_2O, MgO and CaO, its high surface area and porosity [104].

Another example of aged camphor tree ash is using it as a catalyst for biodiesel synthesis. The preparation of the catalyst was performed in the same manner as calcination, but the difference was that camphor tree ash was loaded with potassium carbonate, and the yield obtained by this process was approximately 92%; after recovery of the catalyst, the catalyst was reloaded with potassium carbonate and then calcined. Therefore, the yield was 90% when it was used again [105]. Calcined tectonagrandis leaves [106] are also used for the transesterification of waste cooking oil. CTGL catalysts are active for up to many cycles, and 99% yields are obtained in the first cycle. In the successive cycles, the yield began to decrease, but the lowest yield obtained was near 80%, which is good. To check the reusability of the catalyst, a catalyst is used in the Knoevenagel condensation reaction, and the resulting yield obtained is 97%. In a similar work, banana (*Musa acuminate*) peel ash [107] and trunk ash [108] were used for the production of biodiesel.

Coconut husk is used as a heterogeneous catalyst, but the main component found after calcination is potassium [109]. When the transesterification reaction is performed by using coconut husk as a catalyst (with 5 wt% of jatropha oil), the methanol-to-oil ratio was 12:1 and reaction was performed at room temperature, and the yield obtained was 99%. The only limitation of this reaction is that it is time consuming at room temperature, but if the reaction is performed at 65 °C, then the yield increases and the time of reaction is also converted from hours into minutes. In 2020, Changmai et al. applied orange peel ash as a biogenic catalyst for the conversion of soya bean oil into biodiesel. A high conversion of soya bean oil into biodiesel of 98% was obtained under the optimised reaction conditions, such as methanol:oil ratio of 6:1, catalyst loading of 7 wt% and reaction time of seven hours at room temperature. The high activity of the ash catalyst is attributed to the presence of potassium oxide (K_2O). In the meantime, ironically, the catalyst showed a drastic reduction in catalytic activity on repeated reuse for five successive times [110].

With an aim to increase the recyclability of the biomass-derived catalyst, the same research group has synthesised a novel magnetic nanoparticle-coated orange, *Citrus sinensis* peel (CSP) ash and applied it for biodiesel production from waste cooking oil. A schematic representation of the synthesis of magnetic CSPA ash (CSPA@Fe_3O_4) is shown in Figure 12.4. Interestingly, the magnetic property of the catalyst enables its isolation using an external magnet and greatly increases its recyclability cycles, and accordingly, the catalyst can be reused for nine reaction cycles with a high 91% conversion on the ninth cycle [111].

The biological catalyst derived from the waste crab shell is used for the transesterification of sunflower oil. The crab shell catalyst was prepared by calcination at a temperature of

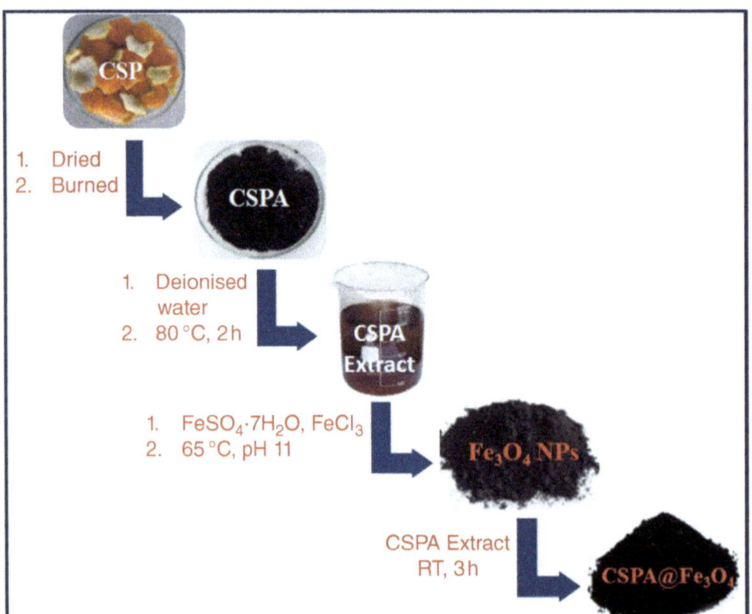

Figure 12.4 Preparation of magnetic *Citrus sinensis* peel ash catalyst. *Source:* From [111] / with permission of Elsevier.

780 °C. The property of this catalyst matches with the commercial CaO, and the property of biodiesel produced by using this biological catalyst matches with all the ASTM standards of diesel [112]. Nowadays, many other methods are used to increase the activity of catalysts, such as the example of crab shell catalysts calcined first and then saturated with Na-ZSM-5. The catalyst obtained CaO/Na-ZSM-5, which has a large surface area and a large pore volume. This catalyst converts neem oil into biodiesel with 95% yield [113].

Calcined crab shell catalysts were used for the conversion of karanja oil into biodiesel. As karanja oil is non-edible oil and catalysts are generated from waste crab shell, both the catalysts decrease the cost of biodiesel generated from these raw materials. By this process, 94% yield was obtained, and the catalyst showed good results until five cycles [114]. Thus, various attempts have been made to continuously improve the yield and make it better fuel for the engine performance.

12.3 Summary

The processing of biomass into the biofuels using suitable catalysts to integrate with plant-based fatty acids acts as an important raw material for the production of nature-friendly sustainable biofuel production. In addition, biodiesel preparation using a suitable reactor, such as microwave, ultrasonic, static mixer, flow reactor and supercritical reactor, with suitable catalysts reduces the reaction time and energy requirement. The greener catalytic processes and sustainable catalysts with capable processing techniques are beneficial for the biofuel production for small to large biofuel producers. All expertise extends their efforts to a wide range of biofuels derived from a variety of industrial, agricultural, forest and other feedstock. The R&D community has achieved a great success in the production of clean, inexpensive fuels from a virtually inexhaustible source of energy.

References

1 Gebremariam, S.N. and Marchetti, J.M. (2018). Economics of biodiesel production: review. *Energy Convers. Manag.* 168: 74–84. https://doi.org/10.1016/j.enconman.2018.05.002.

2 Gavett, S.H., Wood, C.E., Williams, M.A. et al. (2015). Soy biodiesel emissions have reduced inflammatory effects compared to diesel emissions in healthy and allergic mice. *Inhal. Toxicol.* 27 (11): 533–544. https://doi.org/10.3109/08958378.2015.1054966.

3 Ramos, M., Dias, A.P.S., Puna, J.F. et al. (2019). Biodiesel production processes and sustainable raw materials. *Energies 12* (23): 4408. https://doi.org/10.3390/en12234408.

4 Abdelkareem, M.A., Elsaid, K., Wilberforce, T. et al. (2021). Environmental aspects of fuel cells: a review. *Sci. Total Environ.* 752: 141803. https://doi.org/10.1016/j.scitotenv.2020.141803.

5 Changmai, B., Vanlalveni, C., Ingle, A.P. et al. (2020). Widely used catalysts in biodiesel production: a review. *RSC Adv.* 10 (68): 41625–41679. https://doi.org/10.1039/d0ra07931f.

6 Durrett, T.P., Benning, C., and Ohlrogge, J. (2008). Plant triacylglycerols as feedstocks for the production of biofuels. *Plant J.* 54 (4): 593–607. https://doi.org/10.1111/j.1365-313X.2008.03442.x.

7 No, S.Y. (2011). Howvegetable oils and their derivatives affect spray characteristics in ci engines-a review. *At. Sprays* 21 (1): https://doi.org/10.1615/AtomizSpr.v21.i1.60.

8 Marchetti, J.M., Miguel, V.U., and Errazu, A.F. (2007). Possible methods for biodiesel production. *Renew. Sust. Energ. Rev.* 11 (6): 1300–1311. https://doi.org/10.1016/j.rser.2005.08.006.

9 Zabed, H.M., Akter, S., Yun, J. et al. (2019). Recent advances in biological pretreatment of microalgae and lignocellulosic biomass for biofuel production. *Renew. Sust. Energ. Rev.* 105: 105–128. https://doi.org/10.1016/j.rser.2019.01.048.

10 Raghavendra, M.P., Nayaka, S.C., and Gupta, V.K. (2016. https://doi.org/10.1007/978-3-319-43679-1_1). *Microb. Enzymes for Convers. Biomass Bioenergy* 1–26.

11 Chandra, P., Enespa, S.R., and Arora, P.K. (2020). Microbial lipases and their industrial applications: a comprehensive review. *Microb. Cell Factories* 19 (1): 1–42. https://doi.org/10.1186/s12934-020-01428-8.

12 Weber, N., Weitkamp, P., and Mukherjee, K.D. (2001). Steryl and stanyl esters of fatty acids by solvent-free esterification and transesterification in vacuo using lipases from Rhizomucor miehei, Candida antarctica, and Carica papaya. *J. Agric. Food Chem.* 49 (11): 5210–5216. https://doi.org/10.1021/jf0107407.

13 Juangsa, F.B., Prananto, L.A., Oda, T., and Aziz, M. (2018). Highly energy-efficient combination of dehydrogenation of methylcyclohexane and hydrogen-based power generation. *Chem. Eng. Trans.* 226: 31–38. https://doi.org/10.3303/CET1870349.

14 Fjerbaek, L., Christensen, K.V., and Norddahl, B. (2009). A review of the current state of biodiesel production using enzymatic transesterification. *Biotechnol. Bioeng.* 102 (5): 1298–1315. https://doi.org/10.1002/bit.22256.

15 Lam, M.K., Lee, K.T., and Mohamed, A.R. (2010). Homogeneous, heterogeneous and enzymatic catalysis for transesterification of high free fatty acid oil (waste cooking oil) to biodiesel: a review. *Biotechnol. Adv.* 28 (4): 500–518. https://doi.org/10.1016/j.biotechadv.2010.03.002.

16 Sun, C., Qiu, F., Yang, D., and Ye, B. (2014). Preparation of biodiesel from soybean oil catalyzed by Al-Ca hydrotalcite loaded with K2CO3 as heterogeneous solid base catalyst. *Fuel Process. Technol.* 126: 383–391. https://doi.org/10.1016/j.fuproc.2014.05.021.

17 Borah, M.J., Devi, A., Borah, R., and Deka, D. (2019). Synthesis and application of Co doped ZnO as heterogeneous nanocatalyst for biodiesel production from non-edible oil. *Renew. Energy* 133: 512–519. https://doi.org/10.1016/j.renene.2018.10.069.

18 Vasić, K., Podrepšek, G.H., Knez, Ž., and Leitgeb, M. (2020). Biodiesel production using solid acid catalysts based on metal oxides. *Catalysts* 10 (2): 237. https://doi.org/10.3390/catal10020237.

19 Malins, K. (2018). The potential of K3PO4, K2CO3, Na3PO4 and Na2CO3 as reusable alkaline catalysts for practical application in biodiesel production. *Fuel Process. Technol.* 179: 302–312. https://doi.org/10.1016/j.fuproc.2018.07.017.

20 Benedictto, G.P., Legnoverde, M.S., Tara, J.C. et al. (2019). Synthesis of K^+/MgO heterogeneous catalysts derived from $MgCO_3$ for biodiesel production. *Mater. Lett.* 246: 199–202. https://doi.org/10.1016/j.matlet.2019.03.081.

21 Sharma, A., Kodgire, P., and Kachhwaha, S.S. (2019). Biodiesel production from waste cotton-seed cooking oil using microwave-assisted transesterification: optimization and kinetic modeling. *Renew. Sust. Energ. Rev.* 116: 109394. https://doi.org/10.1016/j.rser.2019.109394.

22 Lawan, I., Garba, Z.N., Zhou, W. et al. (2020). Synergies between the microwave reactor and CaO/zeolite catalyst in waste lard biodiesel production. *Renew. Energy* 145: 2550–2560. https://doi.org/10.1016/j.renene.2019.08.008.

23 Pavlović, S.M., Marinković, D.M., Kostić, M.D. et al. (2020). A CaO/zeolite-based catalyst obtained from waste chicken eggshell and coal fly ash for biodiesel production. *Fuel* 267: 117171. https://doi.org/10.1016/j.fuel.2020.117171.

24 Silveira Junior, E.G., Perez, V.H., Reyero, I. et al. (2019). Biodiesel production from heterogeneous catalysts based K_2CO_3 supported on extruded Γ-Al_2O_3. *Fuel* 241: 311–318. https://doi.org/10.1016/j.fuel.2018.12.074.

25 Wang, A., Sudarsanam, P., Xu, Y. et al. (2020). Functionalized magnetic nanosized materials for efficient biodiesel synthesis: Via acid-base/enzyme catalysis. *Green Chem.* 22 (10): 2977–3012. https://doi.org/10.1039/d0gc00924e.

26 Dehghani, S. and Haghighi, M. (2017). Sono-sulfated zirconia nanocatalyst supported on MCM-41 for biodiesel production from sunflower oil: Influence of ultrasound irradiation power on catalytic properties and performance. *Ultrason. Sonochem.* 35: 142–151. https://doi.org/10.1016/j.ultsonch.2016.09.012.

27 Zhang, Y., Xia, X., Duan, M. et al. (2016). Green deep eutectic solvent assisted enzymatic preparation of biodiesel from yellow horn seed oil with microwave irradiation. *J. Mol. Catal. B Enzym.* 123: 35–40. https://doi.org/10.1016/j.molcatb.2015.10.013.

28 Panadare, D.C. and Rathod, V.K. (2016). Microwave assisted enzymatic synthesis of biodiesel with waste cooking oil and dimethyl carbonate. *J. Mol. Catal. B Enzym.* 133: S518–S524. https://doi.org/10.1016/j.molcatb.2017.05.003.

29 Poppe, J.K., Matte, C.R., Fernandez-Lafuente, R. et al. (2018). Transesterification of waste frying oil and soybean oil by combi-lipases under ultrasound-assisted reactions. *Appl. Biochem. Biotechnol.* 186 (3): 576–589. https://doi.org/10.1007/s12010-018-2763-x.

30 Gawas, S.D. and Rathod, V.K. (2018). Enhancement in synthesis of ethyl laurate catalyzed by fermase by combined effect of ultrasound and stage wise addition of ethanol. *Chem. Eng. Process. Process Intensif.* 125: 207–213. https://doi.org/10.1016/j.cep.2017.12.018.

31 Wahidin, S., Idris, A., and Shaleh, S.R.M. (2016). Ionic liquid as a promising biobased green solvent in combination with microwave irradiation for direct biodiesel production. *Bioresour. Technol.* 206: 150–154. https://doi.org/10.1016/j.biortech.2016.01.084.

32 Handayani, P.A., Abdullah, A., and Hadiyanto, H. (2017). Biodiesel production from Nyamplung (Calophyllum inophyllum) oil using ionic liquid as a catalyst and microwave heating system. *Bull. Chem. React. Eng. & Catal.* 12 (2): 293–298. https://doi.org/10.9767/bcrec.12.2.807.293-298.

33 Chang, F. and Zhou, Q. (2020). Facile preparation of β-cyclodextrin-Fe_3O_4 magnetic-brønsted acidic ionic liquid for biodiesel production. *Waste and Biomass Valorization* 11 (5): 1983–1938. https://doi.org/10.1007/s12649-018-0506-4.

34 Gardy, J., Osatiashtiani, A., Céspedes, O. et al. (2018). A magnetically separable SO_4/Fe-Al-TiO_2 solid acid catalyst for biodiesel production from waste cooking oil. *Appl. Catal. B Environ.* 234: 268–278. https://doi.org/10.1016/j.apcatb.2018.04.046.

35 Bucciol, F., Colia, M., Gaudino, E.C., and Cravotto, G. (2020). Enabling technologies and sustainable catalysis in biodiesel preparation. *Catalysts* 10 (9): 988. https://doi.org/10.3390/catal10090988.

36 Laskar, I.B., Rokhum, L., Gupta, R., and Chatterjee, S. (2020). Zinc oxide supported silver nanoparticles as a heterogeneous catalyst for production of biodiesel from palm oil. *Environ. Prog. Sustain. Energy* 39 (3): e13369. https://doi.org/10.1002/ep.13369.

37 Changmai, B., Wheatley, A.E.H., Rano, R. et al. (2021). A magnetically separable acid-functionalized nanocatalyst for biodiesel production. *Fuel* 305: 121576. https://doi.org/10.1016/j.fuel.2021.121576.

38 Estevez, R., Aguado-Deblas, L., Posadillo, A. et al. (2019). Performance and emission quality assessment in a diesel engine of straight castor and sunflower vegetable oils, in diesel/gasoline/oil triple blends. *Energies* 12 (11): 2181. https://doi.org/10.3390/en12112181.

39 Lalthazuala, S., Changmai, B., Kress, T., and Wheatley, A.E.H. (2022). A one-pot route to tunable sugar-derived sulfonated carbon catalysts for sustainable production of biodiesel by fatty acid esterification. *Renew. Energy* 184: 908–919. https://doi.org/10.1016/j.renene.2021.12.001.

40 Su, F. and Guo, Y. (2014). Advancements in solid acid catalysts for biodiesel production. *Green Chem.* 16: 2934–2957. https://doi.org/10.1039/c3gc42333f.

41 Ma, F. and Hanna, M.A. (1999). Biodiesel production: a review. *Bioresour. Technol.* 70 (1): 1–5. https://doi.org/10.1016/S0960-8524(99)00025-5.

42 Bhatti, H.N., Hanif, M.A., Qasim, M., and Rehman, A.-u. (2008). Biodiesel production from waste tallow. *Fuel* 87 (13-14): 2961–2966. https://doi.org/10.1016/j.fuel.2008.04.016.

43 Pruszko, R. (2015). Chapter 20 – biodiesel production. *Bioenergy* 339–359.

44 Luque, R. and Melero, J.A. (2012). Introduction to advanced biodiesel production. *Adv. Biodiesel Prod.* 1–9. https://doi.org/10.1533/9780857095862.1.

45 Diamantopoulos, N. (2015). Comprehensive review on the biodiesel production using solid acid heterogeneous catalysts. *J. Thermodyn. Catal.* 6 (1): 1. https://doi.org/10.4172/2157-7544.1000143.

46 Kumar, R.V., Nachiyar, V.C., Farizah, A.B., and Nityasree, J. (2019). Production and characterization of biodiesel obtained from transesterification of lipid from goat tallow. *J. Environ. Biol.* 40 (4): 601–606. https://doi.org/10.22438/jeb/40/4/MRN-731.

47 Laskar, I.B., Changmai, B., Gupta, R. et al. (2021). A mesoporous polysulfonic acid-formaldehyde polymeric catalyst for biodiesel production from Jatropha curcas oil. *Renew. Energy* 173: 415–421. https://doi.org/10.1016/j.renene.2021.04.004.

48 Marwaha, A., Rosha, P., Mohapatra, S.K. et al. (2018). Waste materials as potential catalysts for biodiesel production: current state and future scope. *Fuel Process. Technol.* 181: 175–186. https://doi.org/10.1016/j.fuproc.2018.09.011.

49 Li, H., Niu, S., Lu, C. et al. (2014). Use of lime mud from paper mill as a heterogeneous catalyst for transesterification. *Sci. China Technol. Sci.* 57 (2): 438–444. https://doi.org/10.1007/s11431-013-5440-x.

50 Hari-Bala, D.X., Guo, Y., Deng, Y. et al. (2006). Multigram scale synthesis and characterization of monodispersed cubic calcium carbonate nanoparticles. *Mater. Lett.* 60 (12): 1515–1518. https://doi.org/10.1016/j.matlet.2005.11.062.

51 Roschat, W., Kacha, M., Yoosuk, B. et al. (2012). Biodiesel production based on heterogeneous process catalyzed by solid waste coral fragment. *Fuel* 98: 194–202. https://doi.org/10.1016/j.fuel.2012.04.009.

52 Liu, W., Yang, J., and Xiao, B. (2009). Application of bayer red mud for iron recovery and building material production from alumosilicate residues. *J. Hazard. Mater.* 161 (1): 474–478. https://doi.org/10.1016/j.jhazmat.2008.03.122.

53 Li, Z., Zhang, J., Li, S. et al. (2021). Feasibility of preparing red mud-based cementitious materials: Synergistic utilization of industrial solid waste, waste heat, and tail gas. *J. Clean. Prod.* 285: 124896. https://doi.org/10.1016/j.jclepro.2020.124896.

54 Ning, G., Zhang, B., Liu, C. et al. (2018). Large-scale consumption and zero-waste recycling method of red mud in steel making process. *Fortschr. Mineral.* 8 (3): 102. https://doi.org/10.3390/min8030102.

55 Wang, M. and Liu, X. (2021). Applications of red mud as an environmental remediation material: a review. *J. Hazard. Mater.* 408: 124420. https://doi.org/10.1016/j.jhazmat.2020.124420.

56 Liu, Q., Xin, R., Li, C. et al. (2013). Application of red mud as a basic catalyst for biodiesel production. *J. Environ. Sci. (China)* 25 (4): 823–829. https://doi.org/10.1016/S1001-0742(12)60067-9.

57 Senthil, M., Visagavel, K., Saravanan, C.G., and Rajendran, K. (2016). Investigations of red mud as a catalyst in Mahua oil biodiesel production and its engine performance. *Fuel Process. Technol.* 149: 7–14. https://doi.org/10.1016/j.fuproc.2016.03.027.

58 Wahyudi, A., Kurniawan, W., and Hinode, H. (2017). Study on deactivation and regeneration of modified red mud catalyst used in biodiesel production. *Green Sustain. Chem.* 7 (4): 247–258. https://doi.org/10.4236/gsc.2017.74019.

59 Sun, R., Li, Y., Zhao, J. et al. (2013). CO_2 capture using carbide slag modified by propionic acid in calcium looping process for hydrogen production. *Int. J. Hydrog. Energy* 38 (31): 13655–13663. https://doi.org/10.1016/j.ijhydene.2013.08.030.

60 Niu, S., Liu, M., Lu, C. et al. (2014). Thermogravimetric analysis of carbide slag. *J. Therm. Anal. Calorim.* 115 (1): 73–79. https://doi.org/10.1007/s10973-013-3268-z.

61 Liu, M., Huang, Y., Liu, Q. et al. (2019). Ferric oxide as a support of carbide slag for effective transesterification of triglycerides in soybean oil. *Energy Convers. Manag.* 198: 111785. https://doi.org/10.1016/j.enconman.2019.111785.

62 Liu, M., Niu, S., Lu, C., and Cheng, S. (2015). An optimization study on transesterification catalyzed by the activated carbide slag through the response surface methodology. *Energy Convers. Manag.* 92: 498–506. https://doi.org/10.1016/j.enconman.2014.12.074.

63 Wang, J., Xing, S., Huang, Y. et al. (2017). Highly stable gasified straw slag as a novel solid base catalyst for the effective synthesis of biodiesel: characteristics and performance. *Appl. Energy* 190: 703–712. https://doi.org/10.1016/j.apenergy.2017.01.004.

64 Kuwahara, Y., Tsuji, K., Ohmichi, T. et al. (2012). Transesterifications using a hydrocalumite synthesized from waste slag: an economical and ecological route for biofuel production. *Cat. Sci. Technol.* 2 (9): 1842–1851. https://doi.org/10.1039/c2cy20113e.

65 Gaide, I., Makareviciene, V., Sendzikiene, E., and Kazancev, K. (2021). Natural rocks–heterogeneous catalysts for oil transesterification in biodiesel synthesis. *Catalysts* 11 (3): 384. https://doi.org/10.3390/catal11030384.

66 Widayat, W., Hadiyanto, H., Wardani, P.W.A. et al. (2020). Preparation of KI/hydroxyapatite catalyst from phosphate rocks and its application for improvement of biodiesel production. *Molecules* 25 (11): 2565. https://doi.org/10.3390/molecules25112565.

67 Volli, V., Purkait, M.K., and Shu, C.M. (2019). Preparation and characterization of animal bone powder impregnated fly ash catalyst for transesterification. *Sci. Total Environ.* 669: 314–321. https://doi.org/10.1016/j.scitotenv.2019.03.080.

68 Babajide, O., Musyoka, N., Petrik, L., and Ameer, F. (2012). Novel zeolite Na-X synthesized from fly ash as a heterogeneous catalyst in biodiesel production. *Catal. Today* 190 (1): 54–60. https://doi.org/10.1016/j.cattod.2012.04.044.

69 Muñoz, R., González, A., Valdebenito, F. et al. (2020). Fly ash as a new versatile acid-base catalyst for biodiesel production. *Renew. Energy* 162: 1931–1939. https://doi.org/10.1016/j.renene.2020.09.099.

70 Yusuff, A.S., Bhonsle, A.K., Trivedi, J. et al. (2021). Synthesis and characterization of coal fly ash supported zinc oxide catalyst for biodiesel production using used cooking oil as feed. *Renew. Energy* 170: 302–314. https://doi.org/10.1016/j.renene.2021.01.101.

71 Arif, H., Achmad, C., and Bachrun, S. (2020). Utilization of modified coal fly ash (CFA) as a catalyst for production of biodiesel from coconut oil: Part 1–characteristics of the catalyst. *Mater. Sci. Forum* 981: 190–195. https://doi.org/10.4028/www.scientific.net/MSF.981.190.

72 Helwani, Z., Ramli, M., Saputra, E. et al. (2020). Impregnation of CaO from eggshell waste with magnetite as a solid catalyst (Fe_3O_4/CaO) for transesterification of palm oil off-grade. *Catalysts* 10 (2): https://doi.org/10.3390/catal10020164.

73 Hamester, M.R.R., Balzer, P.S., and Becker, D. (2012). Characterization of calcium carbonate obtained from oyster and mussel shells and incorporation in polypropylene. *Mater. Res.* 15: 204–208. https://doi.org/10.1590/S1516-14392012005000014.

74 Zhang, Y., Liu, S., and Chen, P. (2013). Study on the properties of calcined waste mussel shell. *Nat. Environ. Pollut. Technol.* 12 (3): 435–438.

75 Hu, S., Wang, Y., and Han, H. (2011). Utilization of waste freshwater mussel shell as an economic catalyst for biodiesel production. *Biomass Bioenergy* 35 (8): 3627–3635. https://doi.org/10.1016/j.biombioe.2011.05.009.

76 Hadiyanto, H., Afianti, A.H., and Navi'A UI, Adetya NP, Widayat W, Sutanto H. (2017). The development of heterogeneous catalyst C/CaO/NaOH from waste of green mussel shell (Perna varidis) for biodiesel synthesis. *J. Environ. Chem. Eng.* 5 (5): 4559–4563. https://doi.org/10.1016/j.jece.2017.08.049.

77 Gupta, A.R. and Rathod, V.K. (2018). Waste cooking oil and waste chicken eggshells derived solid base catalyst for the biodiesel production: optimization and kinetics. *Waste Manag.* 79: 169–178. https://doi.org/10.1016/j.wasman.2018.07.022.

78 Oyelaran, O.A., Ogundana, T.O., Adetayo, O.A. et al. (2021). Waste shells of chicken and guinea fowl eggs as catalysts for the production biodiesel. *Agric. Eng. Int. CIGR J.* 23 (1): 159–167.

79 Niju, S., Begum, M.M.M.S., and Anantharaman, N. (2014). Modification of egg shell and its application in biodiesel production. *J. Saudi Chem. Soc.* 18 (5): 702–706. https://doi.org/10.1016/j.jscs.2014.02.010.

80 Buasri, A. and Loryuenyong, V. (2017). Application of waste materials as a heterogeneous catalyst for biodiesel production from Jatropha Curcas oil via microwave irradiation. *Mater. Today Proc.* 4 (5): 6051–6059. https://doi.org/10.1016/j.matpr.2017.06.093.

81 Cho, Y.B. and Seo, G. (2010). High activity of acid-treated quail eggshell catalysts in the transesterification of palm oil with methanol. *Bioresour. Technol.* 101 (22): 8515–8519. https://doi.org/10.1016/j.biortech.2010.06.082.

82 Hadi, N.N., Ismail, S., MRM, S., and Ghafar, F. (2015). Cockles shell as heterogeneous catalyst for biodiesel production from Jatropha seed oil. *J. Intelek.* 9 (2).

83 Ngadi, N., Sulaiman, S., Rahman, R.A., and Lani, N.S. (2017). Production of biodiesel from palm oil using cockle shell waste as heterogeneous catalyst. *J. Teknol.* 79 (5): https://doi.org/10.11113/jt.v79.4503.

84 Hadi, N.N., Idrus, N.A., Ghafar, F., and Salleh, M.R.M. (2017). Waste cockle shell as natural catalyst for biodiesel production from jatropha oil. *AIP Conf. Proc.* 1901 (1): 110008. https://doi.org/10.1063/1.5010549.

85 Singhasiri, T. and Tantemsapya, N. (2016). The utilization of waste egg and cockle shell as catalysts for biodiesel production from food processing waste oil using stirring and ultrasonic agitation. *Energy Sources, Part A* 38 (21): 3125–3131. https://doi.org/10.1080/15567036.2015.1129375.

86 Buasri, A., Chaiyut, N., Loryuenyong, V. et al. (2013). Calcium oxide derived from waste shells of mussel, cockle, and scallop as the heterogeneous catalyst for biodiesel production. *Sci. World J.* 2013: https://doi.org/10.1155/2013/460923.

87 Boey, P.L., Maniam, G.P., Hamid, S.A., and Ali, D.M.H. (2011). Crab and cockle shells as catalysts for the preparation of methyl esters from low free fatty acid chicken fat. *J. Am. Oil Chem. Soc.* 88 (2): 283–288. https://doi.org/10.1007/s11746-010-1660-4.

88 Kaewdaeng, S., Sintuya, P., and Nirunsin, R. (2017). Biodiesel production using calcium oxide from river snail shell ash as catalyst. *Energy Procedia* 138: 937–942. https://doi.org/10.1016/j.egypro.2017.10.057.

89 Laskar, I.B., Rajkumari, K., Gupta, R. et al. (2018). Waste snail shell derived heterogeneous catalyst for biodiesel production by the transesterification of soybean oil. *RSC Adv.* 8 (36): 20131–20142. https://doi.org/10.1039/c8ra02397b.

90 Liu, H., Shuang, G.H., Jing, W.X. et al. (2016). Mixed and ground KBr-impregnated calcined snail shell and kaolin as solid base catalysts for biodiesel production. Renew. *Energy* 93: 648–657. https://doi.org/10.1016/j.renene.2016.03.017.

91 Krishnamurthy, K.N., Sridhara, S.N., and Ananda Kumar, C.S. (2020). Optimization and kinetic study of biodiesel production from Hydnocarpus wightiana oil and dairy waste scum using snail shell CaO nano catalyst. *Renew. Energy* 146: 280–296. https://doi.org/10.1016/j.renene.2019.06.161.

92 Jairam, S., Kolar, P., Sharma-Shivappa Ratna, R. et al. (2012). KI-impregnated oyster shell as a solid catalyst for soybean oil transesterification. *Bioresour. Technol.* 104: 329–335. https://doi.org/10.1016/j.biortech.2011.10.039.

93 Lin, Y.C., Amesho, K.T.T., Chen, C.E. et al. (2020). A cleaner process for green biodiesel synthesis from waste cooking oil using recycled waste oyster shells as a sustainable base heterogeneous catalyst under the microwave heating system. *Sustain. Chem. Pharm.* 17: 100310. https://doi.org/10.1016/j.scp.2020.100310.

94 Nakatani, N., Takamori, H., Takeda, K., and Sakugawa, H. (2009). Transesterification of soybean oil using combusted oyster shell waste as a catalyst. *Bioresour. Technol.* 100 (3): 1510–1513. https://doi.org/10.1016/j.biortech.2008.09.007.

95 Chakraborty, R., Bepari, S., and Banerjee, A. (2011). Application of calcined waste fish (Labeo rohita) scale as low-cost heterogeneous catalyst for biodiesel synthesis. *Bioresour. Technol.* 102 (3): 3610–3618. https://doi.org/10.1016/j.biortech.2010.10.123.

96 Obadiah, A., Swaroopa, G.A., Kumar, S.V. et al. (2012). Biodiesel production from Palm oil using calcined waste animal bone as catalyst. *Bioresour. Technol.* 116: 512–516. https://doi.org/10.1016/j.biortech.2012.03.112.

97 Ghanei, R., Khalili Dermani, R., Salehi, Y., and Mohammadi, M. (2016). Waste animal bone as support for CaO impregnation in catalytic biodiesel production from vegetable oil. *Waste and Biomass Valorization* 7 (3): 527–532. https://doi.org/10.1007/s12649-015-9473-1.

98 Smith, S.M., Oopathum, C., Weeramongkhonlert, V. et al. (2013). Transesterification of soybean oil using bovine bone waste as new catalyst. *Bioresour. Technol.* 143: 686–690. https://doi.org/10.1016/j.biortech.2013.06.087.

99 Buasri, A., Inkaew, T., Kodephun, L. et al. (2015). Natural hydroxyapatite (NHAp) derived from pork bone as a renewable catalyst for biodiesel production via microwave irradiation. *Key Eng. Mater.* 659: 216–220. https://doi.org/10.4028/www.scientific.net/KEM.659.216.

100 Ansari Samani, M., Hosseinzdeh Samani, B., Lotfalian, A. et al. (2020). The feasibility and optimization of biodiesel production from Celtis australis L. oil using chicken bone catalyst and ultrasonic waves. *Biofuels* 11 (4): 513–521. https://doi.org/10.1080/17597269.2019.1628482.

101 Chen, G., Shan, R., Shi, J. et al. (2015). Biodiesel production from palm oil using active and stable K doped hydroxyapatite catalysts. *Energy Convers. Manag.* 98: 463–469. https://doi.org/10.1016/j.enconman.2015.04.012.

102 Hussain, F., Alshahrani, S., Abbas, M.M. et al. (2021). Review waste animal bones as catalysts for biodiesel production; a mini review. *Catalysts* 11 (5): 630. https://doi.org/10.3390/catal11050630.

103 Sharma, M., Khan, A.A., Puri, S.K., and Tuli, D.K. (2012). Wood ash as a potential heterogeneous catalyst for biodiesel synthesis. *Biomass Bioenergy* 41: 94–106. https://doi.org/10.1016/j.biombioe.2012.02.017.

104 Laskar, I.B., Gupta, R., Chatterjee, S. et al. (2020). Taming waste: Waste Mangifera indica peel as a sustainable catalyst for biodiesel production at room temperature. *Renew. Energy* 161: 207–220. https://doi.org/10.1016/j.renene.2020.07.061.

105 Li, C., Hu, X., Feng, W. et al. (2018). A supported solid base catalyst synthesized from green biomass ash for biodiesel production. *Energy Sources, Part A* 40 (2): 142–147. https://doi.org/10.1080/15567036.2017.1405121.

106 Gohain, M., Laskar, K., Phukon, H. et al. (2020). Towards sustainable biodiesel and chemical production: Multifunctional use of heterogeneous catalyst from littered Tectona grandis leaves. *Waste Manag.* 102: 212–221. https://doi.org/10.1016/j.wasman.2019.10.049.

107 Pathak, G., Das, D., Rajkumari, K., and Rokhum, L. (2018). Exploiting waste: Towards a sustainable production of biodiesel using: Musa acuminata peel ash as a heterogeneous catalyst. *Green Chem.* 20: 2365–2373. https://doi.org/10.1039/c8gc00071a.

108 Rajkumari, K. (2020). A sustainable protocol for production of biodiesel by transesterification of soybean oil using banana trunk ash as a heterogeneous catalyst. *Biomass Convers. Biorefin.* 10 (4): 839–848.

109 Vadery, V., Narayanan, B.N., Ramakrishnan, R.M. et al. (2014). Room temperature production of jatropha biodiesel over coconut husk ash. *Energy* 70: 588–594. https://doi.org/10.1016/j.energy.2014.04.045.

110 Changmai, B., Sudarsanam, P., and Rokhum, L. (2020). Biodiesel production using a renewable mesoporous solid catalyst. *Ind. Crop. Prod.* 145: 111911. https://doi.org/10.1016/j.indcrop.2019.111911.

111 Changmai, B., Rano, R., Vanlalveni, C., and Rokhum, L. (2021). A novel Citrus sinensis peel ash coated magnetic nanoparticles as an easily recoverable solid catalyst for biodiesel production. *Fuel* 286: 119447. https://doi.org/10.1016/j.fuel.2020.119447.

112 Sivakumar, P., Sivakumar, P., Anbarasu, K. et al. (2014). An eco-friendly catalyst derived from waste shell of scylla tranquebarica for biodiesel production. *Int. J. Green Energy* 11 (8): 886–897. https://doi.org/10.1080/15435075.2013.829774.

113 Shankar, V. and Jambulingam, R. (2017). Waste crab shell derived CaO impregnated Na-ZSM-5 as a solid base catalyst for the transesterification of neem oil into biodiesel. *Sustain. Environ. Res.* 27 (6): 273–278. https://doi.org/10.1016/j.serj.2017.06.006.

114 Madhu, D., Chavan, S.B., Singh, V. et al. (2016). An economically viable synthesis of biodiesel from a crude Millettia pinnata oil of Jharkhand, India as feedstock and crab shell derived catalyst. *Bioresour. Technol.* 214: 210–217. https://doi.org/10.1016/j.biortech.2016.04.055.

13

Life Cycle Assessment

13.1 Introduction

The development of industrialisation is related to the production of energy and its utilisation in various economic, scientific and social sectors [1, 2]. The decreasing fossil fuel sources with the increasing greenhouse gas emission with temperature rise have become a critical issue [3, 4]. Hence, it is important to shift and explore alternative choices to overcome the energy crises at the global level and meet the energy demand. Biomass is utilised for the production of biofuels in the form of non-lignocellulosic or lignocellulosic in nature, and others, such as woody, farming manure and by-products, exist in various forms [5, 6]. Technologies used for the production of fuel from the biomass are processes that involve gasification, ignition, pyrolysis, enzymatic hydrolysis routes and the fermentation processes [7, 8]. The environmental effect of conventional fuels can be decreased with the development of biodiesel and biochar from the biofuels [9–11]. The entire production chain of biofuels with its environmental impacts [12] is the evaluation approach identified by life cycle assessment (LCA). The utilisation of potential crops can cause the revolutionary cultivation of crops using the agricultural land to assess the sustainability of biofuels. The conversion of biomass from biological to thermochemical transformation methods involves product and coproducts in the life cycle assessment of the biofuel production chain [13–15].

The development of biofuels in the diversifying energy market reduces the environmental impact and compares conventional fuels. The life cycle assessment tool evaluates the following parameters related to environmental impacts and their production:

1) Energy transformation during complete life cycle
2) Amount of water used in the complete process directly and indirectly
3) Generation of greenhouse gases
4) Amount of resource consumption
5) Extraction of natural resources
6) Release of chemicals in the atmosphere
7) Evaluating the cultivation

Sustainability in Biofuel Production Technology, First Edition. Pratibha S. Agrawal, Pramod N. Belkhode, and Samuel Lalthazuala Rokhum.

8) Amount of oil extracted
9) Manufacturing technique of biodiesel
   The cultivation stage has a significant role in its energy and greenhouse gases with involvement up to 45% and 60%, respectively [16]. The life cycle assessment methodology is useful to numerous important environmentally related aspects of the biofuel life cycle. Various aspects are the emission of substances impacting climate, decrease in the resources, primary energy demand and water consumption. LCA is used to identify environmental burden factors for calculating the output gain in energy on the basis of the amount of energy supplied in biofuels.

13.2 Life Cycle Assessment of Biomass

The life cycle assessment is an important tool as it provides the policy decision to make strategic decisions and improve environmental sustainability. Life cycle assessment involves the following four important stages as per ISO 14040 and IS0 14044:

1) Defining goal and scope
2) Analysis of life cycle inventory
3) Assessment of environmental impacts
4) Interpretation of life cycle [17].

Table 13.1 shows the result of the analysis of 30 life cycle assessments on biofuels. The studies indicated different types of feedstock and biofuel-processing methods with the product obtained [53].

13.3 Feedstock Used

The biodiesel production process affects the economy of individual country depending upon the type of feedstock used for the processing as feedstock prices are mainly responsible for the production cost [54]. In India, there is a maximum use of palm stearin for the production of biodiesel. India produces a very less quantity of biodiesel [55]. Increasing prices of fuel have focused everyone on searching for alternative options for fuel, in which the blending of the biodiesel with the treated waste oil is recommended to overcome the requirement. The impact of processing techniques results in the variation in the prices of biodiesel. The production of biodiesel through various methods affects the economic value as shown in Figure 13.1 for a plant capacity of 8000 tonnes per year of biodiesel production. It shows the variation of prices using vegetable oil and fresh vegetable oil. The variation also affects the use of alkali catalysts as the alkali catalyst process is costlier than other processes [32, 33].

The type of feedstock supplied to the biofuel industries affects the environmental performance. The life cycle assessment is associated with the assessment of the complete biomass supply chain beginning from the feedstock, production process and greenhouse gas emission. The selection of feedstock based on the fertilisers used for the crops results in the variation of crop yield, performance of soil and amount of emission [47].

Table 13.1 Types of feedstock with the corresponding final product.

Feedstock	Method	Outcome	References
Rice and coconut residues	Anaerobic digestion	Electricity	[18]
Sugar cane bagasse	Boiler	Bioenergy	[19, 20]
Biowaste	Anaerobic digestion with gasification	Biomethane	[21, 22]
Microalgal biomass	Transesterification	Biodiesel	[23, 24]
Waste cooking oil	Transesterification	Biodiesel	[23, 25]
Algae harvested	Biochemical processing	Biodiesel	[26, 27]
Cotton straw	Combustion	Electricity	[28, 29]
Biomass and water	Gasification	Hydrogen	[26]
Miscanthus	Anaerobic digestion	Electricity	[30, 31]
Agroforestry residues	Gasification and pyrolysis	Marine fuels	[28]
Rice husk	Combustion	Electricity	[32, 33]
Rice straw	pyrolysis	Methanol	[34]
Woodchips	Boiler	Heat	[30]
Switchgrass	pyrolysis	Gasoline and diesel	[31]
Maize	Anaerobic digestion	Electricity	[35]
Olive	Hydrothermal	Bioenergy	[32]
Rice husk	Alkali water electrolysis	Bio-hydrogen	[33]
Forest residues	pyrolysis	Biodiesel	[36–38]
Rice husk	Torrefaction	pellets	[39, 40]
Coffee cut stems	Enzymatic hydrolysis	Ethanol	[41, 42]
Primary forestry residues	Pyrolysis	Bio-oil	[43]
manure	Anaerobic digestion	Electricity and heat	[44]
Rice straw	Gasification	Methanol	[45]
Sweet sorghum	Combustion	Ethanol	[46]
Corncob	Combustion	Electricity	[47, 48]
Waste date seeds	Esterification	Biodiesel	[49, 50]
Crop residues	Slow pyrolysis	Biofuel and biochar	[41]
Microalgal biomass	Acid catalysis	Biodiesel	[42]
Poultry litter	Pyrolysis and gasification	Biofuel and biochar	[51]
Waste loquat seeds	Transesterification	Biodiesel	[52]

13.4 Purpose of Life Cycle Impact Assessment

The development of any field is carried out by life cycle impact assessment, which covers various influencing parameters with its performance description variables that vary amongst the same substance and impact. The LCA standard ISO 14044 is general in its requirement and helps LCA practitioners make decisions [48].

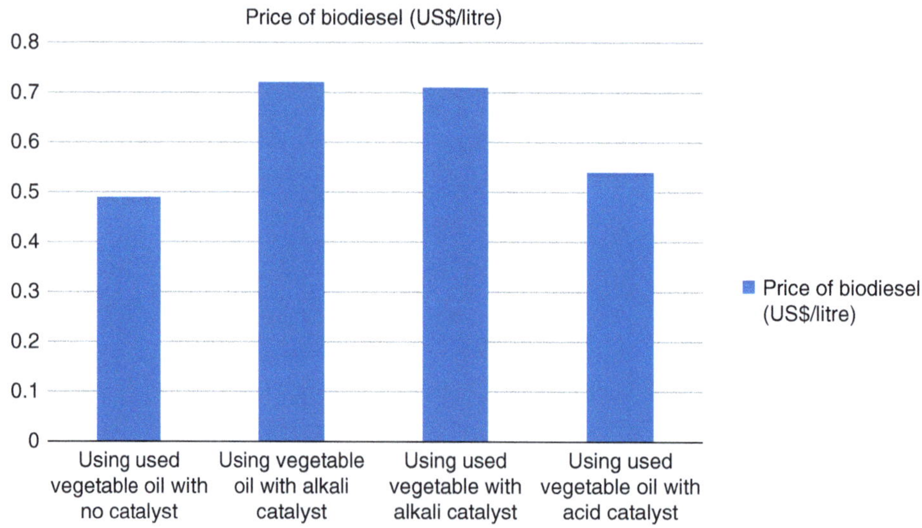

Figure 13.1 Price of biodiesel by using different procedures.

Soya bean biodiesel achieved 76% reduction in greenhouse gas emission without considering induced land use change or 66–72% reduction in overall greenhouse gas emission with induced land use change cases. The life cycle assessment provides a sensitive analysis, which helps to identify the most influencing terms affecting the consumption of fossil fuel and greenhouse gas emissions of biodiesel [49].

Life cycle assessment of fuel used in the vehicle includes a sequence that begins from the production of feedstock to the combustion of fuel and greenhouse gas emission from the combustion stages of biofuels. The benefits of biofuel are quantified in terms of environmental flow, and the energy flow is analysed by the life cycle assessment compared to conventional fossil fuels. Life cycle assessment serves as a principal tool to evaluate greenhouse gas emission and other related factors that impact the environment from a verity of alternative fuel systems [50, 56].

Life cycle assessment is used in various scales for the evaluation of transportation fuel alternatives. Energy security provides a policy for biofuels for sustainability that decreases the net CO_2 emissions. The amount of emission of CO_2 varies as in the case of the burning of biofuels from the conventional fuels [57]. The involvement of CO_2 during biomass feedstock growth results in a net reduction in emission as long as production-related greenhouse gas emission is low, which results in a sustainable agricultural system [58]. The evaluation of transportation fuels with the CO_2 emission with the utilisation of fossil fuels during biofuel production and other production-related greenhouse gas emission is embedded in the life cycle assessment model along with the versions for the suitable changes in the carbon stock to feedstock production.

13.5 Life Cycle for Fossil Fuels

The transportation sector has been using fossil fuel for one hundred years. The life cycle for fossil fuel is shown in Figure 13.2. The production of fuel is more complex than the electrical power system, and the emission of the final product influences the life cycle [59].

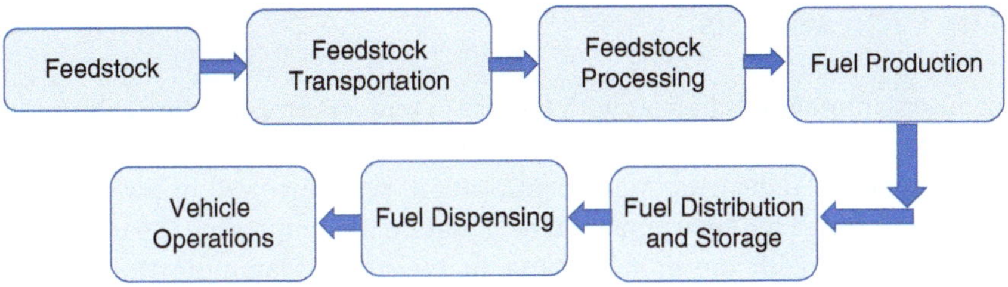

Figure 13.2 Life cycle assessment of fossil fuel.

The evaluation of the environmental performance of alternative transportation fuels is performed with the life cycle assessment considering the industry, government and environmental and consumer decisions. The life cycle assessment study is conducted in consultation with firms, academia, NGOs, industry and government experts to enhance the quality of the method developed and improve the use of biofuels.

13.6 Ethanol Life Cycle

In the transportation sector, ethanol is used in the large level of biofuels, which is primarily made of corn and sugar cane similar to other feedstock used, such as sugar beets, wheat and other starch-bearing crops. Ethanol is produced from the lignocellulosic material using a large number of processes developed with the development of advanced technologies. The process of production of ethanol from the corn is shown in Figure 13.3. The life cycle of ethanol plays a significant role due to variation in the production systems. The selection of different feedstock varies with the emission as it is related to the regional, allocation and process issues. The fertiliser manufacture and land use emission contribute to the variation at a larger scale for the biofuels [59].

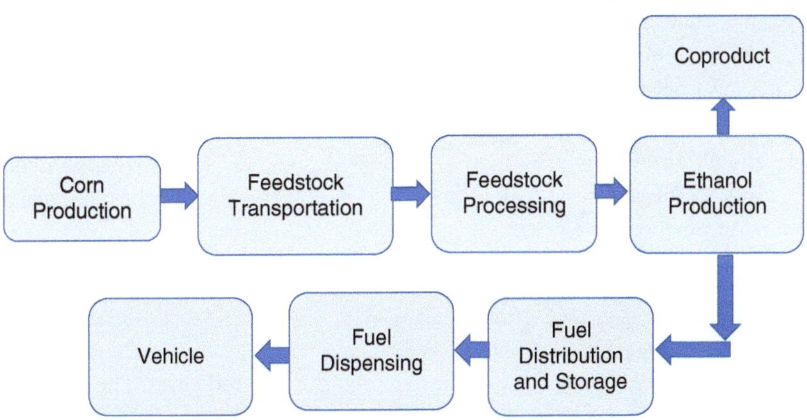

Figure 13.3 Life cycle assessment of ethanol.

13.7 Life Cycle Analysis

Observing the environmental changes with the rise in temperature and the depletion of natural resources, the activities were assessed to reduce the effects on the environment by the government and industries. Many organisations were involved in investigating the environmental issues for environmental degradation and improving the efficiency, reducing process cost with the enhancement in the green marketing through the important tool 'life cycle assessment'. Life cycle assessment evaluates all the stages of a product life with the analysis of the extraction of raw material transportation of material, product disposal, etc., as shown in Figure 13.4. Life cycle assessment is a tool used to evaluate the environmental characteristics related to a product, process by energy analysis, material requirement, evaluation of inputs and corresponding outputs. It includes the product lifespan starting from the raw material to the final manufactured product.

Life cycle assessment involves the following four steps as presented in Figure 13.5:

1) Define goal and scope: Overall view of the product, process and environmental effects for the assessment.
2) Inventory analysis: Energy analysis with the utilisation of water and materials and the release of air emission, solid waste and discharge.
3) Impact Assessment: Effect of energy, water and material along with environment.
4) Clarification: Selection of preferred product, process with the analysis of inventory and impact assessment.

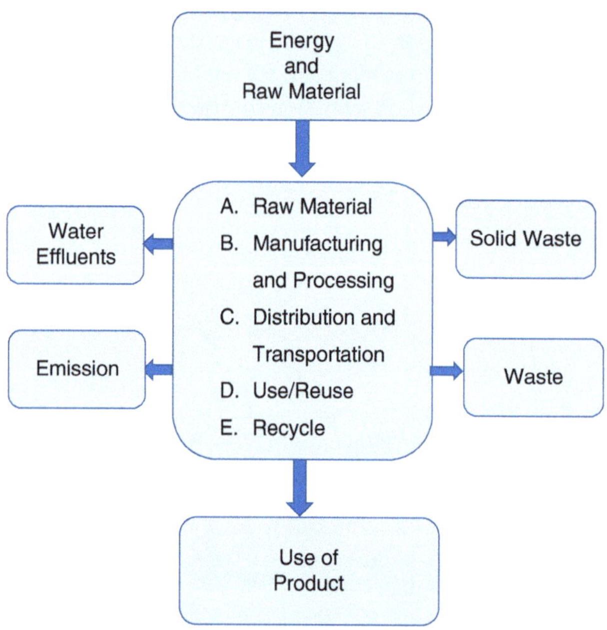

Figure 13.4 Life cycle stages.

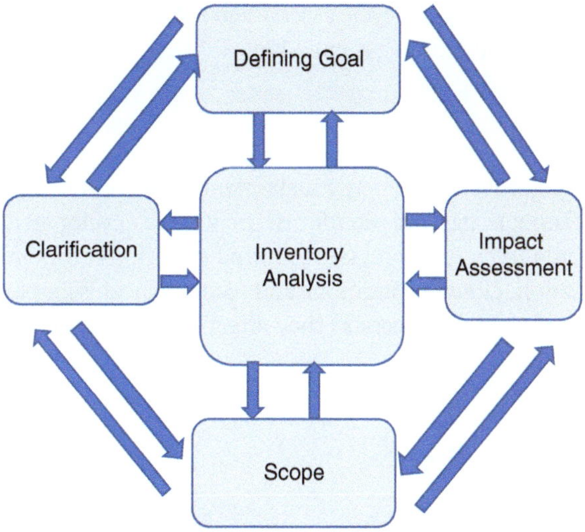

Figure 13.5 Phases of life cycle assessment.

13.8 ISO Life Cycle Assessment Standards

Users were encouraged about the quality product with the attempt of the competition amongst manufacturers and the establishment of concept of life cycle assessment, which was developed in 1980. The ISO 14000 developed a series of life cycle assessment standards [60].

13.8.1 ISO 14040

ISO 14040 of the Environmental Management Series is related to life cycle assessment, particularly the quantitative assessment of various environmental aspects of a product during the complete stages of life cycle. It includes the description of principles and frameworks, which include the analysis of inventory, clarification, financial evaluation, objective and the scope review of life cycle assessment and limitation involved, and the relation between the numerous stages of life cycle assessment.

13.8.2 ISO 14067

ISO 14067 is related to greenhouse gas emission, in which draft indicates the climate change developed by the committee. Various anthropogenic activities were considered as extreme contests for governments and human beings. Greenhouse gas emission initiatives depend on the assessment, monitoring, reporting and verification of greenhouse gas emission.

The objective of ISO 14067 is as follows:

1) To increase the credibility
2) To increase the continuous improvement in product, manufacturing methods and selection of raw materials

3) Greenhouse gas emission management using life cycle assessment tools
4) Customer behaviour in the control of greenhouse gas emission

13.8.3 ISO 13065

ISO 13065 is designed to achieve the objective of bioenergy sustainability means and has a potential role in minimising the greenhouse gas emission for the economic development. Bioenergy is produced by using different types of biomasses followed by various economical operations. The geographical location, climate conditions and technological issues are some of the aspects involved in the production process as they affect the production and supply chain.

13.9 Life Cycle Assessment Benefits

Life cycle assessment facilitates the decision for designing a sustainable product with the aspects of different phases of life cycle assessment. The life cycle assessment benefits in a number of ways as mentioned below:

1) Product Development and Research and Development
   The development of a new product should be as low in emission as possible due to the implementation and regulation of different policies and customer demand. Research and development compares the performance of different raw materials used for the production and the influence of these raw materials on the environmental impact of the end product.
2) Supply Chain Management and Procurement
   It is observed that in most of the industries the supply chain contributes around 75% of the environmental impact that can be reduced within the eco-chain using different assessment tools.
3) Marketing and Sales
   Life cycle assessment plays an important role in the journey of a successful product as it communicates to customers regarding the sustainability of the product.
4) Executive Levsel and Strategic Management
   At this level, a company's decision about the product starting from the raw material selection to the final product through various different processes affects the environmental impact based on the decision to achieve the sustainability.

Overall, a sustainable product means the utilisation of less energy, recycled product and simple processing, which save energy and reduce emission.

13.10 Role of LCA in Public Policies/Regulations

Life cycle assessment's role in the development of public policy is to inform the government and industry about the decision of energy policy, selection of feedstock and lower life cycle environmental impacts. Challenges are involved in the regulatory framework in the application of life cycle analyses. Coproduct allocation systems are developed and designed by some systems to ensure that they provide good results associated with greenhouse gas emissions at the cost of utilisation of land. Environmental regulation information is based on the life cycle basis as it has an important role in different parts of the supply chain related to biofuel production.

13.11 Conclusion

Environmental burdens or impacts of a product are assessed by life cycle assessment (LCA). The identification of life cycle stages, environmental burden factors and energy investment in biofuels, which involves the selection of environmental balancing factors and life cycle comparison, are determined with the life cycle assessment. Life cycle assessment supports determining the concerned technical aspects of the process that need to be optimised to achieve the environmental sustainability. The production processes and future requirements of biofuels are necessary to be identified through the assessment results. The assessment of life cycle stages is evaluated with the consideration of uncertainties. Life cycle assessment is useful to identify the life cycle stages and processes that participate in environmental burdens to determine the energetic return on investment of the amount of energy production of biofuels for environmental factors and to compare the life cycle of environmental problems of products. Finally, it can be concluded that the implementation of best practices brings the changes according to the time for harmony with nature.

References

1 Osman, A.I., Mehta, N., Elgarahy, A.M. et al. (2021). Conversion of biomass to biofuels and life cycle assessment: a review. *Environ. Chem. Lett.* https://doi.org/10.1007/s10311-021-01273-0.

2 Shrivastava, P., Verma, T.N., and Pugazhendhi, A. (2019). An experimental evaluation of engine performance and emisssion characteristics of CI engine operated with Roselle and Karanja biodiesel. *Fuel* https://doi.org/10.1016/j.fuel.2019.115652.

3 Fawzy, S., Osman, A.I., Doran, J., and Rooney, D.W. (2020). Strategies for mitigation of climate change: a review. *Environ. Chem. Lett.* https://doi.org/10.1007/s10311-020-01059-w.

4 Osman, A.I., Hefny, M., Abdel Maksoud, M.I.A. et al. (2021). Recent advances in carbon capture storage and utilisation technologies: a review. *Environ. Chem. Lett.* https://doi.org/10.1007/s10311-020-01133-3.

5 Osman, A.I., Abdelkader, A., Farrell, C. et al. (2019). Reusing, recycling and up-cycling of biomass: a review of practical and kinetic modelling approaches. *Fuel Process. Technol.* https://doi.org/10.1016/j.fuproc.2019.04.026.

6 Kaloudas, D., Pavlova, N., and Penchovsky, R. (2021). Lignocellulose, algal biomass, biofuels and biohydrogen: a review. *Environ. Chem. Lett.* https://doi.org/10.1007/s10311-021-01213-y.

7 Xu, Y., Hu, Y., Peng, Y. et al. (2020). Catalytic pyrolysis and liquefaction behavior of microalgae for bio-oil production. *Bioresour. Technol.* https://doi.org/10.1016/j.biortech.2019.122665.

8 Abou Rjeily, M., Gennequin, C., Pron, H. et al. (2021). Pyrolysis-catalytic upgrading of bio-oil and pyrolysis-catalytic steam reforming of biogas: a review. *Environ. Chem. Lett.* https://doi.org/10.1007/s10311-021-01190-2.

9 Xu, Y., Seshadri, B., Sarkar, B. et al. (2018). Biochar modulates heavy metal toxicity and improves microbial carbon use efficiency in soil. *Sci. Total Environ.* https://doi.org/10.1016/j.scitotenv.2017.11.214.

10 Huang, S., Garg, A., Mei, G. et al. (2020). Experimental study on the hydrological performance of green roofs in the application of novel biochar. *Hydrol. Processes* https://doi.org/10.1002/hyp.13881.

11 Vishnu, D., Dhandapani, B., Kannappan Panchamoorthy, G. et al. (2021). Comparison of surface-engineered superparamagnetic nanosorbents with low-cost adsorbents of cellulose, zeolites and biochar for the removal of organic and inorganic pollutants: a review. *Environ. Chem. Lett.* https://doi.org/10.1007/s10311-021-01201-2.

12 Escobar, J.C., Lora, E.S., Venturini, O.J. et al. (2009). Biofuels: environment, technology and food security. *Renewable Sustainable Energy Rev.* https://doi.org/10.1016/j.rser.2008.08.014.

13 Bilal, M. and Iqbal, H.M.N. (2020). Recent advancements in the life cycle analysis of lignocellulosic biomass. *Curr. Sustain. Energy Reports* 7: 100–107. https://doi.org/10.1007/s40518-020-00153-5.

14 Benavides, P.T., Cronauer, D.C., Adom, F. et al. (2017). The influence of catalysts on biofuel life cycle analysis (LCA). *Sustain. Mater. Technol.* https://doi.org/10.1016/j.susmat.2017.01.002.

15 Martin, E.W., Chester, M.V., and Vergara, S.E. (2015). Attributional and consequential life-cycle assessment in biofuels: a review of recent literature in the context of system boundaries. *Curr. Sustain. Energy Reports* https://doi.org/10.1007/s40518-015-0034-9.

16 Giraldi-Díaz, M.R., De Medina-Salas, L., Castillo-González, E., and De la Cruz-Benavides, M. (2018). Environmental impact associated with the supply chain and production of biodiesel from Jatropha curcas L. through life cycle analysis. *Sustainability* https://doi.org/10.3390/su10051451.

17 Lask, J., Wagner, M., Trindade, L.M., and Lewandowski, I. (2019). Life cycle assessment of ethanol production from miscanthus: a comparison of production pathways at two European sites. *GCB Bioenergy* https://doi.org/10.1111/gcbb.12551.

18 Aberilla, J.M., Gallego-Schmid, A., and Azapagic, A. (2019). Environmental sustainability of small-scale biomass power technologies for agricultural communities in developing countries. *Renew. Energy* https://doi.org/10.1016/j.renene.2019.04.036.

19 Arguelles-Arguelles, A., Amezcua-Allieri, M.A., and Ramírez-Verduzco, L.F. (2021). Life cycle assessment of green diesel production by hydrodeoxygenation of palm oil. *Front. Energy Res.* https://doi.org/10.3389/fenrg.2021.690725.

20 Amezcua-Allieri, M.A., Martínez-Hernández, E., Anaya-Reza, O. et al. (2019). Techno-economic analysis and life cycle assessment for energy generation from sugarcane bagasse: case study for a sugar mill in Mexico. *Food Bioprod. Process.* https://doi.org/10.1016/j.fbp.2019.09.014.

21 Ardolino, F. and Arena, U. (2019). Biowaste-to-biomethane: an LCA study on biogas and syngas roads. *Waste Manage. (Oxford)* https://doi.org/10.1016/j.wasman.2019.02.030.

22 Banerjee, S., Rout, S., Banerjee, S. et al. (2019). Fe2O3 nanocatalyst aided transesterification for biodiesel production from lipid-intact wet microalgal biomass: a biorefinery approach. *Energy Convers. Manage.* https://doi.org/10.1016/j.enconman.2019.05.060.

23 Chung, Z.L., Tan, Y.H., Chan, Y.S. et al. (2019). Life cycle assessment of waste cooking oil for biodiesel production using waste chicken eggshell derived CaO as catalyst via transesterification. *Biocatal. Agric. Biotechnol.* https://doi.org/10.1016/j.bcab.2019.101317.

24 Zhu, Y., Liang, J., Yang, Q. et al. (2019). Water use of a biomass direct-combustion power generation system in China: a combination of life cycle assessment and water footprint analysis. *Renew. Sustain. Energy Rev.* https://doi.org/10.1016/j.rser.2019.109396.

25 DeRose, K., DeMill, C., Davis, R.W., and Quinn, J.C. (2019). Integrated techno economic and life cycle assessment of the conversion of high productivity, low lipid algae to renewable fuels. *Algal Res.* https://doi.org/10.1016/j.algal.2019.101412.

26 Wang, C., Jin, H., Peng, P., and Chen, J. (2019). Thermodynamics and LCA analysis of biomass supercritical water gasification system using external recycle of liquid residual. *Renew. Energy* https://doi.org/10.1016/j.renene.2019.03.129.

27 Wagner, M., Mangold, A., Lask, J. et al. (2019). Economic and environmental performance of miscanthus cultivated on marginal land for biogas production. *GCB Bioenergy* https://doi.org/10.1111/gcbb.12567.

28 Tanzer, S.E., Posada, J., Geraedts, S., and Ramírez, A. (2019). Lignocellulosic marine biofuel: Technoeconomic and environmental assessment for production in Brazil and Sweden. *J. Cleaner Prod.* https://doi.org/10.1016/j.jclepro.2019.117845.

29 Quispe, I., Navia, R., and Kahhat, R. (2019). Life cycle assessment of rice husk as an energy source. A Peruvian case study. *J. Cleaner Prod.* https://doi.org/10.1016/j.jclepro.2018.10.312.

30 Jeswani, H.K., Whiting, A., and Azapagic, A. (2020). Environmental and economic sustainability of biomass heat in the UK. *Energy Technol.* https://doi.org/10.1002/ente.201901044.

31 Lan, K., Ou, L., Park, S. et al. (2020). Life cycle analysis of decentralized preprocessing Systems for Fast Pyrolysis Biorefineries with blended feedstocks in the southeastern United States. *Energy Technol.* https://doi.org/10.1002/ente.201900850.

32 Mendecka, B., Lombardi, L., Micali, F., and De Risi, A. (2020). Energy recovery from olive pomace by hydrothermal carbonization on hypothetical industrial scale: a LCA perspective. *Waste Biomass Valorization* https://doi.org/10.1007/s12649-020-01212-0.

33 Reaño, R.L. (2020). Assessment of environmental impact and energy performance of rice husk utilization in various biohydrogen production pathways. *Bioresour. Technol.* https://doi.org/10.1016/j.biortech.2019.122590.

34 Im-orb, K. and Arpornwichanop, A. (2020). Process and sustainability analyses of the integrated biomass pyrolysis, gasification, and methanol synthesis process for methanol production. *Energy* https://doi.org/10.1016/j.energy.2019.116788.

35 Lask, J., Martínez Guajardo, A., Weik, J. et al. (2020). Comparative environmental and economic life cycle assessment of biogas production from perennial wild plant mixtures and maize (Zea mays L.) in Southwest Germany. *GCB Bioenergy* https://doi.org/10.1111/gcbb.12715.

36 Spatari, S., Larnaudie, V., Mannoh, I. et al. (2020). Environmental, exergetic and economic tradeoffs of catalytic- and fast pyrolysis-to-renewable diesel. *Renew. Energy* https://doi.org/10.1016/j.renene.2020.08.042.
37 Thengane, S.K., Burek, J., Kung, K.S. et al. (2020). Life cycle assessment of rice husk torrefaction and prospects for decentralized facilities at rice mills. *J. Cleaner Prod.* https://doi.org/10.1016/j.jclepro.2020.123177.
38 Aristizábal-Marulanda, V., García-Velásquez, C.A., and Cardona Alzate, C.A. (2021). Environmental assessment of energy-driven biorefineries: the case of the coffee cut-stems (CCS) in Colombia. *Int. J. Life Cycle Assess.* https://doi.org/10.1007/s11367-020-01855-0.
39 Martillo Aseffe, J.A., Martínez González, A., Jaén, R.L., and Silva Lora, E.E. (2021). The corn cob gasification-based renewable energy recovery in the life cycle environmental performance of seed-corn supply chain: an Ecuadorian case study. *Renew. Energy* https://doi.org/10.1016/j.renene.2020.10.053.
40 Al-Mawali, K.S., Osman, A.I., Al-Muhtaseb, A.H. et al. (2021). Life cycle assessment of biodiesel production utilising waste date seed oil and a novel magnetic catalyst: a circular bioeconomy approach. *Renew. Energy* https://doi.org/10.1016/j.renene.2021.02.027.
41 Yang, Q., Mašek, O., Zhao, L. et al. (2021). Country-level potential of carbon sequestration and environmental benefits by utilizing crop residues for biochar implementation. *Appl. Energy* https://doi.org/10.1016/j.apenergy.2020.116275.
42 Saranya, G. and Ramachandra, T.V. (2020). Life cycle assessment of biodiesel from estuarine microalgae. *Energy Convers. Manag.* https://doi.org/10.1016/j.ecmx.2020.100065.
43 Brassard, P., Godbout, S., and Hamelin, L. (2021). Framework for consequential life cycle assessment of pyrolysis biorefineries: a case study for the conversion of primary forestry residues. *Renew. Sustain. Energy Rev.* https://doi.org/10.1016/j.rser.2020.110549.
44 Cusenza, M.A., Longo, S., Guarino, F., and Cellura, M. (2021). Energy and environmental assessment of residual bio-wastes management strategies. *J. Cleaner Prod.* https://doi.org/10.1016/j.jclepro.2020.124815.
45 Da Silva, M.G., Lisbôa, A.C.L., Hoffmann, R. et al. (2021). Greenhouse gas emissions of rice straw-to-methanol chain in Southern Brazil. *J. Environ. Chem. Eng.* https://doi.org/10.1016/j.jece.2021.105202.
46 Fu, J., Yan, X., and Jiang, D. (2021). Assessing the sweet sorghum-based ethanol potential on saline–alkali land with DSSAT model and LCA approach. *Biotechnol. Biofuels* https://doi.org/10.1186/s13068-021-01896-z.
47 *Life-Cycle Assessment of Biorefineries*Dufosse, K., Aoun, W.B., and Gabrielle, B. (2017). Life-Cycle Assessment of agricultural feedstock for biorefineries. In: Elsevier.
48 Hauschild, M.Z., Goedkoop, M., Guinée, J. et al. (2013). Identifying best existing practice for characterization modeling in life cycle impact assessment. *Int. J. Life Cycle Assess.* 18: 683–697.
49 Chena, R., Qin, Z., Han, J. et al. (2018). Life cycle energy and greenhouse gas emission effects of biodiesel in the United States with induced land use change impacts. *Bioresource Technol.* 251: 249–258.
50 Hakizimana, E., Wali, U.G., Sandoval, D., and Venant, K. (2020). Environmental impacts of biomass energy sources in Rwanda. *Energy Environ. Eng.* 7 (3): 62–71. https://doi.org/10.13189/eee.2020.070302.

51 Bora, R.R., Lei, M., Tester, J.W. et al. (2020). Life cycle assessment and Technoeconomic analysis of thermochemical conversion technologies applied to poultry litter with energy and nutrient recovery. *ACS Sustain. Chem. Eng.* https://doi.org/10.1021/acssuschemeng.0c02860.

52 Al-Muhtaseb, A.H., Osman, A.I., Murphin Kumar, P.S. et al. (2021). Circular economy approach of enhanced bifunctional catalytic system of CaO/CeO2 for biodiesel production from waste loquat seed oil with life cycle assessment study. *Energy Convers. Manage.* https://doi.org/10.1016/j.enconman.2021.114040.

53 Rathore, D., Nizami, A.S., Singh, A., and Pant, D. (2016). Key issues in estimating energy and greenhouse gas savings of biofuels: Challenges and perspectives. *Biofuel Res. J.* https://doi.org/10.18331/BRJ2016.3.2.3.

54 De Araújo, C.D.M., De Andrade, C.C., Silva, E.D.S.E., and Dupas, F.A. (2013). Biodiesel production from used cooking oil: a review. *Renew. Sustain. Energy Rev.* 27 (C): 445–452.

55 Barnwal, B.K. and Sharma, M.P. (2005). Prospects of biodiesel production from vegetable oils in India. *Renew. Sustain. Energy Rev.* 9 (4): 363–378. https://doi.org/10.1016/j.rser.2004.05.007.

56 Spoof-Tuomi, K. and Niemi, S. (2020). Environmental and economic evaluation of fuel choices for Short Sea shipping. *Clean Technol.* 2 (1): 34–52. https://doi.org/10.3390/cleantechnol2010004.

57 DeCicco, J. (2012). Biofuels and carbon management. *Clim. Change* 111: 627–640.

58 Marland, G. and Turhollow, A.F. (1991). CO2 emissions from the production and combustion of fuel ethanol from corn. *Energy* https://doi.org/10.1016/0360-5442(91)90004-6.

59 Hossain, N., Zaini, J., and Indra Mahlia, T.M. (2019). Life cycle assessment, energy balance and sensitivity analysis of bioethanol production from microalgae in a tropical country. *Renew. Sustain. Energy Rev.* https://doi.org/10.1016/j.rser.2019.109371.

60 Rezaee, Z. and Elam, R. (2000). Emerging ISO 14000 environmental standards: a step-by-step implementation guide. *Manag. Audit J.* 12 (1/2): 60–67. https://doi.org/10.1108/02686900010304650.

14

Socioeconomic Impact of Biofuel

14.1 Introduction

Global warming has accelerated during the last three decades as a result of the dramatic rise in the environment produced due to fossil fuels [1]. The change in the environment triggered by global warming is a potential cause for events, such as high temperatures, unexpected thunderstorms, cyclones and flooding [2, 3]. The reduction of greenhouse gases and changing climatic behaviour has emerged as a key impetus for renewable fuel research [4]. There is now a troubling challenge to the existence of fossil energy, with the situation of demand exceeding availability [5]. Furthermore, the concerns about environmental pollution as a result of intensive utilisation fossil fuels have prompted experts in the field to look for more environmentally sustainable, possible and renewable fuels [6].

Overabundant fossil energy stocks and a sharp increase in fuel prices have culminated in an ongoing search for potential replacement fuels for internal combustion engines [7, 8]. The demand, in addition to existing petroleum fuels, is rising day by day as the number of automobiles increases [4]. The need for suitable alternative fuels for compression ignition engines is very specific since they are commonly used in several applications [9].

The discussion on the social and economic impact of biofuels is a widely recent discussion topic as it is based on the assessment of biofuel opportunities and costs involved in the production, feedstock and policy documents. The social and economic impact of biofuel production depends upon the efficient producers in export markets and profit analysis [10, 11]. Biodiesel is most effectively used in every country due to it social, environmental and lower costs, so it has no impact on the economy [12]. The production and use of biodiesel were costly for the first time even though the cost was more than that of the diesel due to the selection of feedstock, which was based on the food crops such as sugar, starch or oil content [13]. The cost of biodiesel is affected by various feedstock produced in the geographical area as shown in Figure 14.1 in which various kinds of crops are used as feedstock that are grown in South Africa. The figure shows the quantitative knowledge about the manufacturing cost, outcome and yield of biodiesel [14].

Sustainability in Biofuel Production Technology, First Edition. Pratibha S. Agrawal, Pramod N. Belkhode, and Samuel Lalthazuala Rokhum.

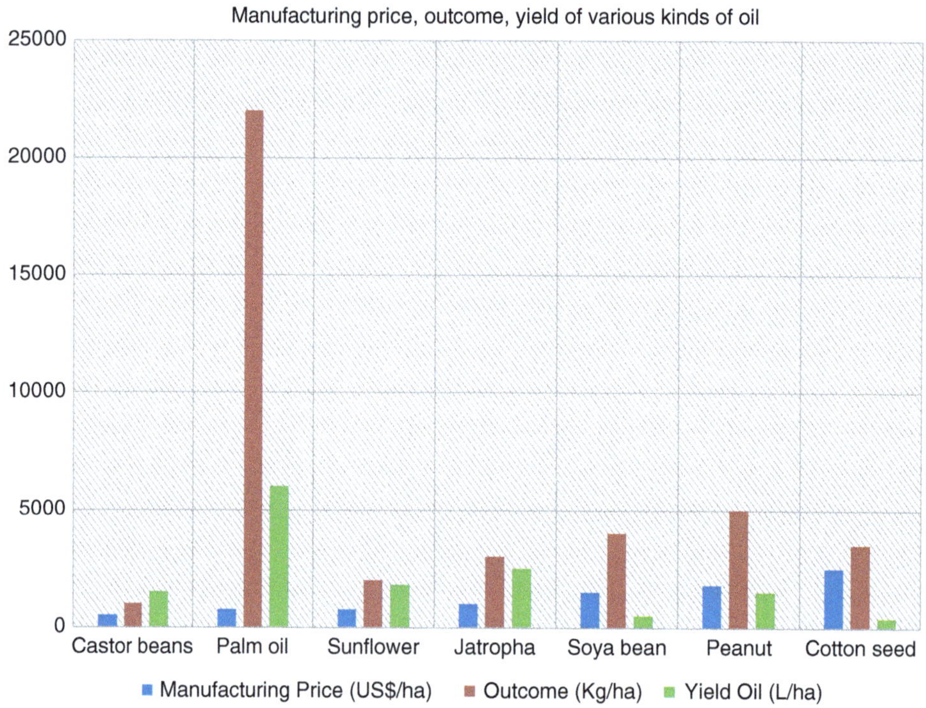

Figure 14.1 Various feedstock used and their performance parameters in South Africa.

Amongst the varieties of edible food crops used to make the biodiesel, the selection is based on the direct effect on food security and food prices.

Agricultural crops used for the production of biofuels, which are the source of energy, are associated with disadvantages in the production of these fuels due to the rise in prices of food items, rise in competition for required land and water and deforestation. The utilisation of agricultural land for the cultivation of biofuels instead of other crops, which lead to the competition between the food production and biofuels, results in the rise of price of the food production [15]. The study of environmental impact assessment is the tool for deciding the maximising opportunities and regional comparative advantages with respect to input demands, resources and energy [16].

Researchers resolved and worked on the other types of feedstock other than food crops and approached the residual and forest, and found out the new route of productions. During second-generation biodiesel, the non-edible oil used for biodiesel production was industrialised [17]. In second-generation biodiesel, the feedstock was non-edible oil [18], such as jatropha [19–21], mahua oil [22], karanja oil [23], neem [24], used/waste cooking oil [25, 26] and others.

Agricultural residues as the feedstock for biodiesel have many advantages. Some of which are as follows:

- It has low cost for the production of biodiesel.
- It reduces the efforts and economy involved in the use of agricultural residues.
- The employability ratio will be increased by adopting this technique on a worldwide level.

14.2 Employment Opportunities in Biofuel Production Industries

Employment opportunities in the rural regions are increasing in biofuel production industries, resulting in growth in the economy by cultivating more crops in the field [27]. There was a 5% rise in GDP in Malaysia due to a rise in the palm oil farming, which created almost 1.4 million jobs [Mpob and Apoc, 'Palm Oil Development']. Continuous improvement in the biodiesel production results in an increase in the agricultural prices of land, which increases the farmers' income [28]. Around one lakh jobs were created in 2011 during first-generation biodiesel in Indonesia [29]. Biofuel policies help to resolve the situation of changes in the prices of crude oil, which affect the foreign exchange and further dependency with other nations [30]. Farmers set up a new market with the development of biodiesel by providing the feedstock to the developed countries and explore the market at the local and international levels [31]. The direct and indirect effects on the gross value added (GVA) and jobs during the starting phase of the biodiesel industry are shown in Figures 14.2 and 14.3. Figure 14.2 shows the temporary effect during the starting period of industries, whereas Figure 14.3 shows the permanent effect, during production, on GVA and jobs [32].

Many job opportunities are created in the biodiesel industries using palm and jatropha feedstock, resulting in the green job generation income. Some countries face the problem for farming of crops due to a shortage of land but has enough sources to build the biodiesel plant. Some countries have enough land but lack a source of biodiesel production. Therefore, the economy of such countries varies depending upon the sources and energy demand.

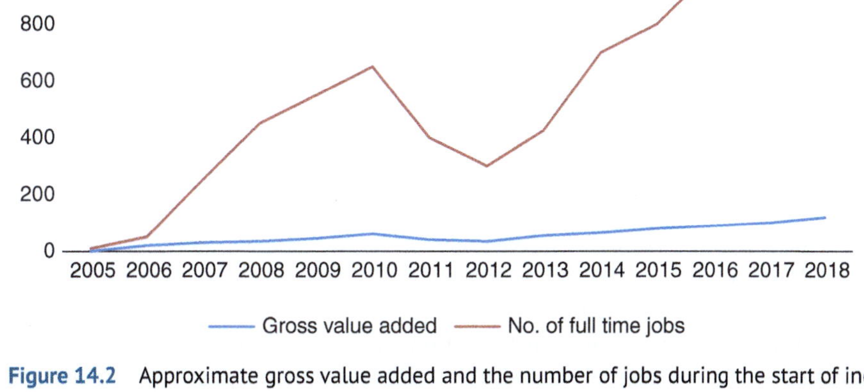

Figure 14.2 Approximate gross value added and the number of jobs during the start of industries.

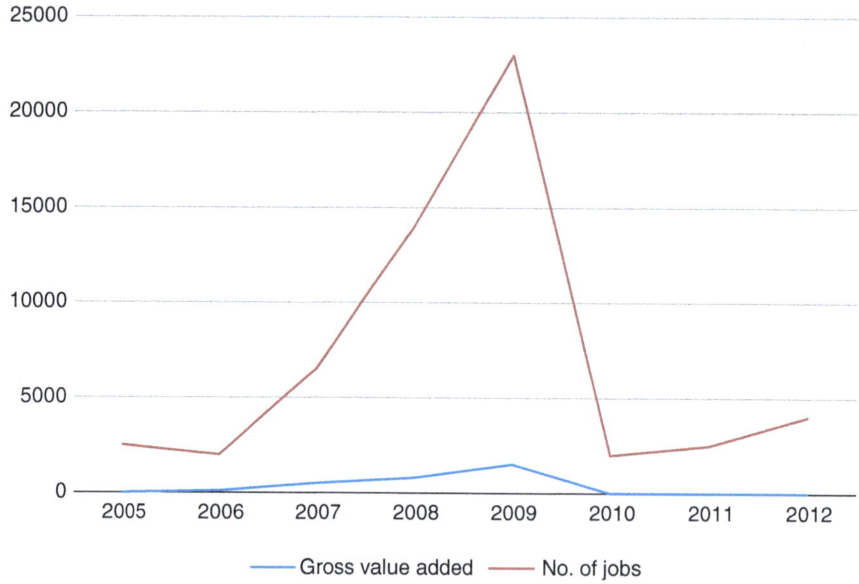

Figure 14.3 Permanent effect on gross value added and the number of jobs in industries.

14.3 Socioeconomic and Environmental Impact

Energy is one of the important factors in the development of the world economy because a shortage of energy development impacts socioeconomic conditions. Awareness amongst the people, suggestions and ideas could contribute to the social acceptance of biofuels [33]. The current surge in biofuel investments and manufacturing capacities is determined by the potential of multiple social, economic, ecological, and geopolitical aids. The socioeconomic and environmental impacts of biofuel vary widely based on the specific condition of the country. The profit and cost for biofuel have a tendency to differ in commodities, landscape and business models. Rural development, employment generation and energy security are significantly involved in the development of a nation. Energy security refers the capacity of a nation to access the energy resources desired to sustain its national power. The economic evaluation of the renewable energy includes the evaluation of the economic costs and advantages to the nation in terms of electricity costs, direct and indirect impacts on jobs, income involved and economic output. Bioenergy supports and benefits the trade market, development of rural civilisation and people's income. The selection of biomasses, conversion process, feedstock selection and location of the feedstock and operations involved in the processing determine the cost of biomass. The amount of alternative energy available results in the cost of biofuel. Trade barriers, price interventions and financial support played an important role in the development of biofuel. Biofuels are strongly bonded to the agricultural markets, and residues obtained from the agricultural supplies affect the food production, whereas next-generation biofuels obtained from lignocellulosic biomass and photosynthetic algae may have rarer direct relations to food production systems. The socioeconomic effect of biofuel includes its impact on food security, supply and accessibility. The first-generation biofuels measured the cost effectiveness, and the second-generation biofuels used the corn and sugar-based feedstock, which reduces the cost by 50% with the indeterminate and different production and conversion processes. Environmental benefit promotes the biofuel due to carbon neutrality means that the amount of carbon released during burning is equal to the amount of carbon utilised in the photosynthesis of plants, which results in a reduction in greenhouse gas emission. Environmental impact includes issues, such as the loss of biodiversity and the degradation of soil and water resource quality. Environmentally being a biofuel-making procedure is projected to have little energy involvement in all steps starting from agriculture to the processing of plants and the manufacturing of biofuels with higher energy balance and better engine performance. Feedstock selection for biofuels affects the ecosystem that influences the biodiversity. Aquatic ecosystem is affected by the feedstock like sugar beet. Soil fertility management is based on the organic and plant nutrients retrieved during land conversion processes related to the feedstock and crop wasted along with hydrological effects. The amount of water required for the irrigation is the same as that required for the food crops that affect the bioenergy. Investment in the biofuel sector is profitable due to the involvement of economic and environmental credibility of coproducts to meet the present demand.

14.4 Biodiesel Industries

Capital investment, manufacturing cost and break-even price of the product are the factors that influence the start-up of any industry [34]. The production of biodiesel is estimated from soya bean oil, including capital cost and operating cost, in which the capital cost of 1 million gallon of biodiesel production was US$ 26312 and the operational cost was US$ 22495.4, resulting in the rise in the price of soya bean oil. Glycerol recovered during the process can be resold in the market [35]. Biodiesel production using waste cooking oil and its comparison with the pure vegetable oil assessed the economic impact due to oil rates in the huge market. Such a large use of oil has forced the government to open more and more WCO collection points for its recycling, which supports biodiesel industries. Malaysian government state practices encourage and provide benefit on taxes [36] as a normal layman is able to process 20–3000 tonnes/month WCO into biodiesel. The production of biodiesel from the waste cooking oil is viable as compared to the production from other sources. The dependency of each nation will be reduced with the generation of biodiesel, and the total investment of plants which include the capital and operating cost determines the biodiesel and the revenue costs that predict the economic analysis of the plant. The installation of biodiesel plants has a positive impact on every nation, but subsidies provided by the government may be barriers that stop the biodiesel industries from moving forward [37]. Biodiesel was sold at the same rate as other fossil fuels in Indonesia during the starting period, and further government took the initiative by reducing the income tax on the investment in the biofuel industries [38]. As per the developed financial and societal growth plans for the biomass energy sector, it was targeted to invest 18 billion Yuan in biofuel industries by successfully utilising non-food raw material [39]. The major goal of the biofuel act 2006 was to increase income and rural area employment by ensuring the availability of feedstock without affecting natural ecosystems, biodiversity and food reserves [40].

14.5 Export and Import of Biodiesel

The consumption of biodiesel decreases with an increase in the production as almost 537 million litres of diesel were saved in 2021 as per the data of southeast region, which may save approximately 50% of diesel due to the replacement of biodiesel. Figure 14.4 [41] shows the analysis of export and import of biodiesel, which clearly shows that import is less as compared to the export of biodiesel in India as most of the biodiesel export to the European Union with the import of biodiesel is almost negligible.

Biodiesel consumption and biodiesel use on the road are shown in detail in Figure 14.5 [41]. Overall, the utility of biodiesel is concerning as it shows an increase in the use of biodiesel up to 4% with a decrease in domestic consumption by 3% in the past 10 years. Government is trying to enhance the implementation of a new policy for increasing the blending of biodiesel in the diesel as the blended biodiesel has a low rate of use.

The use of blended biodiesel is increasing in almost all countries at different percentages. The consumption of biodiesel is less than the production of biodiesel in India. If the consumption of the biodiesel increases in a proper way, it will enhance the biorefineries and beneficial use of biodiesel as compared to other countries.

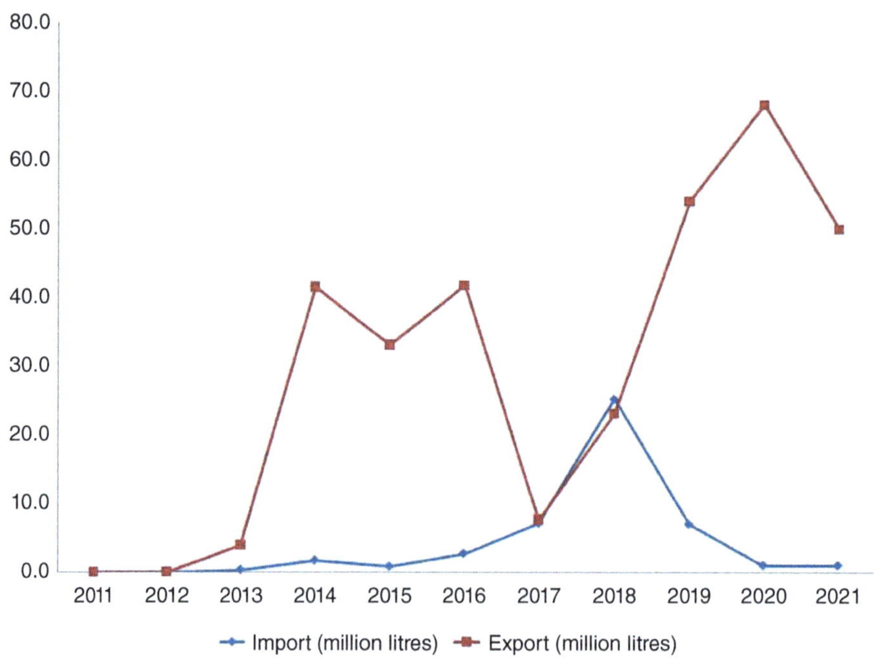

Figure 14.4 Report of export and import of biodiesel.

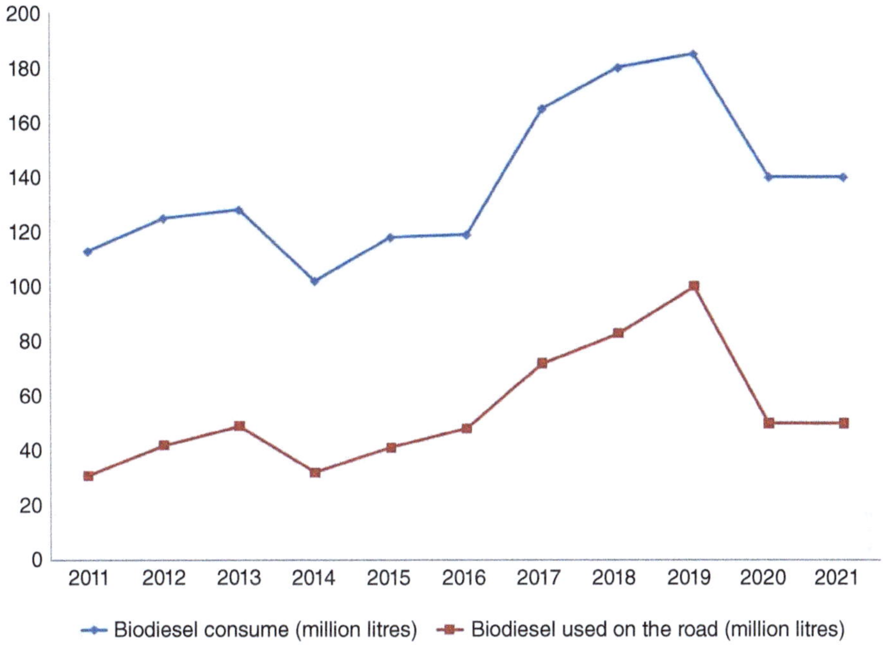

Figure 14.5 Statistical data of biodiesel consumption and biodiesel used on the road.

14.6 Production of Biodiesel

In the recent year, there has been an increase in the production of biodiesel, which results in an increase in the GDP and tariff rates of every country. The implementation of government policies increases the government supports in the biofuel industries in the developing countries [42]. The support of government policies enhanced the biodiesel production from one lakh to ten lakh in 2020 in countries such as Malaysia [43]. This growth resulted in a rapid increase in the economic growth of that country [44]. Single-step transesterification produces the biodiesel with an annual capacity of 150 Kt, a selling price of B10 fuel USD 0.69, a depreciation time of 15 years, and an IRR value of 37% [45]; single-step transesterification produces the biodiesel with an annual production capacity of 614 Kt, with a selling price of B10 fuel USD 0.33, and a production cost of 0.74 million USD [46]. Two-step transesterification produces the biodiesel with an annual production capacity of 784 Kt, a selling price of B10 fuel USD 1.53, a production cost of 0.81 million USD, a depreciation time of 12 years, and an IRR value of 20% [47].

A comparison of the cost analysis of conventional and alternative biodiesel production shows the low cost of capital, raw material and operating cost of the conventional process with a high revenue generation rate as shown in Figure 14.6. Equipment utility and catalyst cost are less, providing major benefits in the conventional process that leads to a difference in economy by the recovery of glycerol. The glycerol by-product produced during

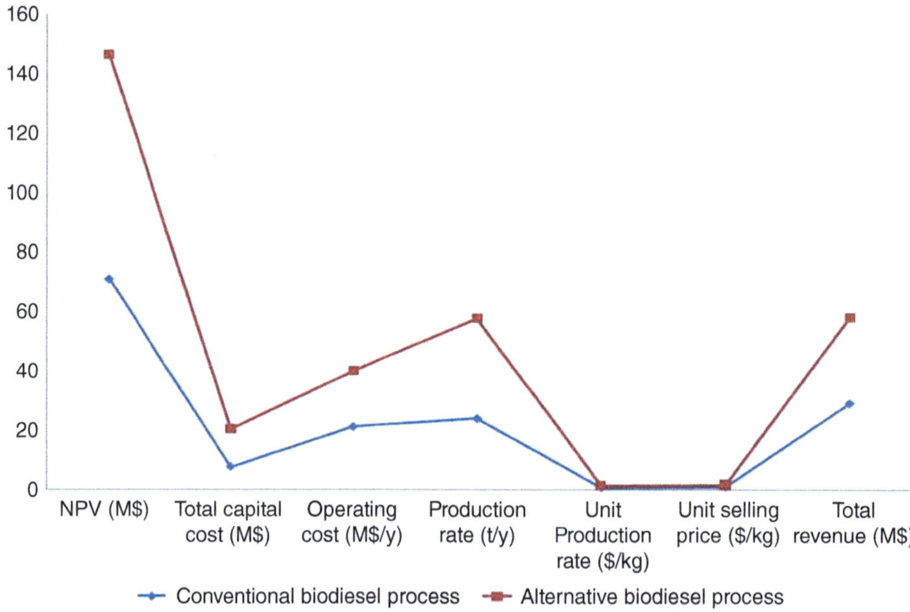

Figure 14.6 Comparison of the cost of conventional and alternative biodiesel production.

transesterification is a carbon source for microalgae that is absorbed and converted into triglycerides, which are again used for biodiesel production. The construction of a large biodiesel plant using feedstock as waste cooking oil of almost 150 000 tonnes per year and its economic analysis of the production of biodiesel are shown in Figure 14.7, which show the result of the past 10 years of the promising profit economy of the UAE country [48].

14.7 Economic Impact of Biofuels

Rural economic development is based on the processing of an agricultural product, which includes the involvement of a new processing plant with the support from various commodity groups and interest of the rural people and higher authorities of the region. New processing plants will generate new employment opportunities and enhance the economy in rural areas. New agricultural processing biofuel projects have received a widespread support and contributed to the society at regional economics. This is proven by the rapid growth of the corn-based ethanol industry for the production of biofuels, which further promoted research on cellulosic biomass conversion and focused on technical issues. Thus, the conversion of biomass into liquid fuels results in the development of rural areas with a high biomass production capacity and provides a platform to compare the economic impact. The finding of the economic globalisation index included various aspects, such as financial, political and societal aspects, and a rise in the production of biofuel level during 2015–2019 (Figure 14.8).

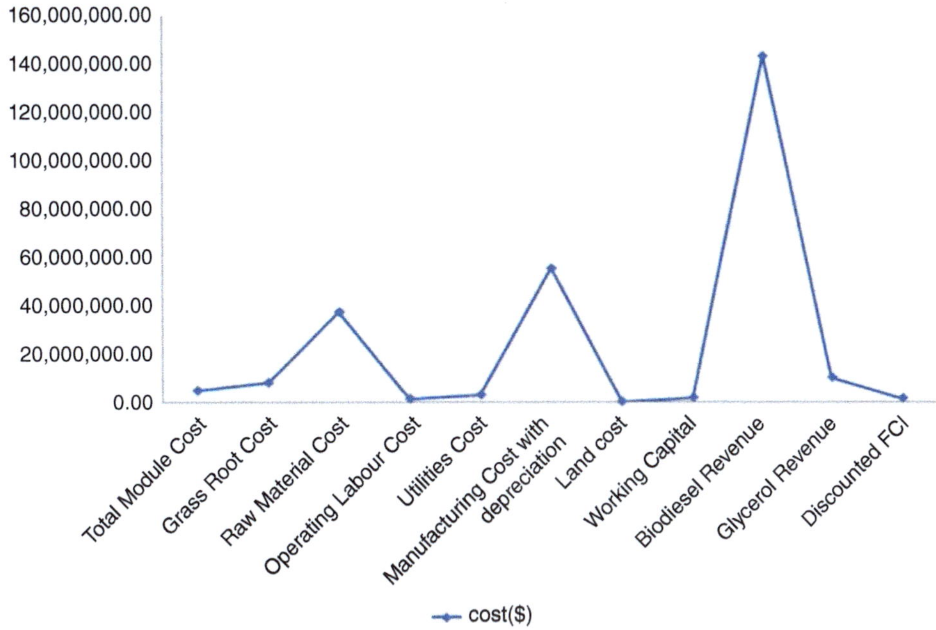

Figure 14.7 Economic analysis of the production of biodiesel in UAE.

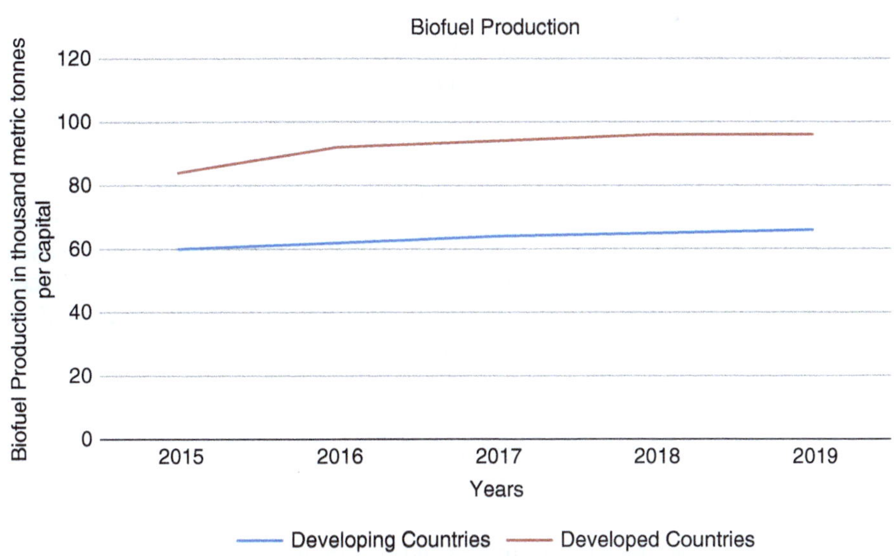

Figure 14.8 Biofuel production (in thousand metric tonnes per capita). *Source:* Adapted from [46, 47].

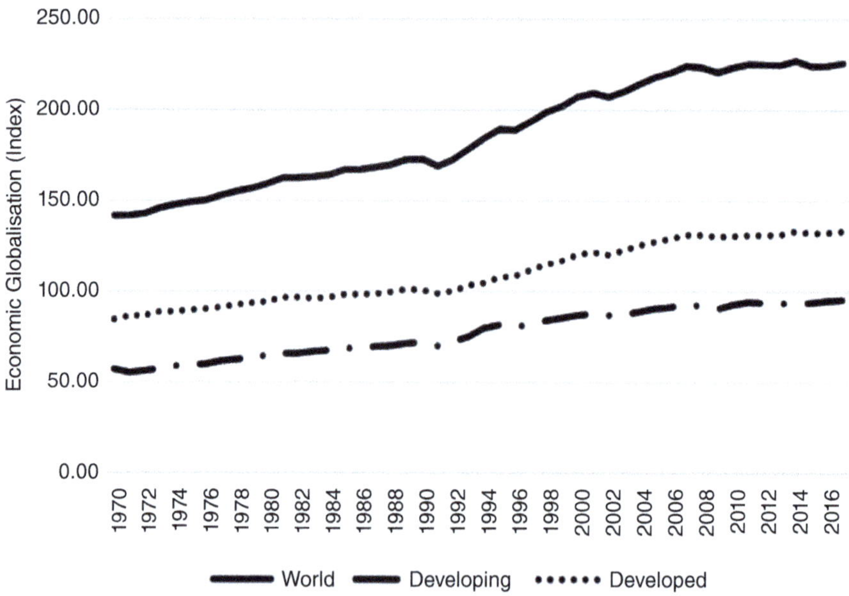

Figure 14.9 World economic globalisation. *Source:* Adapted from [48, 49].

Economic globalisation promotes the development of the biofuels with regard to the developing countries by maintaining the feedstock supply chain and reducing the emission. It helps in the improvement of the feedstock development by providing the knowledge and the availability of trained people and enhancing the skill of labours, crop varieties, agroeconomic practices and farming equipment; the selection of feedstock as per the growth- and location-related climatic conditions would enhance the agricultural practices in terms of agricultural timing and land availability at less resources and cost. The biofuel production depends on the availability of feedstock followed by the conversion processes involved in the production. The advanced technology has enhanced the conversion of biomass into biofuels at large quantities with a lower cost. The problem of advanced technology for the conversion of biomass into biofuels can be overcome with the technology transfer, which results in minimising the cost of production. The economic globalisation helps to maintain the supply chain due to the positive impact on the potential producers and consumers with regard to the biofuels. The world economic globalisation index of the developed and developing countries is increasing as shown in Figure 14.9.

14.8 The Development of Renewable Energy Based on Income

The biofuel demand depends on the environmental condition, and agricultural production means feedstock availability and income of a particular country. Based on the observation of various studies, income is an important factor that will have a positive and/or negative impact on the renewable energy demand. The high income promotes investment in the progress of renewable energy and would help to raise the request for renewable energy by utilising or generating the energy. This results in a decrease in the consumption of non-renewable fuel with an increase in the consumption of renewable energy towards the biofuels, which shows the positive impact of the income. On the contrary, this may affect the demand of renewable energy and divert towards the fossil-based fuel to meet the energy demand. This shows the negative impact of income on the renewable energy. Overall, there is a significant impact of income on the production of biofuels.

14.9 The Development of Renewable Energy Based on Carbon Emission

One of the main reasons for the increasing demand of biofuel renewable energy is the carbon emission. The greenhouse gas is minimised with the liquid biofuels as it is processed from the biomass from the agricultural product, which reduces the dependency of fossil fuel imported from other countries. Renewable energy promotes the sustainability, particularly in the high carbon emission countries.

14.10 Biofuel Impact on the Society

Biofuel policies change the global consequences, such as the requirement of farmland, deforestation and prices of food crops. The impact of biofuels on the society has changed depending on the generation of biofuels and has significant consequences based on the generation of biofuels. One of the significant consequences of the biofuels policy is that it causes a rise in the food prices due to the utilisation of edible crops in the production of biofuels and land utilisation for these crops. This leads to an increase in the food prices and affects the food supply chain. Even the growth of switch grass on the land that was utilised for the growth of edible crops results in a reduction in the growth of edible crops leading to food scarcity. The problem is resolved with the change in the biofuels policies. Atmospheric carbon is reduced by utilising the biofuels as the production of biofuels is from the plants that remove carbon dioxide from the atmosphere, whereas the fossil fuels release the carbon to the atmosphere in large amounts affecting the natural phenomenon. Therefore, the replacement of fossil fuels with biofuels results in the reduction of greenhouse gas emissions. This reduces the dependency of imported fuels as a large amount of foreign currency is spent on oil imports; thus, biofuels are an important option to change the energy matrix for developed and developing countries and boost the economy by creating the jobs, generating the revenue by increasing the biofuel plants at the regional level by selecting suitable feedstock or by utilising the agricultural residues.

Diseases could be a reducing case of asthma, respiratory diseases and other health issues, which results in a reduction in the hospitalisations and medical costs by the utilisation of biofuels instead of fossil fuels. This means that biofuels are more environmentally friendly, and clean fuels that can be used in the diesel engines without modification reduce greenhouse gas emission by approximately 50–75%. Overall, biofuels offer many benefits to the

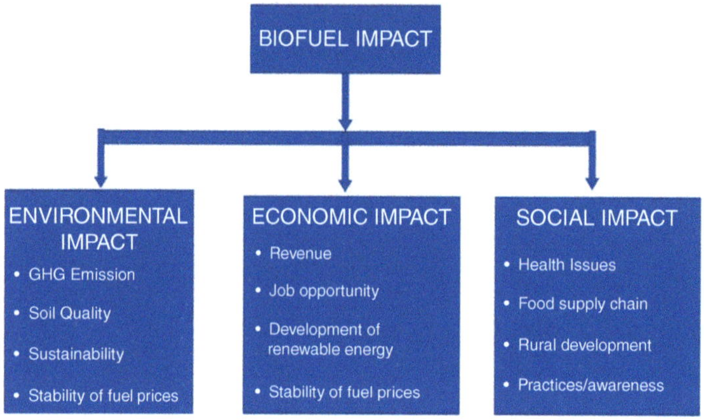

Figure 14.10 Impact of biofuel [50–52].

society in terms of reduction in the greenhouse gases, creation of jobs, rural development and stability in fuel prices (Figure 14.10).

14.11 Barriers in the Production of Biofuels

The production of biofuels is affected due to various barriers involved in each stage during the production of biofuels. The three main barriers, such as feedstock production, conversion and end user, are presented in Figure 14.11 with the expected solution.

The replacement of fossil fuels with the biofuels involves the utilisation of several types of biomasses, which is extended to the rural areas with the extraction of oil from other local sources for power generation applications and diesel engines. The current power supply capacity is inefficient, and emitting high levels of pollutants results in high costs. Sulphur, nitrogen oxides, carbon monoxide and many suspended particulates are the main pollutants released with the burning of fossil fuels, causing the air pollution degradation of ecosystem resulting in the human health problems. Biofuels are the solution to the pollution, good health and environment and social balance with the nature to maintain the harmony in nature.

14.12 Biofuel Desire to Improve the Balance of Trade

Biofuels and trade are two factors that are closely related in terms of carbon emission. Most of the countries decided to replace the fossil fuels with biofuels. Importing and exporting the biofuels either from developing or developed countries significantly affect the trade. In order to maintain the balance between trade objectives, new policies would be more supportive as biofuels can easily be transported and stored, thus promoting the tradable and environmentally preferable product obtained from the agricultural feedstocks.

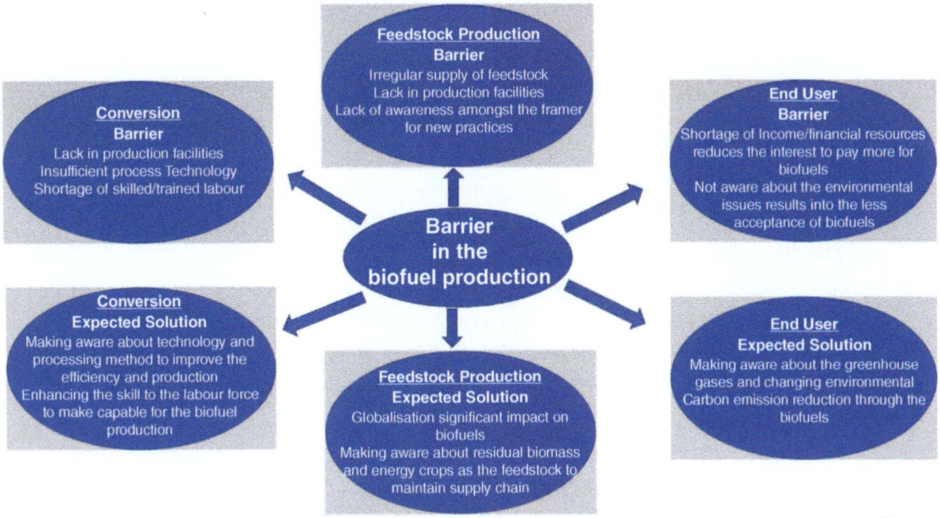

Figure 14.11 Barrier in the biofuel production.

14.13 Conclusion

Reduction in the consumption of the fossil fuel for environmental benefits may hit the market with a decrease in profits, which may not favour the market. The changes occurring from the development of biofuels from the first-generation biofuels to fourth-generation biofuels ensure the solution to all the problems in relation to environment, supply demand of food crops, job creation, feedstock selection, greenhouse gas emission and innovation of new technologies. Energetic benefits ensure the achievement of the market with the help of biofuel policies and innovation in new technologies.

Ethanol produced in first-generation biofuels from the edible items using food crops results in an imbalance in energy, demand supply of food needs, emission of pollutant gases, soil erosion and impact on biodiversity, which created the issue for the national debate. The production of second-generation biofuels, such as bio-ethanol and bio-methanol produced from different feedstock especially from non-food sources such as agricultural waste, results in a lack of feedstock, which is uneconomical for constant energy sources and biofuel generation with cost effectiveness. Algae were used as the feedstock in the third generation, which is more energy dense and eco-friendly but with limitations such as the capital intensity of the process and the cultivation of cultures. Metabolic engineering of algae using the oxygenic photosynthesis microbes was used for the production of biofuels from the biomass in the fourth generation. The projection of biofuels promotes the energy maintenance and reduction in global warming. This further facilitates an increase in the human activities, and the load of fossil fuel consumption would be reduced to renewable eco-friendly fuel carriers. The overall economy of the country is expected to increase by providing opportunities in the employment sector, development in the rural sector and energy security, which enhances benefits to the society.

References

1 Woo, T.H. (2019). Global warming analysis for greenhouse gases impacts comparable to carbon-free nuclear energy using neuro-fuzzy algorithm. *Int. J. Glob. Warm.* 17 (2): 219–233. https://doi.org/10.1504/ijgw.2019.097862.

2 Akasofu, S. (2018). Global warming has been halted: is global warming by the greenhouse gases significant. *Phys. Astron. Int. J.* 2 (5): 445–446. https://doi.org/10.15406/paij.2018.02.00122.

3 Karki, K.B. (2007). Greenhouse gases, global warming and glacier ice melt in Nepal. *J. Agric. Environ.* 8: 1–7. https://doi.org/10.3126/aej.v8i0.721.

4 Simsek, S. (2020). Effects of biodiesel obtained from Canola, sefflower oils and waste oils on the engine performance and exhaust emissions. *Fuel* 265: 117026. https://doi.org/10.1016/j.fuel.2020.117026.

5 Tapanwong, M. (2019). Production of ethyl ester biodiesel from used cooking oil with ethanol and its quick glycerol-biodiesel layer separation using pure glycerol. *Int. J. GEOMATE* 17 (61): 109–114. https://doi.org/10.21660/2019.61.4808.

6 Qasim, M., Ansari, T.M., and Hussain, M. (2017). Combustion, performance, and emission evaluation of a diesel engine with biodiesel like fuel blends derived from a mixture of Pakistani waste canola and waste transformer oils. *Energies* 10 (7): 1023. https://doi.org/10.3390/en10071023.

7 Dong, M., Chang, C.P., Gong, Q., and Chu, Y. (2019). Revisiting global economic activity and crude oil prices: a wavelet analysis. *Econ. Model.* 78: 134–149. https://doi.org/10.1016/j.econmod.2018.08.012.

8 Karagöz, M. (2020). Investigation of performance and emission characteristics of an CI engine fuelled with diesel – waste tire oil – butanol blends. *Fuel* 282: 118872. https://doi.org/10.1016/j.fuel.2020.118872.

9 Ajmal Ahamed, S., Mohamed Riyaz, S., Mahadevan, A. et al. (2020). Preparation and characterization of chemically modified vegetable oils as liquid insulation in transformers. *Int. J. Recent. Technol. Eng.* 8.

10 Aguilar, L.B., Campos, H.M., Leyva, I.R. et al. (2011). Global social and economic impact on the use of biofuels and recomendations for sustainability. *Glob. J. Res. Eng. Automot. Eng.* 11 (5_B): 23–30.

11 Serna, F. and Luis Barrera, H.M. (2011). Impacto Social y Económico en el Uso de Biocombustibles. *J. Technol. Manag. Innov.* 6 (1): 100–114.

12 Chen, M.K. and Chen, X. (2013). Economic, energy security, and greenhouse gas effects of biofuels: implications for policy. *Am. J. Agric. Econ.* 95 (5): 1325–1331.

13 Jokiniemi, T. and Ahokas, J. (2013). A review of production and use of first generation biodiesel in agriculture. *Agron. Res.* 11 (1): 239–248.

14 Yoro, P.T.S. and Yoro, K.O. (2016). Biofuel development initiatives in sub-Saharan Africa: opportunities and challenges. *Climate* 4 (2): 33.

15 Hernández, M. and Hernández, J. (2008). Verdades y mitos de los biocombustibles. *Elementos* 71: 15–18.

16 Stachett, G., Rodrígues, I., Aparecida, I. et al. (2007). Socioenvironmental al impact of biodiesel productionin Brazil. *J. Technol. Manag. Innov.* 2 (2): 46–66.

17 Shikha, K. and Rita, C.Y. (2012). Biodiesel production from non edible-oils: a review. *J. Chem. Pharm. Res.* 4 (9): 4219–4230. https://doi.org/10.32438/wpe.0602149.

18 Atabani, A.E., Silitonga, A.S., Badruddin, I.A. et al. (2012). A comprehensive review on biodiesel as an alternative energy resource and its characteristics. *Renew. Sust. Energ. Rev.* 16 (4): 2070–2093. https://doi.org/10.1016/j.rser.2012.01.003.

19 Laskar, I.B., Changmai, B., Gupta, R. et al. (2021). A mesoporous polysulfonic acid-formaldehyde polymeric catalyst for biodiesel production from Jatropha curcas oil. *Renew. Energy* 173: 415–421. https://doi.org/10.1016/j.renene.2021.04.004.

20 Changmai, B., Wheatley, A.E.H., Rano, R. et al. (2021). A magnetically separable acid-functionalized nanocatalyst for biodiesel production. *Fuel* 305: 121576. https://doi.org/10.1016/j.fuel.2021.121576.

21 Mofijur, M., Masjuki, H.H., Kalam, M.A. et al. (2012). Prospects of biodiesel from Jatropha in Malaysia. *Renew. Sust. Energ. Rev.* 16 (7): 5007–5020. https://doi.org/10.1016/j.rser.2012.05.010.

22 Mahalingam, A., Devarajan, Y., Radhakrishnan, S. et al. (2018). Emissions analysis on mahua oil biodiesel and higher alcohol blends in diesel engine. *Alex. Eng. J.* 57 (4): 2627–2631. https://doi.org/10.1016/j.aej.2017.07.009.

23 Mamilla, V.R., Mallikarjun, M.V., and Rao, G.L.N. (2011). Preparation of biodiesel from Karanja oil. *Int. J. Energy Eng.* 1: 94–100. https://doi.org/10.5963/ijee0102008.

24 Ali, M.H., Mashud, M., Rubel, M.R., and Ahmad, R.H. (2013). Biodiesel from neem oil as an alternative fuel for diesel engine. *Procedia Eng.* 56: 625–630. https://doi.org/10.1016/j.proeng.2013.03.169.

25 Namoco, C.S., Comaling, V.C., and Buna, C.C. (2017). Utilization of used cooking oil as an alternative cooking fuel resource. *ARPN J. Eng. Appl. Sci.* 12: 435–442.

26 Changmai, B., Sudarsanam, P., and Rokhum, L. (2020). Biodiesel production using a renewable mesoporous solid catalyst. *Ind. Crop. Prod.* 145: 111911. https://doi.org/10.1016/j.indcrop.2019.111911.

27 Charles, M.B., Ryan, R., Ryan, N., and RO. (2007). Public policy and biofuels: the way forward. *Energy Policy* 35 (11): 5737–5746.

28 Malik, U.S., Ahmed, M., Sombilla, M.A., and Cueno, S.L. (2009). Biofuels production for smallholder producers in the greater Mekong sub-region. *Appl. Energy* 86: S58–S68.

29 Kumar, S., Shrestha, P., and Salam, P.A. (2013). A review of biofuel policies in the major biofuel producing countries of ASEAN: production, targets, policy drivers and impacts. *Renew. Sust. Energ. Rev.* 26: 822–836.

30 Saravanan, A.P., Pugazhendhi, A., and Mathimani, T. (2020). A comprehensive assessment of biofuel policies in the BRICS nations: implementation, blending target and gaps. *Fuel* 272: 117635.

31 Ewing, M. and Msangi, S. (2009). Biofuels production in developing countries: assessing tradeoffs in welfare and food security. *Environ. Sci. Policy* 12 (4): 520–528.

32 Fernández-Tirado, F. and Parra-López, C. (2020). Economic and social impacts of the biodiesel industry: assessment and policy implications in Spain. Spanish. *J. Agric. Res.* 18 (4): https://doi.org/10.5424/sjar/2020184-16580.

33 Lai, B., Yi, P., Sui, Y., and Zhang, Q. (2021). Energy distribution in EV energy network under energy shortage. *Neurocomputing* 444: 179–188.

34 Tabatabaei, M., Aghbashlo, M., Dehhaghi, M. et al. (2019). Reactor technologies for biodiesel production and processing: a review. *Prog. Energy Combust. Sci.* 74: 239–303.

35 Hoekman, S.K., Broch, A., Robbins, C. et al. (2012). Review of biodiesel composition, properties, and specifications. *Renew. Sust. Energ. Rev.* 16 (1): 143–169.

36 Chin, M. (2011). Biofuels in Malaysia: an analysis of the legal and institutional framework. *Cent. Int. For. Res.* 64.

37 Senatore, A., Dalena, F., Sola, A. et al. (2019). First-generation feedstock for bioenergy production. Second and Third Generation of Feedstocks. *Evol. Biofuels* 35–57.

38 Dillon, H.S., Laan, T., and Dillon, H.S. (2008). Biofuels – At What Cost? Government support for ethanol and biodiesel in Indonesia One of a series of reports addressing subsidies for biofuels in developing countries. In: *The Global Subsidies Initiative (GSI) of the International Institute for Sustainable Development (IISD) Geneva*, Switzerland, 7–34.

39 Sun, J., Xiong, X., Wang, M. et al. (2019). Microalgae biodiesel production in China: a preliminary economic analysis. *Renew. Sust. Energ. Rev.* 104: 296–306. https://doi.org/10.1016/j.rser.2019.01.021.

40 Marquez, G.P.B., Santiañez, W.J.E., Trono, G.C. et al. (2014). Seaweed biomass of the Philippines: sustainable feedstock for biogas production. *Renew. Sust. Energ. Rev.* 38. https://doi.org/10.1016/j.rser.2014.07.056.

41 Proskurina, S., Junginger, M., Heinimö, J., and Vakkilainen, E. (2019). Global biomass trade for energy – part 1: statistical and methodological considerations. *Biofuels Bioprod. Biorefin.* 13 (2): 358–370. https://doi.org/10.1002/bbb.1841.

42 Doku, A. and Di Falco, S. (2012). Biofuels in developing countries: are comparative advantages enough? *Energy Policy* 44: 101–117. https://doi.org/10.1016/j.enpol.2012.01.022.

43 Fevolden, A.M. and Klitkou, A. (2017). A fuel too far? Technology, innovation, and transition in failed biofuel development in Norway. *Energy Res. Soc. Sci.* 23: 125–135. https://doi.org/10.1016/j.erss.2016.10.010.

44 Athukorala, P.C. (2014). Industrialisation through state-MNC partnership: lessons from Malaysia's national car project. *Malaysian J. Econ. Stud.* 51: 113–126.

45 Hussain, M.N., Al Samad, T., and Janajreh, I. (2016). Economic feasibility of biodiesel production from waste cooking oil in the UAE. *Sustain. Cities Soc.* 26: 217–226.

46 Mohammadshirazi, A., Akram, A., Rafiee, S., and Bagheri, K.E. (2014). Energy and cost analyses of biodiesel production from waste cooking oil. *Renew. Sust. Energ. Rev.* 33: 44–49. https://doi.org/10.1016/j.rser.2014.01.067.

47 Vlysidis, A., Binns, M., Webb, C., and Theodoropoulos, C. (2011). A techno-economic analysis of biodiesel biorefineries: assessment of integrated designs for the co-production of fuels and chemicals. *Energy* 36 (8): 4671–4683. https://doi.org/10.1016/j.energy.2011.04.046.

48 Khan, H.M., Iqbal, T., Haider, C.A., and Javaid, A. (2020). Sustainable biodiesel production from waste cooking oil utilizing waste ostrich (Struthio camelus) bones derived heterogeneous catalyst. *Fuel* 277 (1): https://doi.org/10.1016/j.fuel.2020.118091.

49 Fiorese, G., Catenacci, M., and Verdolini, E. (2012). Advanced biofuels: future perspectives from an expert elicitation survey. *SSRN Electron. J.* https://doi.org/10.2139/ssrn.2162834.

50 Jagger, A. (2011). Biofuels for transport in 2050. *Biofuels Bioprod. Biorefin.* 5 (5): 481–485. https://doi.org/10.1002/bbb.330.

51 Gygli, S., Haelg, F., and Potrafke, N.S.J.-E. (2019). The KOF globalisation index – revisited. *Rev. Int. Organ.* 14 (3): 543–574.

52 Subramaniam, Y. and Masron, T.A. (2021). The impact of economic globalization on biofuel in developing countries. *Energy Convers. Manag.* 10: 100064.

Index

Sustainability in Biofuel Production Technology, First Edition. Pratibha S. Agrawal, Pramod N. Belkhode, and Samuel Lalthazuala Rokhum.